Fundamentals of Data Communication Networks

Fundamentals of Data Communication Networks

Oliver C. Ibe

WILEY

This edition first published 2018

Registered Office
John Wiley & Sons, Inc., 111 River Street, Hoboken, NJ 07030, USA

Editorial Office
111 River Street, Hoboken, NJ 07030, USA

For details of our global editorial offices, customer services, and more information about Wiley products visit us at www.wiley.com.

Wiley also publishes its books in a variety of electronic formats and by print-on-demand. Some content that appears in standard print versions of this book may not be available in other formats.

Library of Congress Cataloging-in-Publication Data applied for

ISBN: 9781119436256

Cover design by Wiley
Cover image: © hywards/Shutterstock

Set in 10/12pt WarnockPro by SPi Global, Chennai, India

10 9 8 7 6 5 4 3 2 1

Contents

Preface

There are many books on data communication networks, and so one would ask why write another one. Almost all the books written so far are written for graduate students. The few that are written for an undergraduate audience are aimed at business students who need to understand the buzzwords that they will be encountering in the marketing and sale of data communication equipment. There is no book written for undergraduate electrical engineering students who study the different components of the data communication technologies in isolation. Thus, this book grew out of the following observations:

1. While electrical and computer engineering students study how to build filters in circuit theory class and Fourier analysis, sampling theorem, and modulation in signal and systems, no concerted effort is made to combine these topics into one system that they can relate to.
2. In this information age, most undergraduate electrical and computer engineering students do not know what an IP address is, how to design a data network, or how domain name system (DNS) works.
3. While every student uses a mobile device, most undergraduate engineering students in general and undergraduate electrical and computer engineering students in particular do not know what 2G, 3G, and 4G networks are. To them these are mere buzzwords.
4. Data security is a very important issue and still many undergraduate electrical and computer engineering students do not understand the basic concepts of data security.

This book seeks to address these issues and grew out of the lecture notes for a class with a similar title as this book that I have been teaching in the Department of Electrical and Computer Engineering at the University of Massachusetts Lowell. Most of the students are juniors and seniors who have taken signals and systems, calculus II, and circuit theory I.

The book is organized along the seven-layer framework of the open systems interconnection (OSI) model, which is a conceptual hierarchical model that specifies the communication functions of each layer in a communication system. The lowest layer of the hierarchy is called the physical layer (or Layer 1), while the highest layer is called the application layer (or Layer 7). Specifically, Chapter 1 gives an overview of data communication networks, including how they are classified, and a discussion of the OSI model. Chapter 2 discusses the lowest layer of the OSI model, which is the physical layer (or Layer 1). The different topics covered in signals and systems as well as circuit theory are discussed, including Fourier series, Fourier transform, the different multiplexing and modulation schemes, sampling theorem, different media types, and channel impairments.

Chapters 3–5 discuss topics related to the data link layer or Layer 2. Chapter 3 discusses data link layer protocols. Chapter 4 discusses techniques called multiple access schemes that are used to access data communication networks, while Chapter 5 discusses local area networks that use Layer 2 protocols.

Chapters 6 and 7 discuss the network layer or Layer 3. Chapter 6 discusses IP addressing, while Chapter 7 discusses how information is sent from source to destination, a process that is called routing. Different routing protocols are discussed, where a protocol is a defined set of rules that ensure an effective communication.

Chapters 8 and 9 discuss the transport layer or Layer 4. Chapter 8 discusses two protocols that were defined in the early days of the Internet. These are the transmission control protocol (TCP) and the user datagram protocol (UDP). These two protocols were defined when the Internet, which is a data communication network, was expected to be used mainly for non-real-time applications. Thus, they are optimized for file transfers. With the Internet being used for both traditional non-real-time applications and new real-time applications such as streaming video and voice over IP, two new transport-layer protocols have been defined to deal with this new environment. These protocols are the stream control transmission protocol (SCTP) and the datagram congestion control protocol (DCCP), and they are discussed in Chapter 9.

Chapter 10 discusses two protocols that are used in the highest layer called the application layer. These are the dynamic host configuration protocol (DHCP) and the DNS.

Chapter 11 provides an introduction to mobile communication networks. It discusses the different generations of the mobile communication network called 1G, 2G, 3G, 4G, and 5G networks, where 1G stands for first-generation network, 2G stands for second-generation network, and so on.

Finally, Chapter 12 gives an introduction to network security. It discusses the security threats that are encountered in a data communication network and the steps used to overcome them. There are exercises at the end of each chapter.

We have been able to cover these chapters in one semester, but the instructor is free to skip some chapters that they believe that the students are not equipped to handle. However, it is strongly recommended that Chapters 1 through 9 be covered, while Chapters 10–12 will be covered when time permits.

May 2017

Oliver C. Ibe
Lowell, Massachusetts

Acknowledgments

I would like to thank my editor at Wiley, Brett Kurzman, for his whole-hearted support of this book from the beginning. I also want to thank the anonymous reviewers for their useful comments that helped to improve the presentation of the material in the book. I would like to thank my colleagues in the Electrical and Computer Engineering (ECE) Department at UMass Lowell for encouraging me to develop the course. The course is taught every semester with an enrollment of more than 50 students each semester. I would like to thank all my students who have taken so much interest in the course, thereby justifying the effort that has gone into preparing the material at the undergraduate level. I would like to thank the Dean of the Francis College of Engineering, Dr. Joseph Hartman, for his keen interest in my career development and for appointing me the acting chair of ECE at the time the decision to develop this course was made. Finally, I would like to thank my wife Christie for her support and encouragement, and our children Chidinma, Ogechi, Amanze, and Ugonna for the joy they bring into my life.

1

Overview of Data Communication Networks

1.1 Introduction

In a very broad sense, a network is any interconnected group of people or devices that are capable of sharing meaningful information with one another. In the telecommunication sense, a data communication network is a collection of two or more computing devices that are interconnected to enable them to share data. Data communication networking arose in response to the need to share data in a timely manner. Data sharing and information dissemination are critical to the success of any business. Thus, data communication networks are important to all contemporary organizations.

As discussed earlier, data communication networks deal with the transfer of data between two points. Data originates at the *source* and is finally delivered to the destination, which is also called a *sink*. Sometimes, the source and destination are interconnected by one link; at other times, the data must traverse multiple links to reach the destination. A typical communication environment includes multiple sources and sinks that are interconnected by communication links to build a network. Thus, a communication network is essentially an arrangement of hardware and software that allows users to exchange information.

1.2 Data Communication Network Model

A communication model is necessary to enable us to introduce the main elements of a communication system as well as to define some of the terminology used in the remainder of this book. A communication system consists of the following:

- A *source* that generates the information.
- A *source encoder* that converts the information into an electrical form called *message signal* $m(t)$.

Fundamentals of Data Communication Networks, First Edition. Oliver C. Ibe.

- A *transmitter* that is used to convert the message signal into a form acceptable to the channel.
- The *channel* is the path or link that connects the transmitter and the receiver; it can be metallic, optical fiber, or air.
- A *receiver* performs an inverse function of that of the transmitter to recover the message signal.
- A *source decoder* converts the electrical signal back to a form acceptable to the receiver.
- A *sink* is the user of the information generated by the source.

The model is illustrated in Figure 1.1.

Note that information flow can be bidirectional because what is a source at one time can be a sink at another time. Thus, Figure 1.1 shows the basic blocks used to process information as it flows in one direction.

The simplest data communication network consists of a source that is directly connected to a sink, as shown in Figure 1.2.

In a more complex network, the two communicating nodes are interconnected by a complex structure, which is usually represented by a cloud as shown in Figure 1.3.

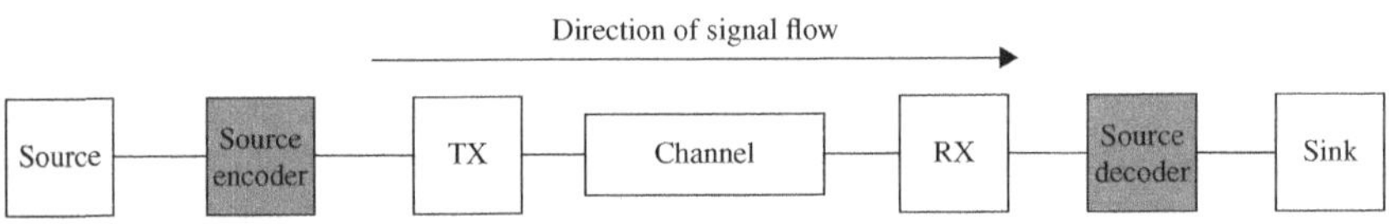

Figure 1.1 A Data Communication Network Model.

Figure 1.2 A Simple Data Communication Network.

Figure 1.3 Representation of a Complex Network Structure.

1.3 Classification of Data Communication Networks

There are different ways to classify a data communication network, which are as follows:

- Transmission method
- Data flow direction
- Network topology
- Geographical coverage
- Transmission medium
- Data transfer technique
- Network access technique
- Media sharing technique.

In this section, we describe each of these methods.

1.3.1 Transmission Method

Data transmission method can be classified in two fundamental ways: *asynchronous* and *synchronous* transmissions. Asynchronous transmission is used when data is transmitted as individual characters. In this method, each character is preceded by one start bit and one or two stop bits that are used by the receiver for synchronization purposes. The need for synchronization arises from the fact that the interval between characters is random, which means the receiver that has been idle for some time needs to know when data is coming in.

Synchronous transmission is used to transmit large blocks of data at a time. In this scheme, data is usually organized in frames and each frame is preceded by a *flag* that consists of a few bits, and terminated by another flag. It is more efficient than asynchronous transmission because the overhead is smaller on a character-by-character basis. Figure 1.4 illustrates the difference between an asynchronous transmission scheme and a synchronous transmission scheme.

1.3.2 Data Flow Direction

Three ways are used to characterize the direction of data flow: *simplex, half duplex,* and *full duplex.* In a simplex transmission, data can only flow in one direction, which is usually from the source to the sink. This is illustrated in Figure 1.5.

In a half-duplex transmission (HDX) data can flow in both directions, but never simultaneously. It first flows in one direction, and then in the other direction. Thus, one station is the source and the other is the sink. Then the roles are interchanged such that the previous source becomes the sink and the previous sink becomes the source, and so on. This is illustrated in Figure 1.6.

In a full-duplex transmission (FDX), data can flow in both directions simultaneously. It can be viewed as a pair of simplex lines between the source and

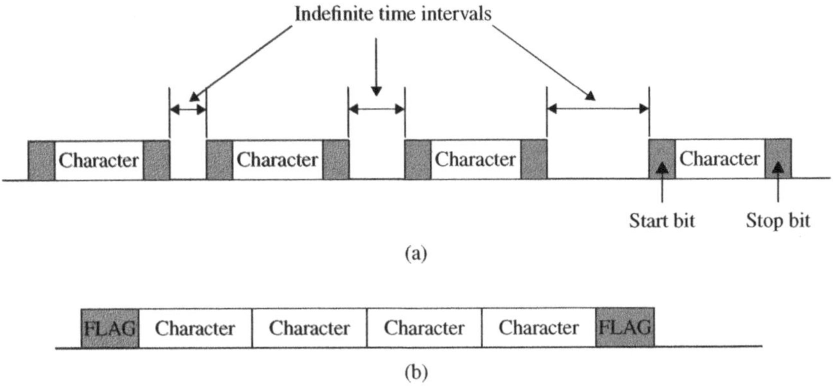

Figure 1.4 Asynchronous versus Synchronous Transmission. (a) Character-oriented asynchronous transmission and (b) frame-oriented synchronous transmission.

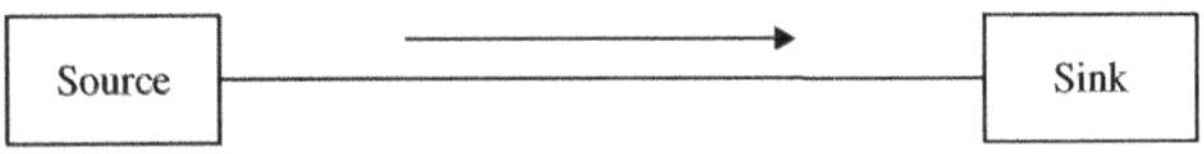

Figure 1.5 Simplex Transmission.

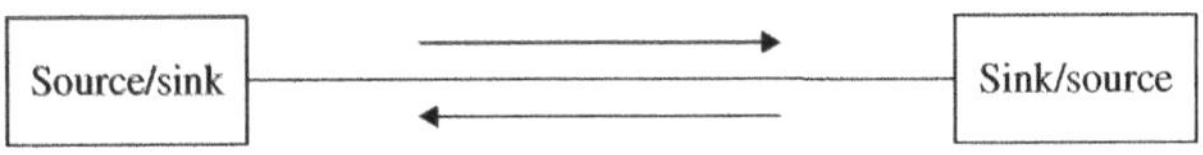

Figure 1.6 Half- Duplex Transmission.

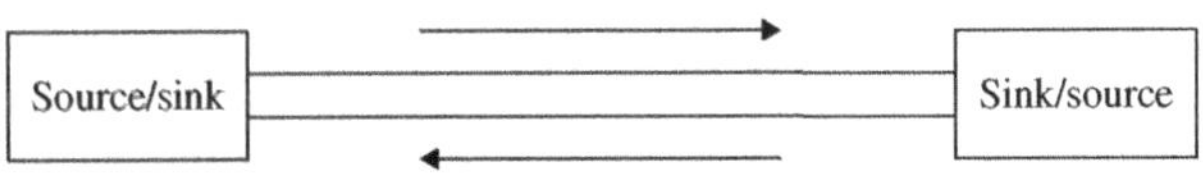

Figure 1.7 Full- Duplex Transmission.

sink with one line going from the source to the sink and the other going from the sink to the source. This is illustrated in Figure 1.7.

1.3.3 Network Topology

Network topology refers to the different geometrical configurations that can be used to build a network. Different topologies exist and include the following:

- Point-to-point (P2P)
- Point-to-multipoint
- Multidrop
- Bus
- Ring (or loop)
- Star

- Tree
- Mesh.

In the *P2P topology*, a link permanently connects two nodes or network devices. The P2P topology is illustrated in Figure 1.8. Note that the link interconnecting two nodes can be either a wireless (or air) or a wired connection.

In the *point-to-multipoint topology*, one node is connected to multiple nodes, each in a P2P manner, as illustrated in Figure 1.9.

In the *multidrop* topology, all nodes are interconnected by a single link with one node that is the master node and the other nodes are secondary or slave nodes. The master node usually controls access to the link and is located at one end of the link as illustrated in Figure 1.10.

The *bus* topology is similar to the multidrop topology with the exception that there is no master–slave relationship; all nodes are peers. The topology is illustrated in Figure 1.11.

Figure 1.8 Point-to-Point Topology.

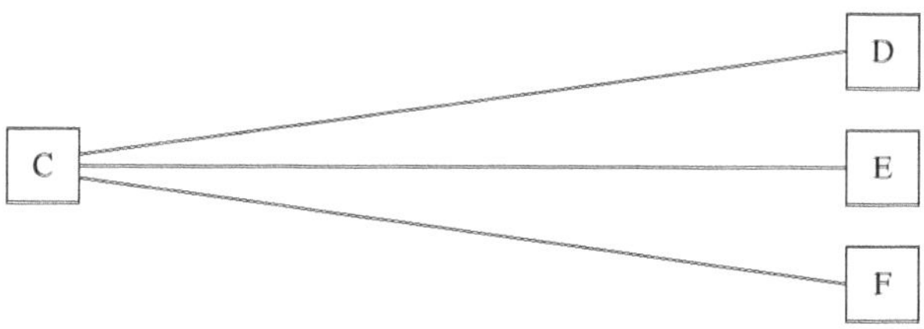

Figure 1.9 Point-to-Multipoint Topology.

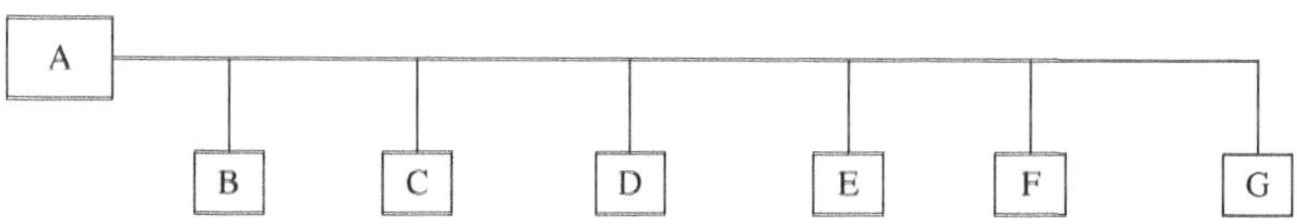

Figure 1.10 Mulitdrop Topology.

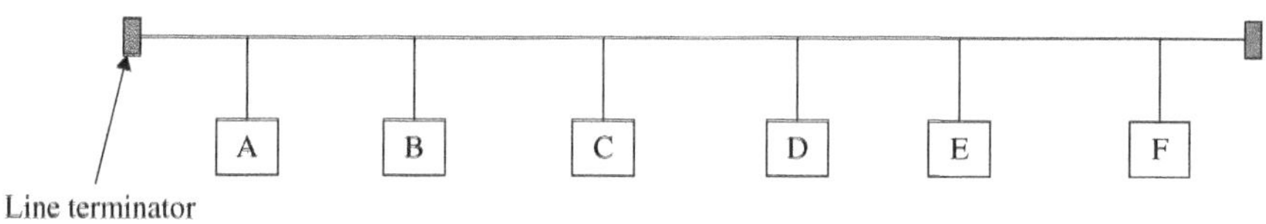

Figure 1.11 Bus Topology.

The *line terminator* in the figure is used to prevent a signal that comes to the end of a transmission line from bouncing back and corrupting other signals on the line. This "bouncing back" is called *signal reflection* and it has the tendency to interfere with and, therefore, corrupt the data on the line.

In *ring* topology, the nodes are connected serially in a P2P manner with the last node connected to the first node to form a loop. This is illustrated in Figure 1.12.

A *star* topology is a topology in which each node is connected in a P2P manner to a central node, called a *hub*. This is illustrated in Figure 1.13. Note that the star topology is similar to the point-to-multipoint topology. The difference between the two is that in the star topology, the hub is a passive device that does not control access to the network, while in the point-to-multipoint topology, the central node is an active device that controls communication in the network.

A *tree* topology is formed by connecting multiple buses together to form a system of branching links with no closed loop. It has a special node called the *headend* from which information flows to the other nodes. The topology is illustrated in Figure 1.14.

In the *mesh* topology, the network nodes are interconnected in an arbitrary manner. Generally, users are connected to only a subset of the nodes and another set of internal nodes provides a switching facility that moves data

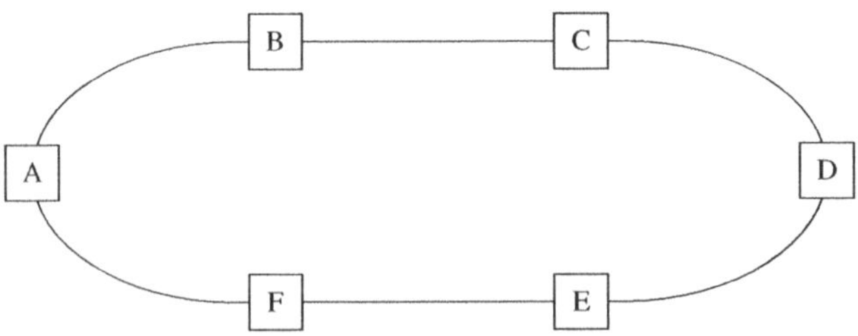

Figure 1.12 Ring Topology.

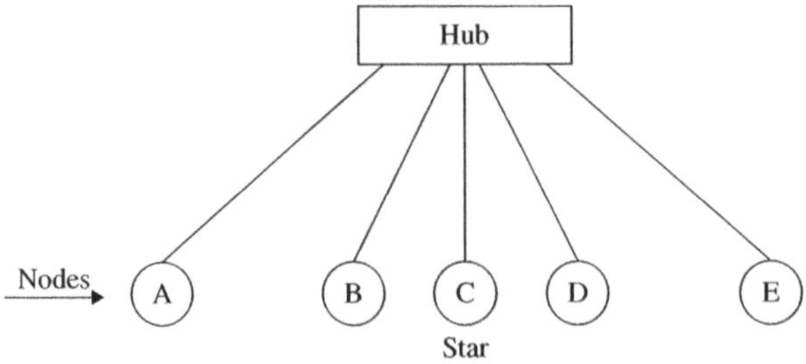

Figure 1.13 Star Topology.

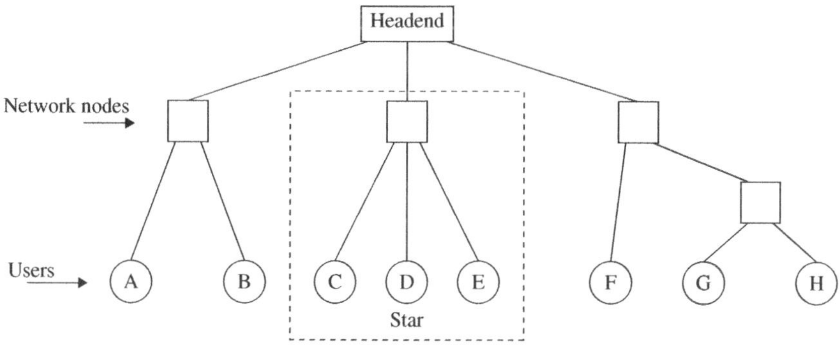

Figure 1.14 Tree Topology.

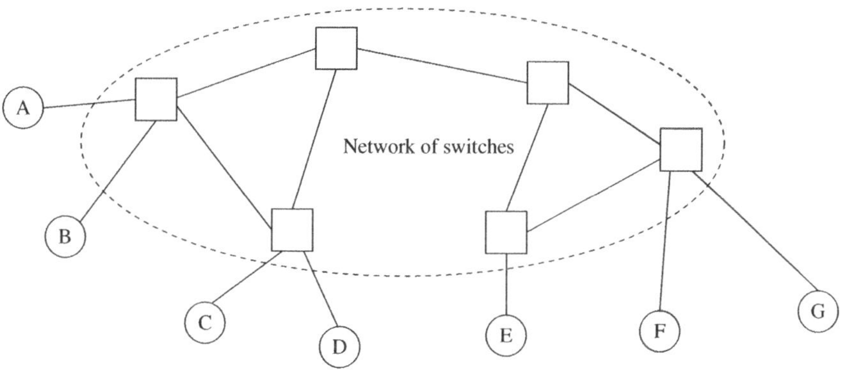

Figure 1.15 Mesh Topology.

from one node to another until it reaches its destination. An example of the mesh topology is shown in Figure 1.15.

1.3.4 Geographical Coverage

Networks are sometimes classified according to their geographical coverage. The following are examples of networks that are based on their geographical coverage:

- *Personal area networks* (PANs) are networks that interconnect devices that are within the reach of an individual, usually, within a range of 10 m. These devices are usually cell phones, tablets, and laptops.
- *Local area networks* (LANs) cover small geographical areas, typically one building, a floor, or a campus. Examples include the Ethernet and token ring networks.

- *Metropolitan area networks* (MANs) interconnect LANs in a campus or metropolitan area. Example includes the fiber distributed data interface (FDDI).
- *Wide area networks* (WANs) cover much larger areas such as a country (e.g., the public switched telephone network, or PSTN), or the globe (e.g., the Internet).

1.3.5 Transmission Medium

Data communication networks can also be classified according to the type of medium over which the signal propagates. In this case, there are two types of transmissions: *guided transmission* and *wireless transmission*; wireless transmission is also called *unguided transmission.*

In guided transmission, a physical path is provided along which the signal propagates. Guided transmission includes the twisted pair, coaxial cable, and optical fiber. In wireless transmission, the medium over which the signal propagates is mostly the air. Such networks use radio transmission.

1.3.6 Data Transfer Technique

There are two ways in which data can be transferred from source to destination: *switching* and *broadcasting.*

In a *switched network,* data is transferred from source to destination through a series of intermediate switching nodes. Data passes through a subset of the network nodes. There are two types of switched networks: *circuit-switched* networks and *packet-switched* networks. *Circuit switching* involves establishing a path from source to destination before the commencement of communication. The path is dedicated to the source–destination pair for the duration of communication session.

Packet switching involves organizing data in blocks called *packets* that are sent in a *store-and-forward* manner without prior establishment of communication path. By store-and-forward we mean that when a node receives a packet, it stores the packet and checks it for errors. If the packet is found to have an error, it is discarded. If it is found to be error-free, it is then scheduled for transmission to the next node on its way to the destination. There are two types of packet switching: *virtual circuit switching* and *datagram service.* Virtual circuit switching uses the same path for all packets belonging to the same session. Datagram service can use different paths for the different packets in a session.

In a *broadcast network,* a transmission from a source is received by all nodes in the network. Thus, unlike a switched network where data passes through only a subset of the nodes in the network, a broadcast network generally ensures that all the nodes in the network see the transmitted data.

1.3.7 Network Access Technique

There are two network access techniques that are closely related to the transfer technique used: *switched network access* and *broadcast network access*. Switched network access uses either circuit switching or packet switching. *Circuit switching* involves three phases:

- *Call setup phase* is used to establish communication path between the source and the sink.
- *Data transfer phase* is used to transmit the data after the path has been established.
- *Call teardown phase* is used to clear, tear down, or delete the communication path after the communication has been completed.

Packet switching sends packets of data into the network to be routed in a store-and-forward manner without prior establishment of the communication path.

Broadcast network access uses two access methods: *random access* in which users contend for control of the channel and *controlled access* where no contention is allowed. Controlled access uses one of two polling schemes: *centralized polling* (or *roll-call polling*) and *distributed polling*.

Centralized polling is used in master–slave systems where the master (or controller) uses a round-robin scheme to invite each station to transmit its data. Thus, no station can transmit until it is explicitly invited by the controller to transmit. *Distributed polling* generally uses a *token passing scheme* in peer-to-peer systems to control access. When a station receives the token, it transmits its data and after that it passes the token to the next node in a logical order. If a station that has no data to transmit receives the token, it simply passes the token to the next station in the logical order.

1.3.8 Media Sharing Technique

Some transmission media provide more capacity than one user can use. The utilization of such media can be increased by allowing multiple users to transmit their data simultaneously, or close to simultaneously. Three methods exist for sharing such media among the users, which are as follows:

(a) *Frequency-division multiplexing* (FDM) where the frequency spectrum of the medium is partitioned into multiple frequency blocks called *channels* that are assigned to users who can use these channels simultaneously without interference from each other.

(b) *Time-division multiplexing* (TDM) where transmission time is divided into nonoverlapping time slots that are assigned to users. Transmissions are staggered using a round-robin method to schedule the transmissions. Specifically, when it is time for a user to transmit, he uses the entire transmission medium for the duration of the time slot and then relinquishes

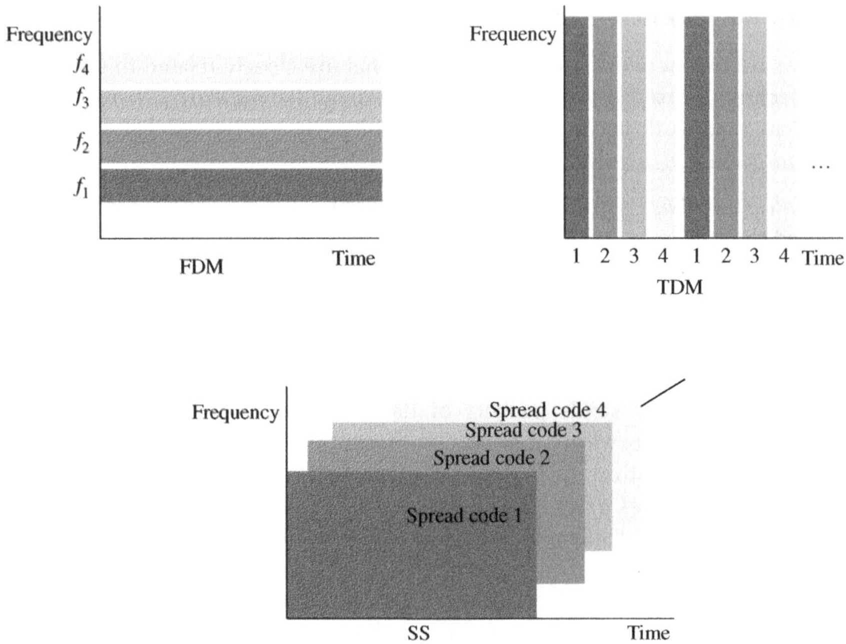

Figure 1.16 Media Shring Schemes.

control of the medium for the next user in the logical order. There are two types of TDM:

- *Synchronous TDM* where a time slot is dedicated to user whether or not the user has data to transmit.
- *Statistical multiplexing* (also called *asynchronous TDM*) where a user gets a slot only when he has data to transmit.

(c) *Spread spectrum* (SS) where the output signal, which appears like noise, occupies more bandwidth than original signal. There two types of SS:

- *Direct sequence* SS, where the frequency spectrum of a data signal is spread using a code that is uncorrelated with that signal and the codes are unique for every user and uncorrelated with other codes.
- *Frequency hopping* SS, where the transmitted frequency is pseudorandomly changed at a rate called the "hopping rate." The hopping pattern (i.e., the fixed order of frequencies that the user hops into) assigned to a user constitutes the channel for that user.

The different media sharing schemes are illustrated in Figure 1.16. They can be summarized as follows. In FDM, a user is assigned a part of the frequency spectrum that they use all the time. In TDM, the user is allowed to use the entire frequency spectrum but only part of the time. Finally, in SS, each user can use the entire spectrum all the time as long as they use a code that is uncorrelated to other codes to transmit their data.

1.4 Data Network Architecture

Because data communication deals with exchange of data messages between computers, transferring a message from one computer to another is not a trivial task as all conditions have to be anticipated and the necessary course of action exhaustively specified. To simplify intercomputer communication, the International Standards Organization (ISO) proposed a seven-layer architectural model called the *Open Systems Interconnection* (OSI) *Reference Model* for implementing data communication between cooperating systems. Each layer deals with a specific data communication function and provides services for the layer immediately above it while using the services of the layer immediately below it, except for the lowest layer that has no layer below it and the uppermost layer that has no layer above it.

One advantage of the model is that the implementation of one layer can change with technology without affecting the implementation of other layers as long as it provides the same services to the immediate upper layer as before. For example, we can change the network interface card (NIC) on a laptop from wired NIC to wireless NIC without affecting other layers; only the layer that is concerned with direct connection to the medium is affected.

1.4.1 The OSI Protocol Reference Model

In this book, we use the term "protocol" very often. A *protocol* is a set of rules that ensure the effective exchange of information. The seven-layer reference model is shown in Table 1.1.

- *Physical layer* defines the electrical and mechanical standards and signaling required to establish, maintain, and terminate connections. It deals with issues such as size and shape of connectors, signal strength, bit representation, and bit synchronization.

Table 1.1 The OSI model.

Layer 7	Application layer
Layer 6	Presentation layer
Layer 5	Session layer
Layer 4	Transport layer
Layer 3	Network layer
Layer 2	Data link layer
Layer 1	Physical layer

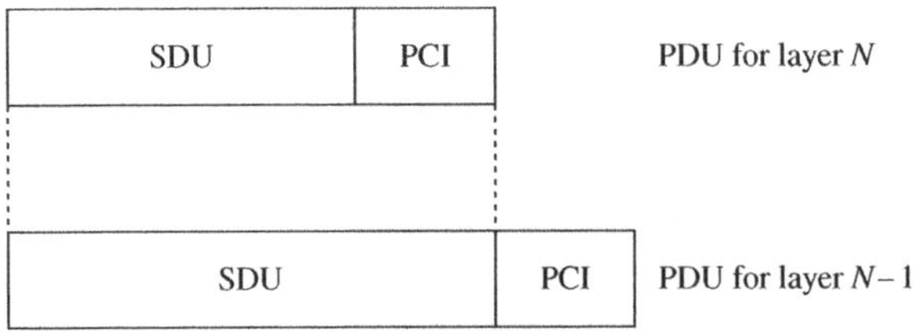

Figure 1.17 PDU Generation Process.

- *Data link layer* is responsible for organizing data in frames and for detecting errors that occur in a frame.
- *Network layer* is responsible for routing data to its destination and for providing unique addresses in the network.
- *Transport layer* is responsible for reliable end-to-end data transfer.
- *Session layer* is responsible for establishing, maintaining, and terminating sessions between applications.
- *Presentation layer* is responsible for translating data in a form that can be understood by the receiver.
- *Application layer* is responsible for providing services to end-user applications that lie outside the scope of the OSI model. It defines procedures by which end-user applications access network services.

At each layer, the protocol for that layer creates a *protocol data unit* (PDU) for transmission that includes the header information required by that protocol and the data to be transmitted. The header is called the *protocol control information* (PCI), and is the information exchanged between entities at a given layer. The PDU of layer N is passed down to layer $N-1$ where it becomes a unit called the *service data unit* (SDU) of that layer. Layer $N-1$ adds its PCI to the SDU to create its PDU that is passed down to layer $N-2$, and so on. The concept of PDU generation is illustrated in Figure 1.17.

Figure 1.18 shows how different layers add their overheads to an application until it gets down to the physical layer where it is forwarded to the next node on its way to the destination. Each layer, with the exception of the physical layer, appends its PCI that is manifested in the form of a layer header (LH). In addition to a header, layer 2 also appends a layer trailer (LT) to create a frame that is passed on to the physical layer. The physical layer interprets the frame as a sequence of zeros and ones that need to be transmitted over the medium.

1.4.2 The Internet Architecture

Internet architecture is also layered but has only four layers (or three, if the so-called *user process* layer is not included). The architecture is shown in Figure 1.19, where it is compared to the OSI model.

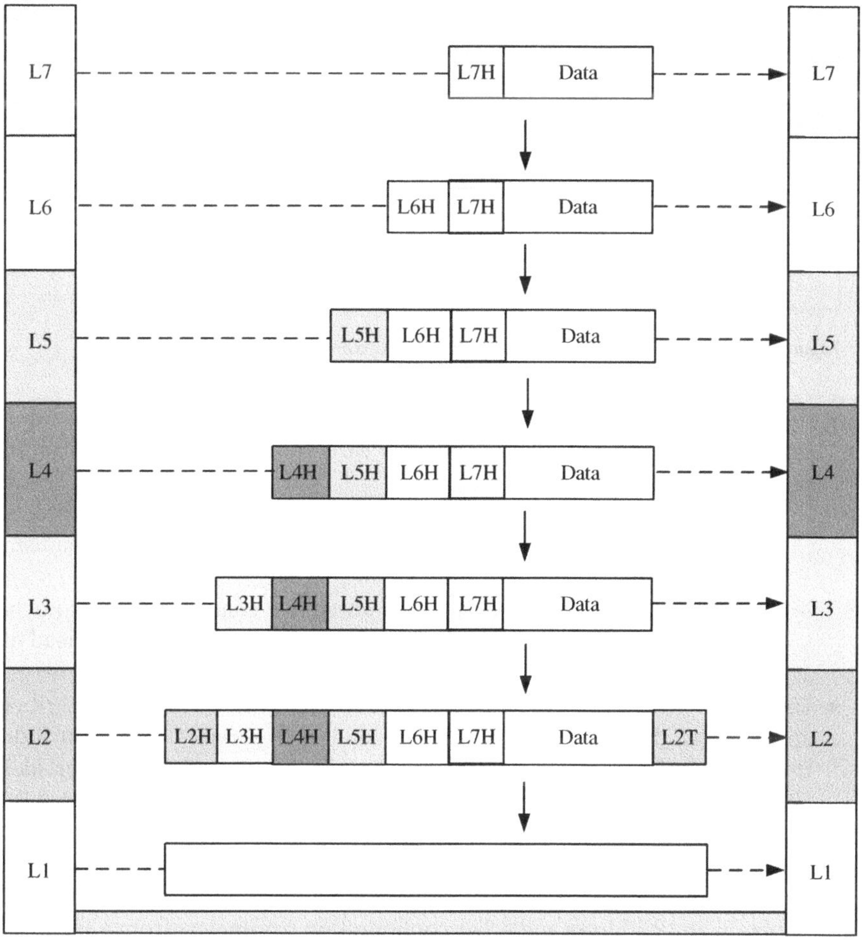

Figure 1.18 Composite Frame Construction Process.

The layers are as follows:

- *Network access layer* is concerned with the functions performed by the physical and data link layers.
  - No specific protocols are defined for this layer; it is expected that the network will rely on the data link and physical layers of the appropriate networks.
- *Internet layer* is the top part of the network layer.
  - The *Internet Protocol* (IP) defined for this layer is a simple connectionless datagram protocol that provides no error recovery and no delivery guarantee.

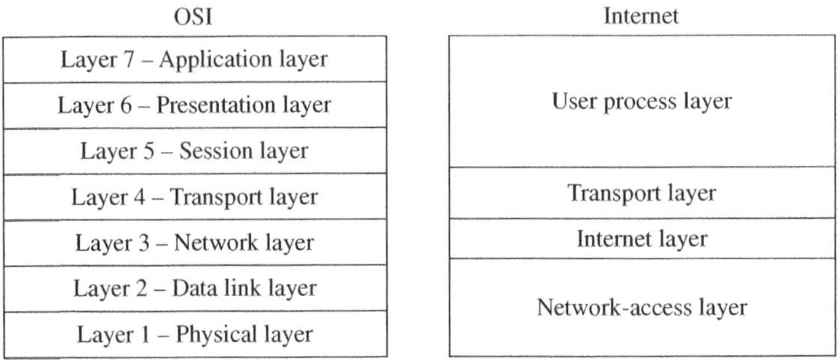

Figure 1.19 Comparison of the Internet Architecture and the OSI Model.

- *Transport layer* provides reliable data transfer between two communicating systems. Two protocols were defined in early days of data communication networks in this layer:
  - *Transmission Control Protocol* (TCP) is a connection-oriented protocol that provides reliable end-to-end transfer and uses window mechanism for flow control.
  - *User Datagram Protocol* (UDP) is an unreliable connectionless protocol designed for applications that do not need the reliability and overhead of TCP.
- *User process layer* (sometimes called *application layer*) describes applications that are used to provide end-user services. These include the file transfer protocol (FTP), a protocol that is used to transfer files between a client and a server; the simple mail transfer protocol (SMTP), a protocol that is used to send and receive emails; the hypertext transfer protocol (HTTP), a protocol that is used to surf the web; and the dynamic host configuration protocol (DHCP), a protocol that is used by a client host to obtain an IP address and other configuration parameters from a server.

1.5 Summary

Many of the basic concepts covered in this book are discussed in this chapter. More detailed discussions on these and other topics are done in the other chapters of the book.

Exercises

1 What is the difference between half-duplex and full-duplex modes of communication?

2 What does circuit switching mean?

3 What does packet switching mean?

4 Name the two types of packet switching.

5 Name two protocols used in the transport layer.

6 What is the name of the layer 6 of the OSI protocol architecture and what is its function?

7 What is the name of the layer 3 of the OSI protocol architecture and what are its functions?

8 Name three techniques used to share a transmission medium.

9 Name one advantage of hierarchical network architecture.

2

Physical Layer

2.1 Introduction

In this chapter, we consider the basic functions performed by the physical layer. In particular, we consider the following:

- Signal classification
- Fourier analysis: Fourier series and Fourier transform
- Modulation and demodulation
- Sampling theorem
- Analog-to-digital conversion
- Channel sharing schemes
- Modems
- Guided and unguided transmission media
- Channel impairments.

2.2 Classification of Signals

Message signals are classified as either analog (or continuous-time) or digital (or discrete-time). Analog (or continuous-time) signals, which include speech, audio, and video, have an infinite number of values. Digital signals are predominantly binary in nature and thus are represented by two values: 0 and 1. These two values are called *binary digits*, or *bits*.

Signals found in communication systems are complex waveforms. However, in many instances, these waveforms can be analyzed as one or more sine waves of the form:

$$c(t) = A \sin(2\pi ft + \varphi) \tag{2.1}$$

where A is the *amplitude* of the sinusoid, f the *frequency* in Hertz (Hz), and φ the *phase* in radians.

Fundamentals of Data Communication Networks, First Edition. Oliver C. Ibe.

As mentioned in Chapter 1, a transmitter is required to convert signals into a form suitable for transmission over a channel. A key operation is to change the signal's frequency range to match the frequency range, or *bandwidth* (BW), offered by the channel. In some systems, the signal's frequency content (or *frequency spectrum*) may be applied directly to a channel; this is known as *baseband* operation. In other cases, the signal's frequency components are usually different from those that the channel can carry. For these cases, the baseband frequencies must be translated to much higher frequencies by a process called *modulation*. Such signals that are located in frequency bands that are not native to them are referred to as *broadband* signals.

2.3 Periodic Signals

A periodic signal is a signal that completes a pattern within a time called a *period* and repeats that pattern over identical subsequent periods. Thus, a periodic signal repeats itself after a time interval called a period. The completion of a full pattern is called a *cycle*. Thus, a period is defined as the amount of time (in seconds) required to complete one full cycle. It can also be defined as the smallest amount of time it takes for the signal to repeat itself.

Specifically, let a signal be represented by $x(t)$. If $x(t + T) = x(t)$ for all t, then the signal is periodic with period T. Generally, the period of a signal is the smallest value of T that satisfies this condition. The *frequency*, f, of a periodic signal is the number of complete cycles that can occur per second. The frequency of a periodic signal is related to its period as follows:

$$f = \frac{1}{T} \tag{2.2}$$

Figure 2.1 is an illustration of examples of periodic signals.

2.4 Fourier Analysis of Periodic Signals

Given a signal, it is important to understand the range of frequencies that it contains because we need to know whether the medium over which it is being transmitted can pass the signal without distorting it. According to the French mathematician Jean-Baptiste Joseph Fourier, any periodic signal $g(t)$ with period T can be represented mathematically as the sum of an infinite number of sines and cosines as follows:

$$g(t) = a_0 + \sum_{n=1}^{\infty} a_n \cos(nwt) + \sum_{n=1}^{\infty} b_n \sin(nwt) \tag{2.3}$$

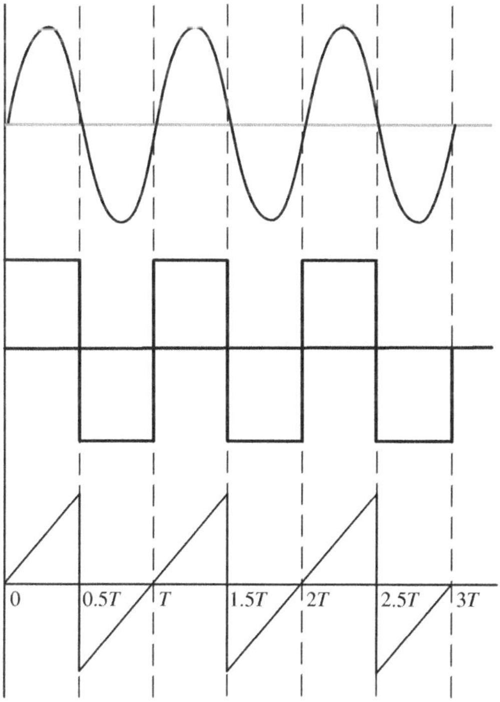

Figure 2.1 Examples of Periodic Signals.

where $w = 2\pi f = 2\pi/T, f$ is called the fundamental frequency and nf, for $n > 1$, is the nth *harmonic* and

$$a_0 = \frac{1}{T}\int_0^T g(t)dt = \frac{1}{T}\int_{-T/2}^{T/2} g(t)dt \tag{2.4}$$

$$a_n = \frac{2}{T}\int_0^T g(t)\cos(nwt)dt = \frac{2}{T}\int_{-T/2}^{T/2} g(t)\cos(nwt)dt \quad n = 1, 2, \ldots \tag{2.5}$$

$$b_n = \frac{2}{T}\int_0^T g(t)\sin(nwt)dt = \frac{2}{T}\int_{-T/2}^{T/2} g(t)\sin(nwt)dt \quad n = 1, 2, \ldots \tag{2.6}$$

Example 2.1 Consider the pulse train shown in Figure 2.2 whose frequency components we need to understand, with $T = 2$. It is required to find the parameters $a_0, a_n, b_n,\ \ n \geq 1$.

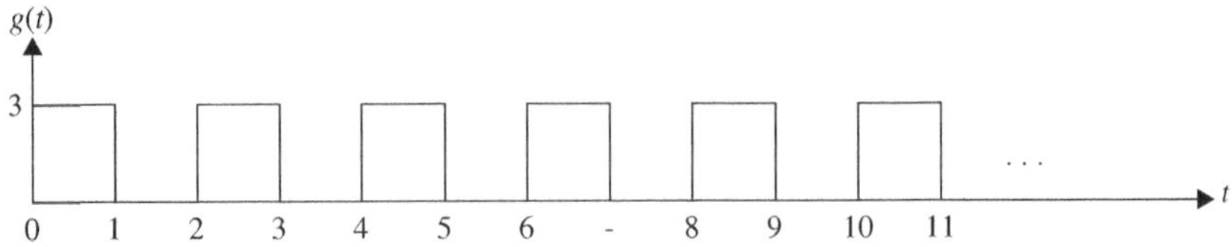

Figure 2.2 An Example.

Solution:

$$a_0 = \frac{1}{T}\int_0^T g(t)dt = \frac{1}{2}\int_0^1 3dt = \frac{3}{2}$$

$$a_n = \frac{2}{T}\int_0^T g(t)\cos(nwt)dt = \frac{2}{T}\int_0^T g(t)\cos(2n\pi ft)dt$$

$$= \frac{2}{T}\int_0^T g(t)\cos\left(\frac{2n\pi t}{T}\right)dt$$

$$= \int_0^1 3\cos(n\pi\, t)dt = \frac{3}{n\pi}\{\sin(n\pi t)\}_0^1 = \frac{3\sin(n\pi)}{n\pi} = 0$$

$$b_n = \frac{2}{T}\int_0^T g(t)\sin(nwt)dt = \frac{2}{T}\int_0^T g(t)\sin(2n\pi ft)dt$$

$$= \frac{2}{T}\int_0^T g(t)\sin\left(\frac{2n\pi t}{T}\right)dt$$

$$= \int_0^1 3\sin(n\pi\, t)dt = \frac{3}{n\pi}\{-\cos(n\pi t)\}_0^1 = \frac{3}{n\pi}\{1-(-1)^n\}$$

$$= \begin{cases} \dfrac{6}{n\pi} & n \text{ odd} \\ 0 & n \text{ even} \end{cases}$$

Thus:

$$g(t) = \frac{3}{2} + \frac{6}{\pi}\left\{\sin(wt) + \frac{1}{3}\sin(3wt) + \frac{1}{5}\sin(5wt) + \cdots\right\}$$

The importance of Fourier analysis of a signal is that it enables us to determine the frequency spectrum of the signal. Thus, if the components of the nth harmonic are very negligible, then the nth and higher harmonics of the signal can be filtered off without adversely affecting the quality of the signal.

2.4.1 Reconstructing a Function from its Fourier Series

Consider the periodic signal waveform shown in Figure 2.3.

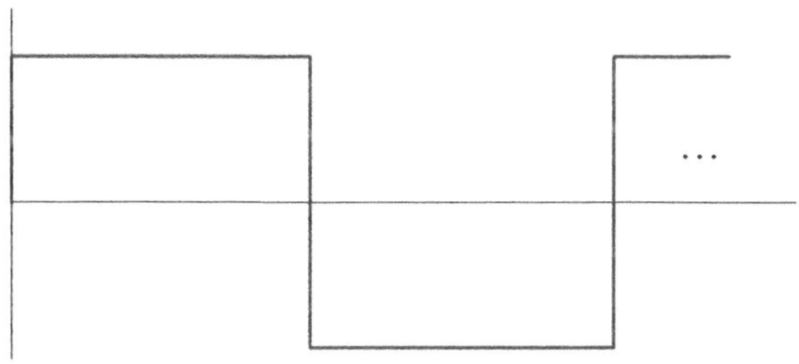

Figure 2.3 A Periodic Signal.

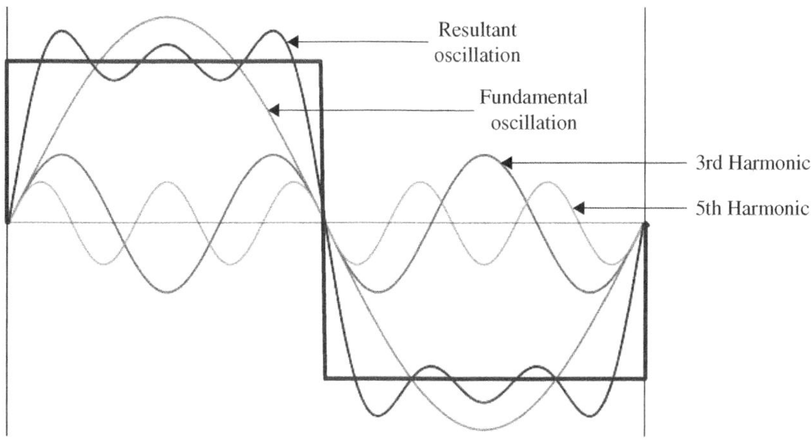

Figure 2.4 Reconstruction of Signal with Fundamental, Third, and Fifth Harmonics.

It can be shown that:

$$g(t) = \frac{4}{\pi}\left\{\sin(\omega t) + \frac{1}{3}\sin(3\omega t) + \frac{1}{5}\sin(5\omega t) + \frac{1}{7}\sin(7\omega t) + \cdots\right\}$$

Figure 2.4 shows the result of adding the signal at the fundamental frequency, the third harmonic and fifth harmonic.

From the figure, we observe that the Fourier series does not converge to the original waveform even as the number of terms increases. The approximated function shows large amounts of ripples at points of discontinuity. This is known as the *Gibbs phenomenon.*

2.4.2 Fourier Analysis of Even and Odd Functions

A function $f(x)$ is defined to be an *even function* if $f(-x) = f(x)$. For example, the function $f(x) = x^2$ is an even function because $f(-x) = (-x)^2 = x^2 = f(x)$. Similarly, a function $f(x)$ is defined to be an *odd function* if $f(-x) = -f(x)$. For example, the function $f(x) = x$ is an odd function because $f(-x) = -x = -f(x)$.

The Fourier series for odd and even functions are very interesting. Specifically:

1. If $x(t)$ is an even function, then $b_n = 0$ and the Fourier series will be the sum of cosine functions only.
2. If $x(t)$ is an odd function, then $a_n = 0$ and the Fourier series will be the sum of sine functions only.

For example, the function,

$$x(t) = \begin{cases} 1 & 0 \le t \le 1 \\ -1 & -1 \le t \le 0 \end{cases}$$

is an odd function. It can be shown that if $x(t)$ is a periodic function, then the Fourier series is given by:

$$x(t) = \frac{4}{\pi}\left\{\sin(wt) + \frac{1}{3}\sin(3wt) + \frac{1}{5}\sin(5wt) + \cdots\right\}$$

which shows that $a_n = 0$. Similarly, the function,

$$y(t) = \begin{cases} 1 & -0.5 \le t \le 0.5 \\ -1 & 0.5 \le t \le 1.5 \end{cases}$$

is an even function. It can be shown that if $y(t)$ is a periodic function the Fourier series is given by:

$$y(t) = \frac{4}{\pi}\left\{\cos(wt) - \frac{1}{3}\cos(3wt) + \frac{1}{5}\cos(5wt) - \frac{1}{7}\cos(7wt) + \cdots\right\}$$

which shows that $b_n = 0$.

2.4.3 Parseval's Theorem

Consider a signal (or function) that has a Fourier series given by:

$$g(t) = a_0 + \sum_{n=1}^{\infty} a_n \cos(nwt) + \sum_{n=1}^{\infty} b_n \sin(nwt)$$

The Parseval's theorem provides a useful way to relate the Fourier coefficients to the function they describe. Specifically, it states that

$$\frac{1}{T}\int_0^T |g(t)|^2 dt = a_0^2 + \frac{1}{2}\sum_{n=1}^{\infty}(a_n^2 + b_n^2) \tag{2.7}$$

The integral on the left-hand side is the average power of the signal. Thus, the theorem says that the sum of the squares of the Fourier coefficients is equal to the average value of $|g(t)|^2$ over one period, or the average power.

2.4.4 Complex Form of Fourier Series

We can obtain the Fourier series in a different form by noting that

$$\cos(nwt) = \frac{e^{inwt} + e^{-inwt}}{2}, \quad \sin(nwt) = \frac{e^{inwt} - e^{-inwt}}{2i}$$

Thus:

$$\begin{aligned} a_n \cos(nwt) + b_n \sin(nwt) &= \frac{a_n}{2}(e^{inwt} + e^{-inwt}) + \frac{b_n}{2i}(e^{inwt} - e^{-inwt}) \\ &= \frac{a_n}{2}(e^{inwt} + e^{-inwt}) - \frac{ib_n}{2}(e^{inwt} - e^{-inwt}) \\ &= \frac{1}{2}(a_n - ib_n)e^{inwt} + \frac{1}{2}(a_n + ib_n)e^{-inwt} \\ &= c_n e^{inwt} + c_{-n} e^{-inwt} \quad n > 0 \end{aligned}$$

This gives:

$$x(t) = \sum_{n=-\infty}^{\infty} c_n e^{inwt}, \quad c_0 = \frac{1}{2}a_0 \tag{2.8}$$

where c_n is given as:

$$c_n = \frac{1}{T}\int_0^T x(t)e^{-inwt}dt \quad w = \frac{2\pi}{T} = 2\pi f \tag{2.9}$$

2.5 Fourier Transform of Nonperiodic Signals

The Fourier analysis enables us to perform a *spectral analysis* of periodic signals, where spectral analysis deals with determining the distribution of power over frequency. For nonperiodic signals, we can obtain the spectral components using the Fourier transform. Thus, Fourier transform maps a time series into the series of frequencies (their amplitudes and phases) that make up the time series. Inverse Fourier transform maps the series of frequencies (their amplitudes and phases) back into the corresponding time series. The two functions are inverses of each other.

In communications theory the signals are usually voltages, and Fourier transform is essential to understanding how an arbitrary aperiodic signal behaves when it passes through filters, amplifiers, and communications channels.

For a signal $x(t)$, its Fourier transform, $X(w)$, is given by:

$$X(w) = \int_{-\infty}^{\infty} x(t)e^{-jwt}dt \tag{2.10}$$

Conversely, given $X(w)$ we can recover $x(t)$ by means of the inverse Fourier transform as follows:

$$x(t) = \frac{1}{2\pi}\int_{-\infty}^{\infty} X(w)e^{jwt}dw \tag{2.11}$$

Table 2.1 Fourier Transform Pairs.

$x(t)$	$X(w)$
$e^{-a\|t\|}, \quad a > 0$	$\frac{2a}{a^2 + w^2}$
$e^{-at}, \quad a > 0, \ t \geq 0$	$\frac{1}{a + jw}$
1	$2\pi\delta(w)$
$\delta(t)$	1
e^{jw_0t}	$2\pi\delta(w - w_0)$
$\begin{cases} 1 & -T/2 < t < T/2 \\ 0 & \text{otherwise} \end{cases}$	$T\frac{\sin(wT/2)}{(wT/2)}$
$x(at)$	$\frac{1}{\|a\|}X\left(\frac{w}{a}\right)$
$x(t - \tau)$	$e^{-jw\tau}X(w)$
$\cos(w_0t)$	$\pi\{\delta(w - w_0) + \delta(w + w_0)\}$
$\sin(w_0t)$	$-j\pi\{\delta(w - w_0) - \delta(w + w_0)\}$

Thus, $X(w)$ provides a complete description of $x(t)$, and vice versa. This relationship is sometimes expressed as follows: $X(w) \leftrightarrow x(t)$. Some common Fourier transform pairs are given in Table 2.1.

2.6 Filters

An electrical filter is a circuit that can be designed to modify, reshape, or reject all unwanted frequencies of an electrical signal and accept or pass only those signals wanted by the circuit designer. In other words, it "filters out" unwanted signals, and an ideal filter will separate and pass sinusoidal input signals based on their frequency. Filters are classified according to the frequency range of signals that they allow to pass through them while blocking or "attenuating" the rest. The most commonly used filter designs are as follows:

- *Low-pass filter*, which only allows low-frequency signals from 0 Hz to its upper cutoff frequency f_h Hz to pass while blocking those at higher frequencies.
- *High-pass filter*, which only allows high-frequency signals from its cutoff frequency f_l to infinity to pass through while blocking those at any lower frequency.
- *Band-pass filter*, which allows signals falling within a certain frequency band defined between two frequencies to pass through while blocking both the lower and higher frequencies on either side of this frequency band.

- *Band-stop filter*, which passes all frequencies with the exception of those within a specified range of frequencies called the stop band, which are greatly attenuated. It operates in a way that is directly opposite to the way the band-pass filter operates.

These filters are illustrated in Figure 2.5.

The circuit to realize a band-pass filter is essentially a combination of a high pass filter tuned to f_l and a low-pass filter tuned to f_h. Note the reversal; that is, high pass for the lower frequency and low pass for the higher frequency. This is illustrated in Figure 2.6.

Similarly, to obtain a band-stop filter we connect a low-pass filter and high-pass filter in parallel. The signal will take both routes, but the low-frequency components will only make it through the low-pass filter and

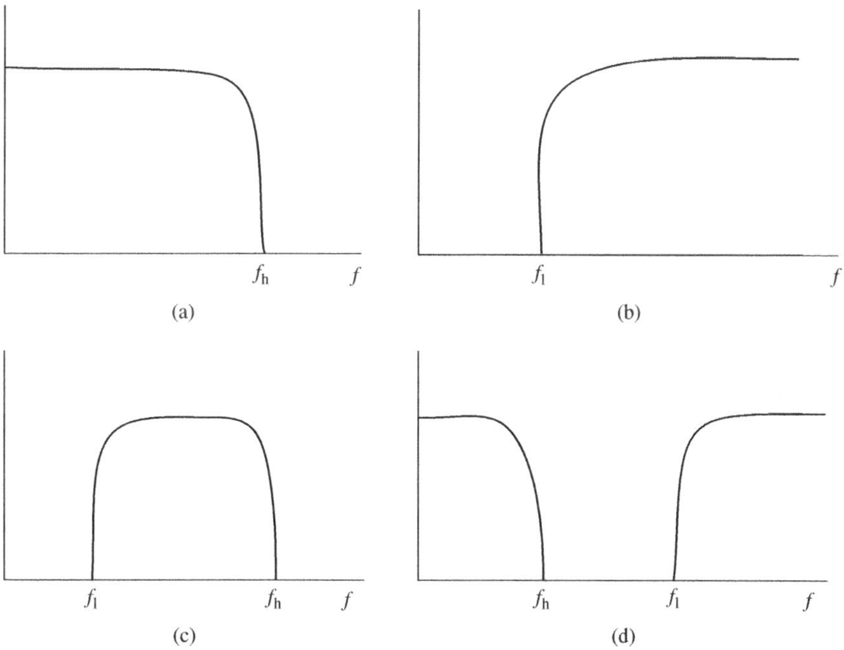

Figure 2.5 Illustration of Different Types of Filters. (a) Low-Pass Filter; (b) High-Pass Filter; (c) Band-Pass Filter; and (d) Band-Stop Filter.

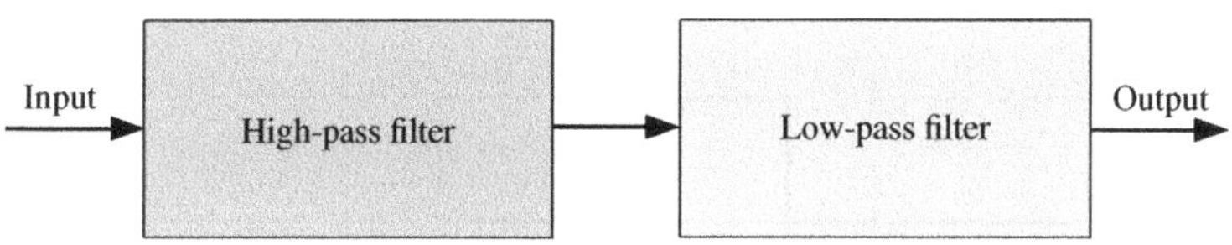

Figure 2.6 Band-pass Filter Configuration.

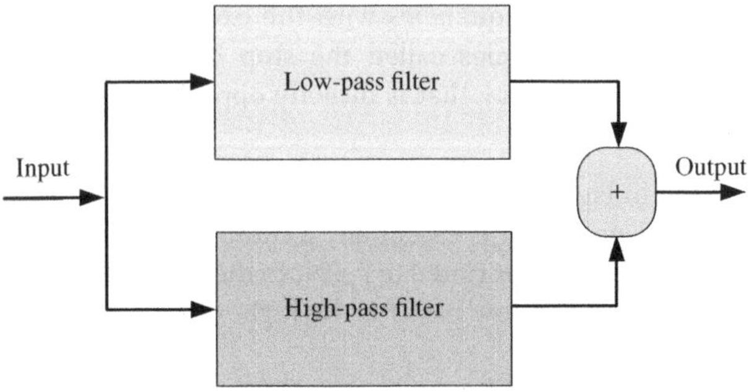

Figure 2.7 Band-stop Filter Configuration.

the high-frequency (HF) components will only make it through the high-pass filter. The cutoff frequencies of the two filters define the stop band; that is, the stop band is in the range $[f_h, f_l]$. This configuration is shown in Figure 2.7.

2.7 Line Coding

Line coding is the process of converting digital data to digital signals. It is the representation of the digital signal to be transmitted by a digital waveform. There are three primary groups of line coding schemes: *unipolar*, *polar*, and *bipolar*. Before we describe these schemes, we first define the concept of a self-clocking signal.

A self-clocking signal is one that can be decoded without the need for a separate clock signal or other source of synchronization. This is usually done by including embedded synchronization information within the signal, and adding constraints on the coding of the data payload such that false synchronization can easily be detected.

A unipolar encoding scheme involves the transmission of only a single nonzero voltage level (+V for a 1, and 0 volts for a 0). An example of a unipolar coding scheme is shown in Figure 2.8.

In the polar encoding scheme, binary 1s and 0s are represented by equal positive and negative levels. There are three basic types of polar encoding schemes:

- *Non-return-to-zero* (NRZ), in which binary 1s and 0s are represented by equal positive and negative levels. Specifically, binary ones are represented

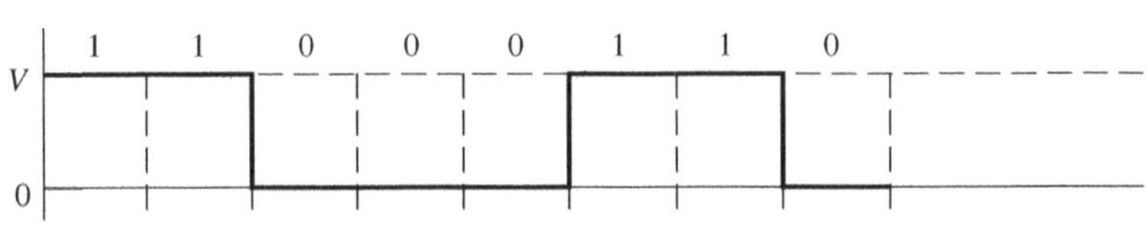

Figure 2.8 Example of Unipolar Encoding.

by a positive voltage while binary zeros are represented by a negative voltage, with no other neutral or rest condition. Thus, the binary low and high states, represented by numerals 0 and 1, are transmitted by specific and constant DC (direct-current) voltages.

- *Return-to-zero* (RZ), which uses two voltage levels for binary 1 and binary 0 (+V and –V respectively) and a rest state of zero voltage. Specifically, bit times are divided into two halves. For a binary 1, the first half of the bit time is +V and the second half is zero voltage. Similarly, for a binary 0, the first half of the bit time is –V and the second half is zero voltage. Thus, the signal state is determined by the voltage in the first half of each data bit time. The signal returns to a rest state called the *zero state* during the second half of each bit time.
- *Manchester*, which is essentially a hybrid scheme that encompasses the features of NRZ and RZ. Specifically, similar to RZ the duration of each bit is divided into two halves. Thus, there is a mandatory transition in the middle of each bit. A "one" is positive in the first half and negative in the second half – a high to low transition. Similarly, a "zero" is negative in the first half and positive in the second half – a low to high transition. Unlike RZ and like NRZ, there is no resting state.

The different polar encoding schemes are illustrated in Figure 2.9. Note that in RZ the signal is self-clocking, which means that a separate clock does not need to be sent alongside the signal but suffers from using twice the bandwidth

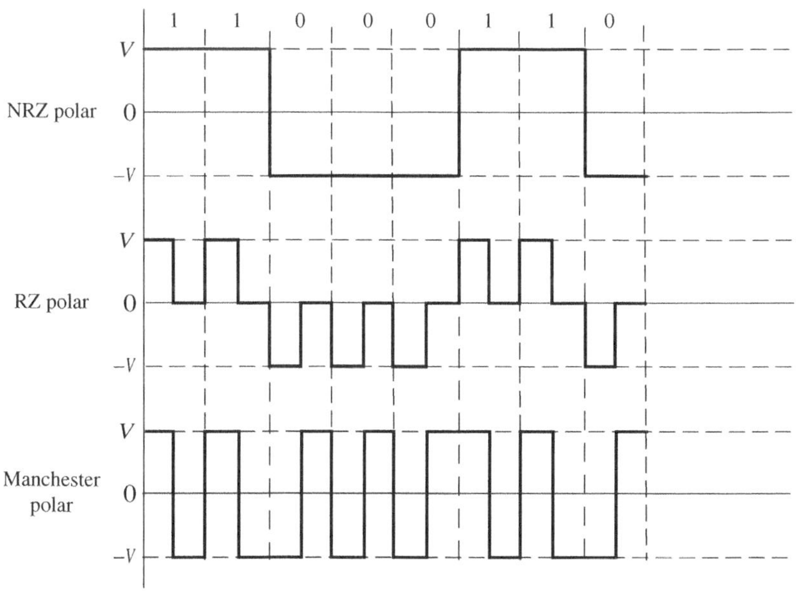

Figure 2.9 Examples of Polar Encoding Schemes.

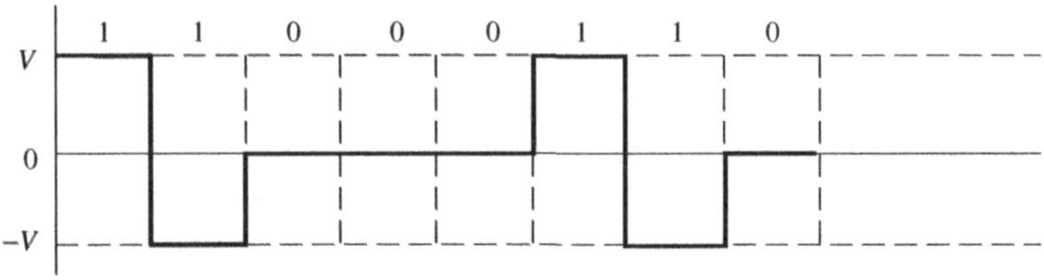

Figure 2.10 Example of Bipolar Encoding.

to achieve the same data rate as compared to NRZ format. NRZ is not inherently a self-clocking signal, so some additional synchronization technique must be used to avoid bit slips; for example, through run lengths of significant size. (A run of data is a sequence in which the same data value occurs consecutively; for example, the sequence CAAAAAB has a run of 5A's.)

In bipolar encoding, binary 1s are represented by alternating positive or negative values, and the binary 0 is represented by a zero level. Thus, bipolar encoding uses three signal levels: +V, 0, −V. This is illustrated in Figure 2.10. This type of bipolar encoding is called *alternating mark inversion* (AMI).

2.8 Modulation

The goal of any communication system is to transfer information from one place to another. This information is usually embedded in low-frequency signals that inherently cannot travel very far from the source without being heavily attenuated. In order to get low-frequency signals to longer distances, a higher frequency signal called a *carrier wave* is to "carry" them to these locations. A major issue then is how to "add" a low-frequency signal to a carrier wave. The solution is to change some characteristic of the carrier wave in sympathy with the amplitude of the low-frequency signal. The process is called *modulation.*

According to the Cambridge Dictionary, to modulate means to "vary the strength or amount of something." Thus, to modulate is simply to "change." In telecommunication, the purpose of modulation is to shift an information bearing signal that is centered around a very low frequency to a signal that is centered about a much higher frequency. Modulation is extremely necessary in communication systems for the following reasons:

(i) *Practical antenna length*: From the theory of radio propagation, it is known that in order to transmit a wave effectively, the length of the transmitting antenna should be approximately equal to the wavelength of the wave. We know that the product of the wavelength of a signal and the frequency of the signal is equal to the velocity of propagation in the transmission medium. Thus, for any signal that is to be transmitted over the air at a given frequency f, we have that

$$\text{Wavelength} = \frac{\text{velocity}}{\text{frequency}} = \frac{3 \times 10^8}{f(\text{Hz})}\ \text{m}$$

Audio frequencies range from 20 Hz to 20 kHz. This means that if they are transmitted directly into space, the length of the transmitting antenna required would be extremely large. For example, to transmit a frequency of 20 kHz directly into space, we would need an antenna whose length is on the order of

$$\frac{3 \times 10^8}{20 \times 10^3} = 1.5 \times 10^4 = 15{,}000\ \text{m}$$

This would not be a practically feasible antenna. On the other hand, if we "change" the frequency to 10 MHz, we will need an antenna length on the order of 30 m, which can be easily constructed. If we "change" the frequency to 100 MHz, we will need an antenna length on the order of 3 m. If we "change" the frequency to the gigahertz range (i.e., on the order of 10^9 Hz), we will need antenna lengths on the order of centimeters. Thus, as the frequency increases, the length of the antenna becomes smaller.

(ii) *Operating range*: The energy of a signal depends on its frequency: the greater the frequency of the signal, the greater the energy it possesses. Because audio signal frequencies are small, these signals cannot be transmitted over long distances. The only practical solution is to modulate a high-frequency carrier wave with an audio signal, which permits the transmission to occur at the high frequency of the carrier wave.

(iii) *Wireless communication*: Radio transmission is a wireless process. As we noted earlier, at audio frequencies, wireless transmission is not practicable because not only is the efficiency of the transmission poor, but also it requires antennas that are impractical to construct. Thus, for wireless communication, efficient transmission of electrical energy and practical antennas are possible only at high frequencies. For this reason, all radio communication systems involve modulation.

(iv) *Mutual interference*: If all audio signals from different sources are transmitted as baseband signals, they would all interfere with each other since they occupy the same frequency band. If modulation is done, each signal will occupy a different frequency band and all the signals can be transmitted simultaneously without any interference.

At the source, the goal is to impress the signal to be transmitted on a carrier wave in such a way that it can be recovered at the destination. As stated earlier, the process of impressing the signal on a carrier wave is known as modulation. After the signal has been impressed on the carrier wave, the wave becomes a *modulated carrier*. At the receiver at the destination, the modulated carrier wave is *demodulated* to extract the original baseband signal.

A large number of radio transmitters usually transmit at the same time. For a receiver to pick up only one wanted signal and to reject the rest, it is necessary to assign a carrier with a known frequency to each transmitter, modulate this carrier with the signal, and then design the receiver to pick up only that known carrier frequency and reject the rest, using appropriate filtering methods. The same concept is used in carrying a large number of telephone conversations over a single pair of wires or optical fiber.

Consider the signal:

$$c(t) = a \ \sin(2\pi ft + \varphi)$$

where

a = amplitude of the carrier
f = frequency of the carrier
φ = phase of the carrier

The quantity $2\pi f$ is usually defined as the *angular frequency* and denoted by w. Thus, we may also write:

$$c(t) = a \sin(wt + \varphi)$$

In order to modulate a carrier wave, we have to change one or more of the three basic characteristics of the wave, namely:

- altering the amplitude of the carrier sine wave,
- altering the frequency of the carrier sine wave, and
- altering the phase of the carrier sine wave.

These three methods of modulating the carrier wave, respectively, lead to the following three basic modulation schemes:

- Amplitude modulation (AM)
- Frequency modulation (FM)
- Phase modulation (PM).

Frequency modulation and phase modulation are also together called *angle modulation.* A signal that alters one or more of these characteristics of the carrier is the information that is being transmitted, which is called the *modulating signal.*

2.8.1 Trigonometric Refresher Course

$$\cos(A - B) = \cos A \cos B + \sin A \sin B \tag{2.12}$$

$$\cos(A + B) = \cos A \cos B - \sin A \sin B \tag{2.13}$$

Adding the two equations, we obtain the following identity:

$$\cos A \cos B = \frac{1}{2}\{\cos(A - B) + \cos(A + B)\}$$

Subtracting Eq. (2.13) from Eq. (2.12), we obtain the following identity:

$$\sin A \sin B = \frac{1}{2}\{\cos(A - B) - \cos(A + B)\}$$

When $A = B$, we obtain:

$$\cos(2A) = \cos^2 A - \sin^2 A \qquad (2.14)$$
$$1 = \cos^2 A + \sin^2 A \qquad (2.15)$$

which are well-known trigonometric identities.

2.8.2 Amplitude Modulation

In amplitude modulation, the information to be transmitted is used to vary the amplitude of the carrier. Amplitude modulation is implemented by mixing the carrier wave in a nonlinear device with the modulating signal. This produces upper and lower sidebands, which are the sum and difference frequencies of the carrier wave and modulating signal. Let the carrier signal be represented by:

$$c(t) = a_c \cos(w_c t)$$

Let the modulating signal be represented by:

$$m(t) = a_m \cos(w_m t)$$

The amplitude modulation process is implemented as shown in Figure 2.11, where the modulating signal is mixed with the carrier wave in a multiplier circuit, and the carrier wave is later added to the mixed signal.

Thus, the modulated carrier is given by:

$$\begin{aligned} s(t) &= m(t)c(t) + c(t) = c(t)\{1 + m(t)\} \\ &= a_c \cos(w_c t)\{1 + a_m \cos(w_m t)\} = a_c \cos(w_c t)\{1 + \mu a_c \cos(w_m t)\} \\ &= a_c \cos(w_c t) + \frac{a_c^2 \mu}{2}\{\cos(w_c - w_m) + \cos(w_c + w_m)\} \end{aligned}$$

$$\mu = \frac{a_m}{a_c} \equiv \text{modulation index} \qquad (2.16)$$

Multiplying the modulation index by 100 gives the *percentage of modulation*. Thus, we observe a unique property of AM, which is that the *envelope* of the

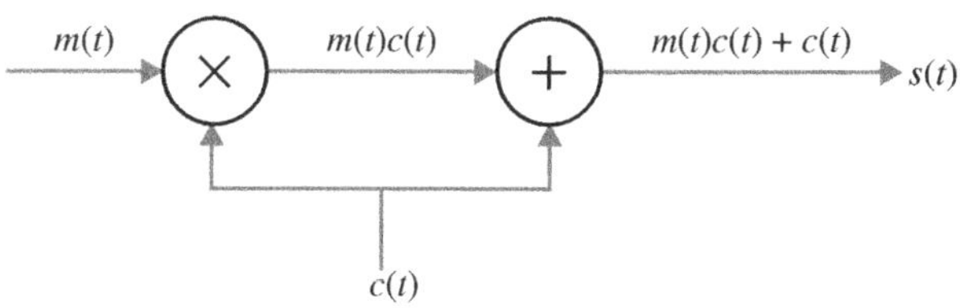

Figure 2.11 The Modulation Process.

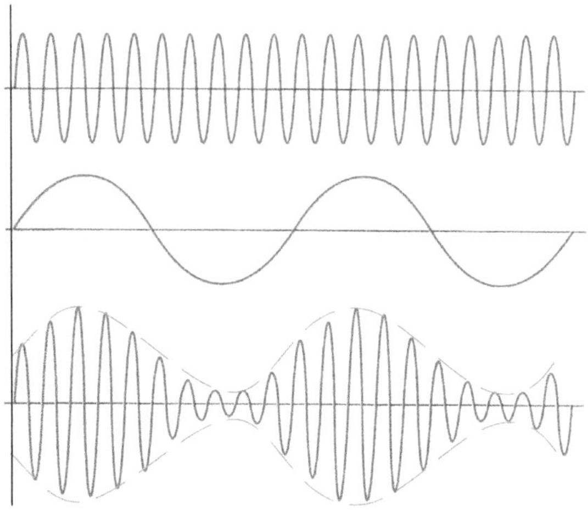

Figure 2.12 Illustration of Amplitude Modulation Process.

modulated carrier has the same shape as the modulating signal. Figure 2.12 illustrates the shape of the amplitude modulated carrier.

We can also compute the modulation index as follows. The largest and smallest values of the modulated wave are given, respectively, by:

$$a_{max} = a_c + a_m$$
$$a_{min} = a_c - a_m$$

From these we obtain the amplitudes of the carrier signal and the modulating signal as follows:

$$2a_c = a_{max} + a_{min} \Rightarrow a_c = \frac{1}{2}\{a_{max} + a_{min}\}$$
$$2a_m = a_{max} - a_{min} \Rightarrow a_m = \frac{1}{2}\{a_{max} - a_{min}\}$$

Thus, the modulation index is given by:

$$\mu = \frac{a_m}{a_c} = \frac{\frac{1}{2}\{a_{max} - a_{min}\}}{\frac{1}{2}\{a_{max} + a_{min}\}} = \frac{a_{max} - a_{min}}{a_{max} + a_{min}} \tag{2.17}$$

In the frequency domain, the modulated carrier has three frequency components:

- The lower-sideband (LSB) frequency $(f_c - f_m)$
- The carrier frequency f_c
- The upper-sideband (USB) frequency $(f_c + f_m)$.

The bandwidth (BW) of the modulated wave is equal to the difference between the upper sideband frequency and lower sideband frequency. That is,

$$BW = (f_c + f_m) - (f_c - f_m) = 2f_m \tag{2.18}$$

We demodulate the signal at the receiving end by multiplying the modulated carrier by the unmodulated carrier, as follows:

$$\begin{aligned}
s(t)c(t) &= a_c \cos\left\{2\pi f_c t\right\} a_c \cos(2\pi f_c t) \\
&+ \frac{\mu a_c^2}{2} \cos\left\{2\pi(f_c - f_m)t\right\} a_c \cos(2\pi f_c t) \\
&+ \frac{\mu a_c^2}{2} \cos\left\{2\pi(f_c + f_m)t\right\} a_c \cos(2\pi f_c t) \\
&= \frac{a_c^2}{2}\left[1 + \cos\left\{4\pi f_c t\right\}\right] + \frac{\mu a_c^3}{4}\left[\cos\left\{-2\pi f_m t\right\} + \cos\left\{2\pi(2f_c - f_m)t\right\}\right] \\
&+ \frac{\mu a_c^3}{4}\left[\cos\left\{2\pi f_m t\right\} + \cos\left\{2\pi(2f_c + f_m)t\right\}\right] \\
&= \frac{\mu a_c^3}{2} \cos\left\{2\pi f_m t\right\} + \frac{a_c^2}{2}\left[1 + \cos\left\{4\pi f_c t\right\}\right] \\
&+ \frac{\mu a_c^3}{4}\left[\cos\left\{2\pi(2f_c - f_m)t\right\} + \cos\left\{2\pi(2f_c + f_m)t\right\}\right]
\end{aligned}$$

Thus, the frequency components that are present after the demodulation process are f_m, $2f_c$, $2f_c - f_m$, and $2f_c + f_m$. Since we have chosen $f_c >> f_m$, it is obvious that the other three frequency components are much greater than f_m. This means that we can use a low-pass filter to recover the modulating signal. Figure 2.13 is a summary of the amplitude modulation and demodulation process.

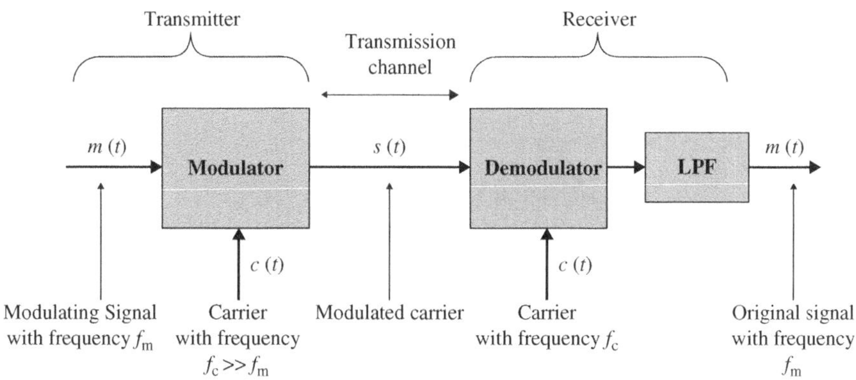

Figure 2.13 Summary of the Modulation and Demodulation Process.

2.8.2.1 Overmodulation and Distortion

The modulation index should be a number between 0 and 1. If the amplitude of the modulating voltage is greater than the carrier voltage, the modulation index, μ, will be greater than 1, causing distortion of the modulated waveform. If the distortion is great enough, the intelligence signal becomes unintelligible. The modulated carrier in Figure 2.12 is said to be *undermodulated* and, therefore, does not suffer any signal distortion; in this case, $\mu < 1$. This is similar to Figure 2.14(a). In Figure 2.14 (b), the modulated carrier is said to be *perfectly* (or *critically*) modulated because $\mu = 1$. Finally, in Figure 2.14(c), the carrier is said to be *overmodulated*, which means that $\mu > 1$. As can be seen from Figure 2.14(c), the lower part of the envelope overshoots the origin, and the upper part of the envelope overshoots the origin, thereby causing the signal to be distorted.

2.8.2.2 Single-Sideband Suppressed-Carrier Amplitude Modulation

In radio transmission, the AM signal is amplified by a power amplifier and fed to the antenna with a characteristic impedance that is ideally, but not necessarily, almost pure resistance. The AM signal is really a composite of several signal voltages, namely, the carrier and the two sidebands, and each of these signals produces power in the antenna. The total transmitted power P_T is simply the sum of the carrier power P_c and the power in the two sidebands P_{USB} and P_{LSB}:

$$P_T = P_c + P_{LSB} + P_{USB}$$

To see how the power in an AM signal is distributed, we use V_c and V_m as the peak voltage values of the carrier and modulating cosine waves. Thus, the

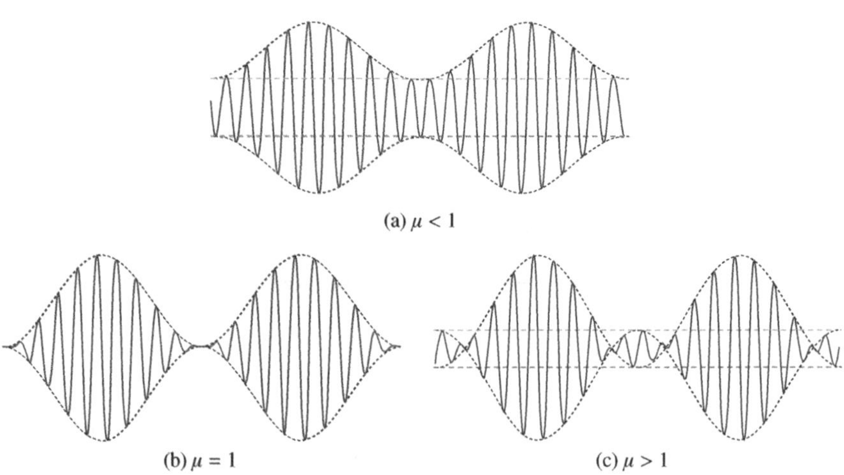

Figure 2.14 Illustrations of Undermodulation, Critical (or Perfect) Modulation, and Overmodulation.

modulated carrier wave is given by:

$$v_{AM} = V_c \cos(2\pi f_c t) + \frac{1}{2} V_m \cos\{2\pi(f_c - f_m)t\} + \frac{1}{2} V_m \cos\{2\pi(f_c + f_m)t\}$$

From our discussion, we recall that the second and third terms on the right-hand side are the lower and upper sidebands, respectively. For power calculations, we use the root mean square (RMS) values for the voltages. We can convert from peak values to RMS values by multiplying the peak value by $\frac{1}{\sqrt{2}} = 0.707$. Thus, the modulated carrier wave is given in terms of RMS values as follows:

$$v_{AM} = \frac{V_c}{\sqrt{2}} \sin(2\pi f_c t) + \frac{V_m}{2\sqrt{2}} \cos\{2\pi(f_c - f_m)t\} + \frac{V_m}{2\sqrt{2}} \cos\{2\pi(f_c + f_m)t\}$$

It is well known that the output power P when V is the output RMS voltage and R is the resistive part of the impedance of the antenna is given by $P = V^2/R$. Thus, the power in the AM wave is given by:

$$P_T = \frac{\left(V_c/\sqrt{2}\right)^2}{R} + \frac{\left(V_m/2\sqrt{2}\right)^2}{R} + \frac{\left(V_m/2\sqrt{2}\right)^2}{R} = \frac{V_c^2}{2R} + \frac{V_m^2}{8R} + \frac{V_m^2}{8R}$$

Now, we know that $V_m = \mu V_c$, where μ is the modulation index. Thus, we obtain:

$$P_T = \frac{V_c^2}{2R} + \frac{V_m^2}{8R} + \frac{V_m^2}{8R} = \frac{V_c^2}{2R} + \frac{\mu^2 V_c^2}{8R} + \frac{\mu^2 V_c^2}{8R} = \frac{V_c^2}{2R}\left\{1 + \frac{\mu^2}{4} + \frac{\mu^2}{4}\right\}$$

$$= P_c\left\{1 + \frac{\mu^2}{2}\right\} \tag{2.19}$$

In practice, we want to avoid overmodulation, which means that we operate with $\mu \le 1$. This means that the carrier power, which contains no information, constitutes two-thirds or more of the total transmitted power. This is a major drawback in conventional AM.

Another problem in conventional AM is that it uses twice the bandwidth (upper and lower side are identical). Figure 2.15 shows the frequency components of the modulated carrier and their relative amplitudes.

With single sideband with suppressed carrier, we filter out the carrier and one of the sidebands. Assume that we filter out the USB and thus the transmitted modulated carrier is:

$$s_{SC}(t) = \frac{\mu a_c}{2} \cos\{2\pi(f_c - f_m)t\}$$

The generation of the SSB is illustrated in Figure 2.16.

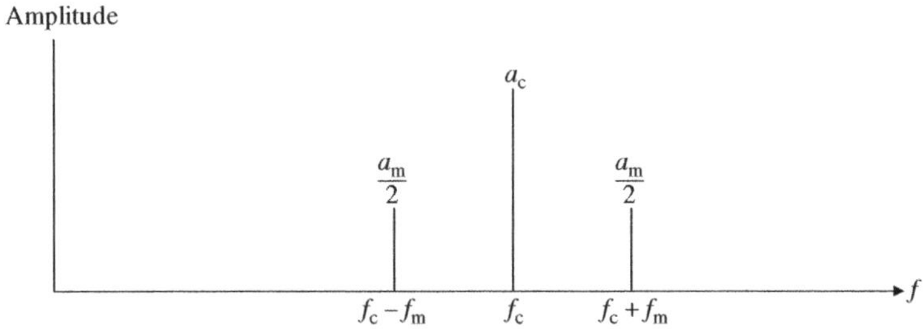

Figure 2.15 Frequency Components of the Modulated Carrier.

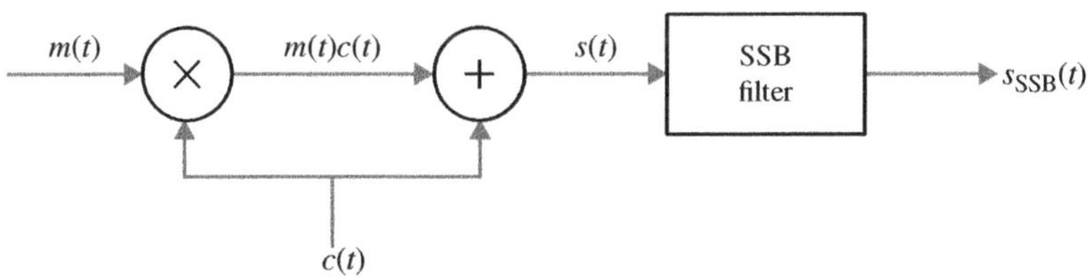

Figure 2.16 Realization of the SSB.

At the receiver, we multiply the received signal by the carrier signal to obtain:

$$\begin{aligned} s_{SC}(t)c(t) &= \frac{\mu a_c}{2}\cos\{2\pi(f_c - f_m)t\}a_c\cos(2\pi f_c t) \\ &= \frac{\mu a_c^2}{4}\left[\cos\{-2\pi f_m t\} + \cos\{2\pi(2f_c - f_m)t\}\right] \\ &= \frac{\mu a_c^2}{4}\left[\cos\{2\pi f_m \mathrm{t}\} + \cos\{2\pi(2f_c - f_m)t\}\right] \end{aligned}$$

Thus, we can use a low-pass filter to recover the modulating signal.

2.8.3 Frequency Modulation

As previously stated frequency modulation and phase modulation are referred to as *angle modulation*. Consider the signal,

$$c(t) = a_c\cos\{\theta_i(t)\}$$

a_c is the carrier amplitude and $\theta_i(t)$ is the phase angle. The instantaneous frequency is given by:

$$f_i(t) = \frac{1}{2\pi}\frac{d\theta_i(t)}{dt}$$

Thus, given the instantaneous frequency we can obtain $\theta_i(t)$ as follows:

$$\theta_i(t) = 2\pi\int_0^t f_i(\tau)d\tau$$

In frequency modulation, we let the instantaneous frequency of the carrier wave vary in sympathy with the modulating signal as follows:

$$f_i(t) = f_c + k_f m(t) \Rightarrow \theta_i(t) = 2\pi f_c t + 2\pi k_f \int_0^t m(\tau)d\tau$$

k_f is called the *frequency sensitivity* or the *frequency deviation* and represents the maximum shift of $f_i(t)$ relative to f_c. Thus, the frequency-modulated (FM) signal is given by:

$$s_{FM}(t) = a_c \cos\{\theta_i(t)\} = a_c \cos\left\{ 2\pi f_c t + 2\pi k_f \int_0^t m(\tau)d\tau \right\} \quad (2.20)$$

Figure 2.17 is an illustration of the frequency modulation process. As the amplitude of the modulating signal increases the frequency of the carrier increases also, and as the amplitude of the modulating signal decreases the frequency of the carrier decreases.

An FM signal is called a *constant envelope signal* because the amplitude of the carrier signal is kept constant while the frequency of the carrier is changed according to the amplitude of the modulating message signal. Frequency modulation is the most popular analog modulation technique. Because an FM signal is a constant envelope signal, it has a better noise immunity than AM. This is why FM radios have a better performance than AM radios.

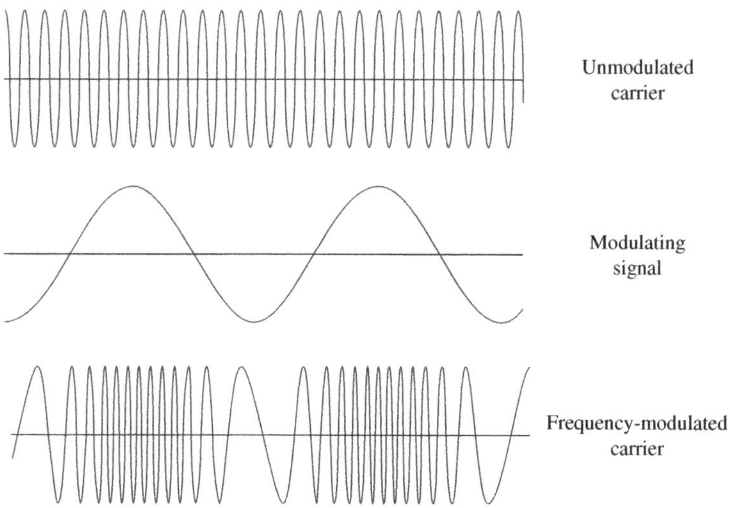

Figure 2.17 Illustration of the Frequency Modulation Process.

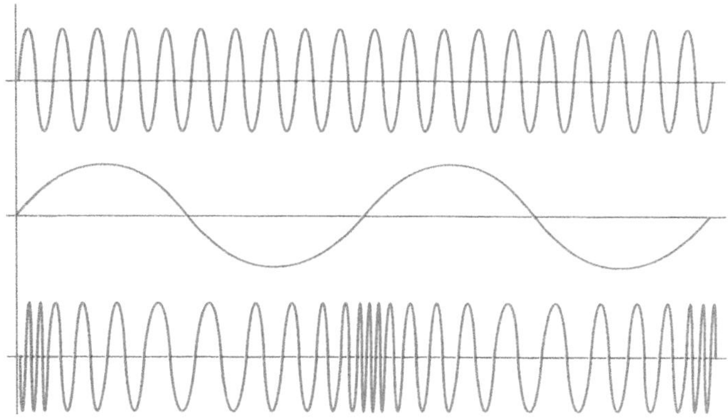

Figure 2.18 Illustration of the Phase Modulation Process.

2.8.4 Phase Modulation

As stated earlier, phase modulation and frequency modulation are called angle modulation. In phase modulation, the phase angle varies with the instantaneous value of the amplitude of the modulating signal; that is,

$$
\begin{aligned}
c(t) &= a_c \cos\{2\pi f_c t + \varphi(t)\} \\
\varphi(t) &= k_p m(t) \\
s_{PM}(t) &= a_c \cos\{2\pi f_c t + k_p m(t)\} \\
k_p &= \text{phase sensitivity or phase deviation}, \quad k_p \leq 180^\circ
\end{aligned} \tag{2.21}
$$

As in the case of frequency modulation, a phase-modulated signal is a constant envelope signal that is immune to noise. However, phase modulation requires a more complex demodulation scheme. As illustrated in Figure 2.18, when the signal is making a positive phase transition, the frequency of the signal increases; and when it makes a negative phase transition, the frequency of the signal decreases. So, while FM and PM are similar in the sense that the frequency of the carrier changes with the state of the signal, changes associated with FM are connected with the amplitude of the modulating signal while changes associated with PM are connected with the transitions from positive to negative values and vice versa.

2.9 Sampling Theorem

Sampling is a method of converting continuous-time signals into discrete-time signals. It is widely used in the analysis of discrete-time systems. A common way to represent the sampling of a continuous-time signal at regular intervals

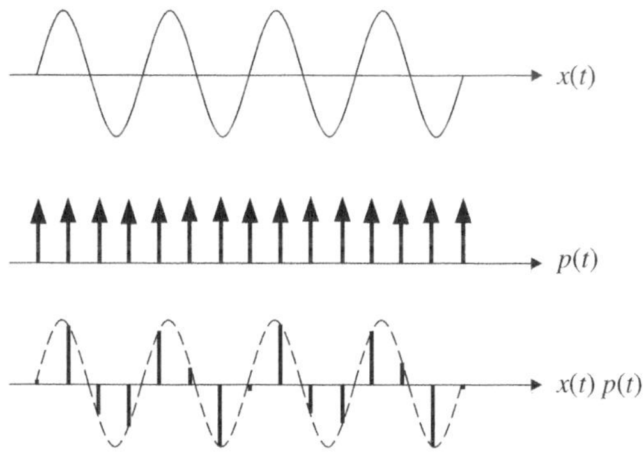

Figure 2.19 The Sampling Process.

is through the use of a periodic *impulse train* signal, $p(t)$, multiplied by the continuous-time signal:

$$x_p(t) = x(t)p(t)$$

$$p(t) = \sum_{n=-\infty}^{\infty} \delta(t - nT) \tag{2.22}$$

T is the sampling period and $w_S = 2\pi/T$ is the sampling frequency; alternatively, $f_S = 1/T$. This is known as *impulse train sampling*. Note $x_p(t)$ is still a continuous time signal. It can be observed that sampling is a form of amplitude modulation where the pulse train is the carrier wave and the signal to be sampled is the modulating signal. The sampling process is illustrated in Figure 2.19.

2.9.1 Analyzing Impulse Train Sampling

Let us consider the effect that this sampling has on the frequency decomposition (Fourier transform) of the impulse train signal $x_p(t)$. By definition,

$$x_p(t) = x(t)p(t) = \sum_{n=-\infty}^{\infty} x(t)\delta(t - nT) = \sum_{n=-\infty}^{\infty} x(nT)\delta(t - nT)$$

The signal $p(t)$ is periodic and its Fourier transform is given by:

$$P(\omega) = \frac{2\pi}{T} \sum_{k=-\infty}^{\infty} \delta(\omega - k\omega_s)$$

One property of the Fourier transform is that if $x_p(t) = x(t)p(t)$, then:

$$X_p(\omega) = \frac{1}{2\pi} \int_{-\infty}^{\infty} X(\theta)P(\omega - \theta)d\theta$$

That is, the Fourier transform of the product of two functions is the convolution of the Fourier transforms of the two functions. Thus, $X_p(w)$ is the

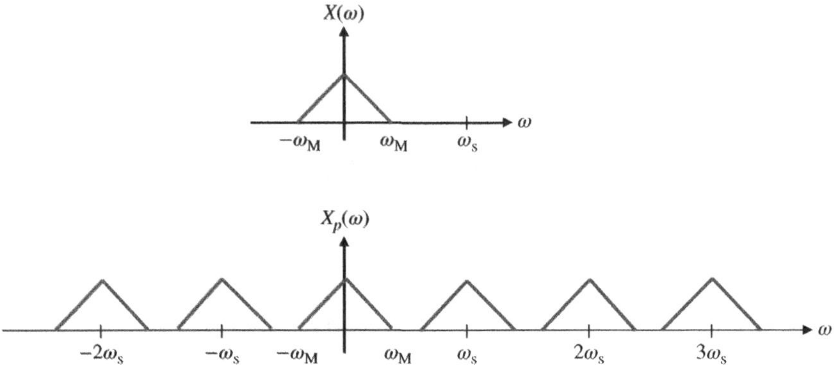

Figure 2.20 Spectral Profile of $X_p(w)$.

convolution of $X(w)$ and $P(w)$ scaled by $1/2\pi$, which is usually denoted by:

$$X_p(\omega) = \frac{1}{2\pi} X(w) * P(\omega)$$

Substituting for $P(w)$, we obtain the following:

$$\begin{aligned} X_p(\omega) &= \frac{1}{T}\int_{-\infty}^{\infty} X(\theta) \sum_{k=-\infty}^{\infty} \delta((\omega - k\omega_s) - \theta)d\theta \\ &= \frac{1}{T}\sum_{k=-\infty}^{\infty} \int_{-\infty}^{\infty} X(\theta)\delta((\omega - k\omega_s) - \theta)d\theta \\ &= \frac{1}{T}\sum_{k=-\infty}^{\infty} X(\omega - k\omega_s) \end{aligned} \tag{2.23}$$

Therefore, $X_p(w)$ is a periodic function of w, consisting of a superposition of shifted replicas of $X(w)$, scaled by $1/T$. This is illustrated in Figure 2.20.

2.9.2 Reconstruction of the Continuous-Time Signal

When the sampling frequency w_S is greater than twice the band-limited frequency w_M, there is no overlap of the spectrum $X(w)$. If this is true, the original signal $x(t)$ can be recovered from $x_p(t)$ by passing the latter through a low-pass filter. This is illustrated in Figure 2.21, where the signal is said to be *oversampled*.

When the sampling frequency w_S is less than twice the band-limited frequency w_M, there is an overlap of the replicas in the spectrum of $X(w)$. In this case, the original signal $x(t)$ cannot be accurately recovered from the impulse sampled $x_p(t)$ because there is distortion caused by interference from adjacent copies of the signal. This is illustrated in Figure 2.22, where the signal is said to be *undersampled*.

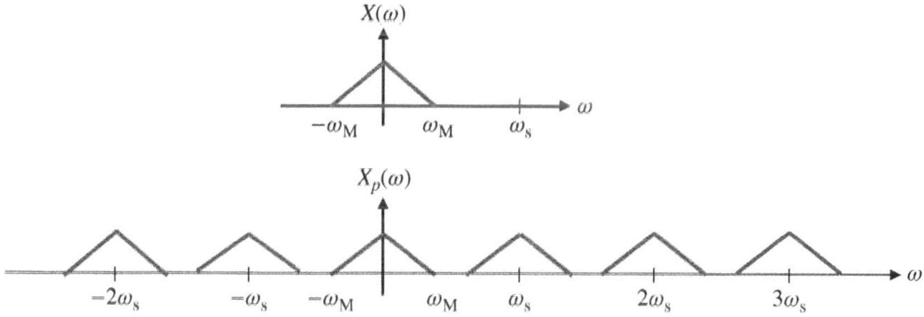

Figure 2.21 Illustration of Oversampling.

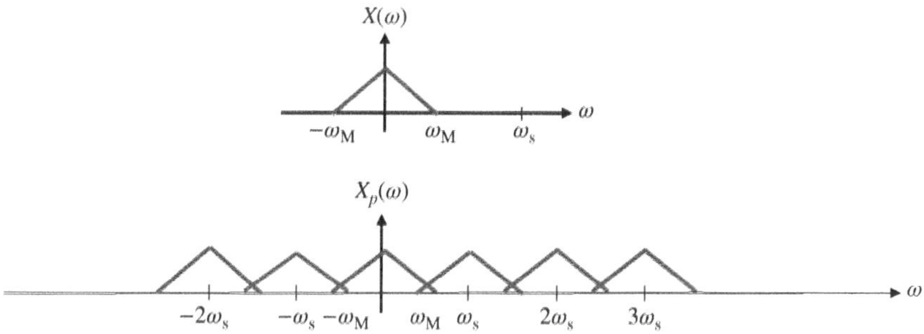

Figure 2.22 Illustration of Undersampling.

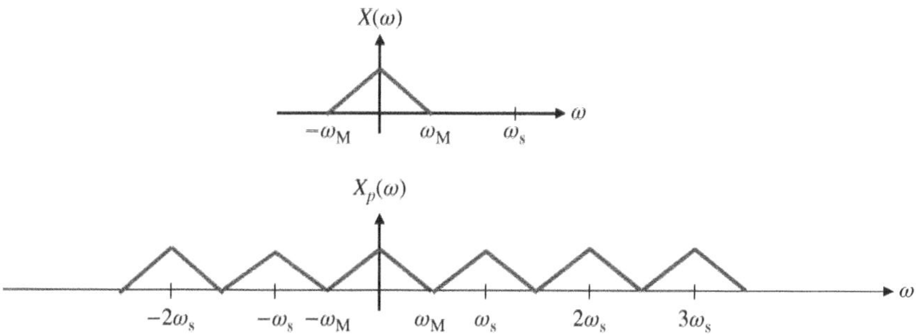

Figure 2.23 Illustration of Perfect Sampling.

When the sampling frequency w_S is exactly twice the band-limited frequency w_M, the signals are just touching each other. In this case, the original signal $x(t)$ can still be recovered from the impulse sampled $x_p(t)$, by passing it through a low-pass filter. This is the case in Figure 2.23, where the signal is said to be *critically* or *perfectly sampled*; that is, where one element ends the other begins with no overlap.

2.9.3 Statement of the Sampling Theorem

Let $x(t)$ be a band-limited (or frequency-limited) signal, which means that $X(w) = 0$ for $|w| > w_M$. Then, $x(t)$ can be uniquely determined by its samples $\{x(nT)\}$ when the sampling frequency satisfies the condition $w_S \geq 2w_M$, where $w_S = 2\pi f_S = 2\pi/T$. The quantity $2w_M$ is called the *Nyquist rate*; it represents the smallest frequency at which the signal can be sampled and be reproduced from its samples. When $w_S < 2w_M$, we obtain what is called ***aliasing***, which means that the reproduced signal takes on an identity that is different from that of the original signal.

In summary, the sampling theorem states that a continuous-time band-limited signal can be represented perfectly by its samples at uniform intervals of T seconds if $T = 1/f_S$ is small enough, which also means that $w_S = 2\pi f_S$ is large enough. The minimum sampling rate for the signal to be perfectly recovered from its samples is twice the highest frequency of the signal and is called the Nyquist rate.

For example, voice has a frequency range 300–3400 Hz, but we usually round the upper range to 4000 Hz. Thus, according to the sampling theorem, the smallest sampling rate (i.e., the Nyquist rate) for voice is $2 \times 4000 = 8000$ samples/second.

2.9.4 Proof of the Sampling Theorem

Let $\delta_T(t)$ be a periodic impulse train defined by:

$$\delta_T(t) = \sum_{n=\infty}^{\infty} \delta(t - nT)$$

The Fourier transform of $\delta_T(t)$ is given by:

$$\Im\{\delta_T(t)\} = w_0 \sum_{k=\infty}^{\infty} \delta(w - kw_0) = \frac{2\pi}{T} \sum_{k=\infty}^{\infty} \delta\left(w - \frac{2\pi k}{T}\right)$$

The Fourier transform pair is illustrated in Figure 2.24.

Consider a band-limited signal $m(t)$ whose bandwidth is f_m; let $w_m = 2f_m$. The signal $m(t)$ and the spectrum of its Fourier transform $M(w)$ are shown in Figure 2.25.

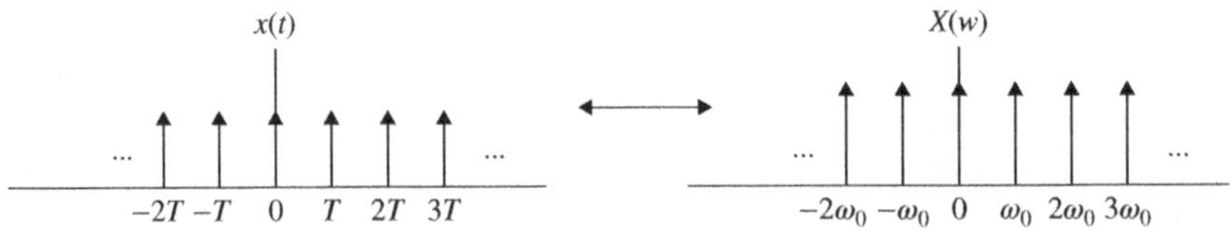

Figure 2.24 The Fourier Transform Pair of Impulse Train.

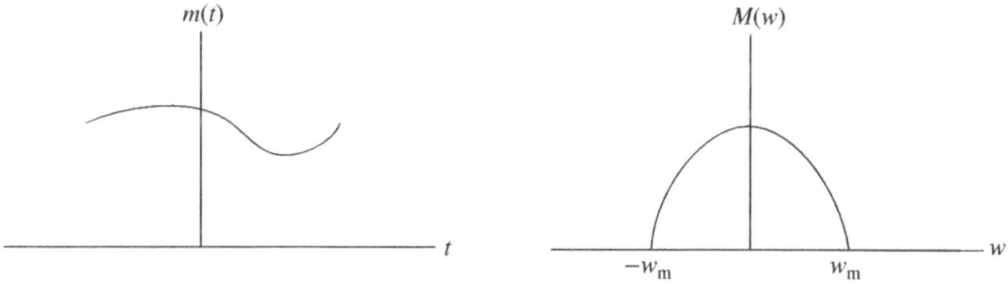

Figure 2.25 The Fourier Transform Pair of Arbitrary Band-limited Signal.

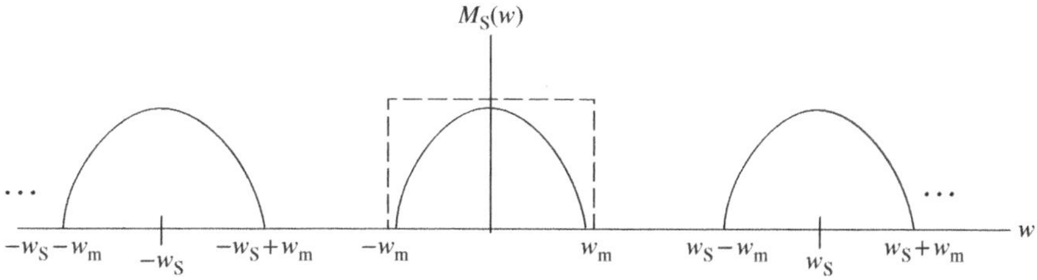

Figure 2.26 Separation of Signal Copies that Can be Perfectly Recovered.

Let $\delta_T(t)$ be the sampling signal for $m(t)$ with the sampling frequency $f_S = 1/T$. Let the signal $m_S(t)$ be the sampled signal. Then the Fourier transform $M_S(w)$ of $m_S(t)$ is given by:

$$m_s(t) = m(t)\delta_T(t)$$

$$M_s(w) = \Im\{m(t)\delta_T(t)\} = \Im\left\{ m(t)\sum_{n=-\infty}^{\infty}\delta(t-nT)\right\}$$

$$= \frac{1}{T}M(w) * \sum_{n=-\infty}^{\infty}\delta(w-nw_s) = \frac{1}{T}\sum_{n=-\infty}^{\infty}M(w-nw_s)$$

Thus, the Fourier transform of the sampled signal consists of an infinite number of copies of the Fourier transform of the original signal, as shown in Figure 2.26.

To recover the original signal, we use a low-pass filter with a bandwidth of $2w_m$. If $w_S - w_m \geq w_m$, which implies that $w_S \geq 2w_m$, then we can completely recover the original signal. If $w_S - w_m < w_m$, which implies that $w_S < 2w_m$, then *aliasing* will occur, where aliasing is the phenomenon whereby the copies of the original signal spectrum interfere with each other causing the reconstructed signal to differ from the original signal. Aliasing is illustrated in Figure 2.27. The sampling rate $w_S = 2w_m$ is called the Nyquist rate.

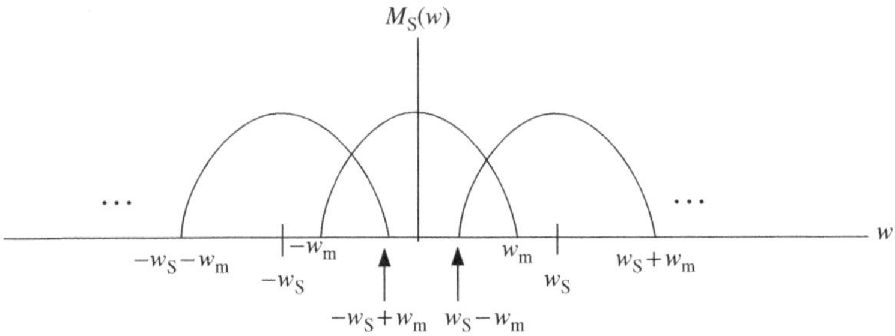

Figure 2.27 Illustration of Aliasing.

To avoid aliasing, a signal should either be critically sampled (i.e., $f_S = 2f_m$, which is the Nyquist rate) or oversampled whereby f_S is greater than the Nyquist rate.

2.10 Analog-to-Digital Conversion: From PAM to PCM

Analog-to-digital conversion is a method used to convert an analog signal, such as voice and video, into a digital signal. The first step in the process is to sample the analog signal uniformly at specified intervals in keeping with the sampling theorem. The sample values are transmitted by pulses whose amplitudes are proportional to those of the message signal at the sampling instant. The sampling process essentially converts analog amplitudes to discrete levels and is a type of modulation called *pulse amplitude modulation* (PAM).

PAM is one of the three *pulse modulation* schemes; the other two are *pulse width modulation* (PWM) and *pulse position modulation* (PPM). Although a PAM signal has discrete values, it is still an analog signal that is not suitable for transmission over long distances because noise can cause the amplitude of the signal to change along the transmission path. Figure 2.28 illustrates the sampling process.

2.10.1 Pulse Code Modulation

PAM is the most prevalent method of converting analog signals to digital pulses. It is frequently used as an intermediate step in the *pulse code modulation* (PCM). The next step after PAM is to ***quantize*** the digital pulses, which means to approximate the amplitude value of a PAM pulse to the nearest integer on a predefined set of permitted integers. The maximum integer level is usually defined as $L = 2^k$, where $k = 1, 2, 3, \ldots$. This enables the PAM pulses to be represented as k-bit code words. For example, when $k = 5$, $L = 32$, and each PAM pulse will have a value between 0 and 32. For voice signals, we usually have $L = 256$, which means that each voice sample is represented by $k = 8$ bits.

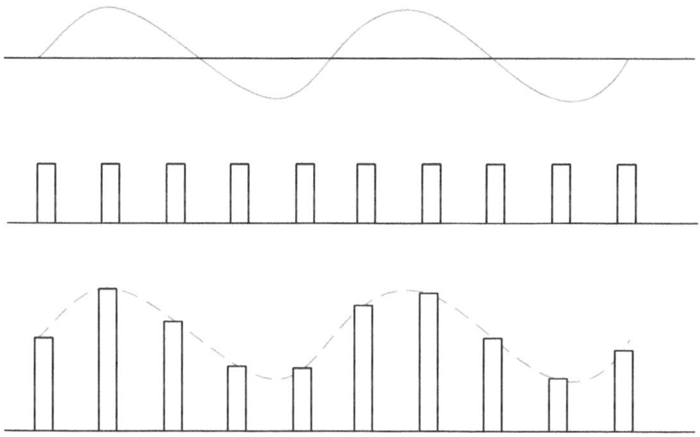

Figure 2.28 Illustration of the Sampling Process.

2.10.2 Quantization Noise

By definition, noise is an unwanted signal. Consider a PCM scheme that has $L = 32$, which as we stated earlier, means that each sample will be represented by a 5-bit code word since $32 = 2^5$. Assume that a PAM sample has a value of 14.5, which lies between the integer values 14 and 15. The value may be rounded up to 15 or down to 14; noninteger values are not permitted. Whichever way we choose to approximate 14.5, an error in the sample value has been introduced; the value used to represent the sample is different from its true value.

When a PAM value is not at exactly one of the 2^k integer values, the error introduced in approximating it to the closest integer value is called *quantization noise*. One way to reduce quantization noise is to make L as large as possible; but this means that the code words will be large and bandwidth required to transmit the signal will also be large.

To understand the bandwidth requirement, we consider the following case. As we discussed earlier, voice has a frequency range 300–3400 Hz, but we usually round the upper range up to 4000 Hz and the lower range down to 0 Hz, which means that the bandwidth of the voice signal is 4000 Hz. Thus, according to the sampling theorem, the smallest sampling rate (i.e., the Nyquist rate) for voice is $2 \times 4000 = 8000$ samples/second. We usually represent each sample by 8 bits, which means that $L = 2^8 = 256$ levels. By processing the voice at 8 bits/sample and 8000 samples/second, we are transmitting the voice signal at the rate of 8 bits/sample × 8000 samples/second = 64,000 bits/s = 64 kb/s.

If $L = 2^9 = 512$, which means that we represent each sample by 9 bits, the transmission rate will be 9 bits/sample × 8000 samples/second = 72,000 bits/second = 72 kb/s.

Thus, while we are likely to reduce quantization noise using 512 levels, increasing the bit representation of a sample by 1 causes the transmission rate to increase by 8 kb/s.

2.11 Basic Digital Modulation Schemes

When the modulating signal is a digital signal, we obtain the following schemes:

- *Amplitude-shift keying* (ASK)
- *Frequency-shift keying* (FSK)
- *Phase-shift keying* (PSK).

2.11.1 Amplitude-Shift Keying

ASK describes the technique whereby the carrier wave $c(t)$ is multiplied by the digital signal $m(t)$. Mathematically, the modulated carrier signal $s(t)$ is given by:

$$s(t) = m(t)c(t) = m(t)\sin(2\pi ft + \varphi)$$

where the amplitude of the carrier has been set to $a = 1$. If we set $\varphi = 0$, then, ASK means that:

when $m(t)$ is binary 1, $s(t) = \sin(2\pi ft)$;
when $m(t)$ is binary 0, $s(t) = 0$.

The ASK scheme is illustrated in Figure 2.29.

The type of ASK defined so far is called *on–off keying*, which acts like a switch that uses the presence of a carrier to indicate a binary one and the absence of

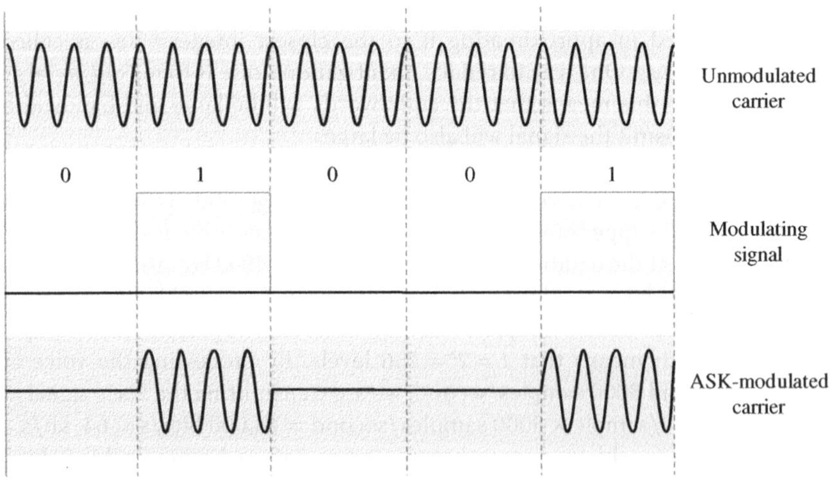

Figure 2.29 ASK Scheme.

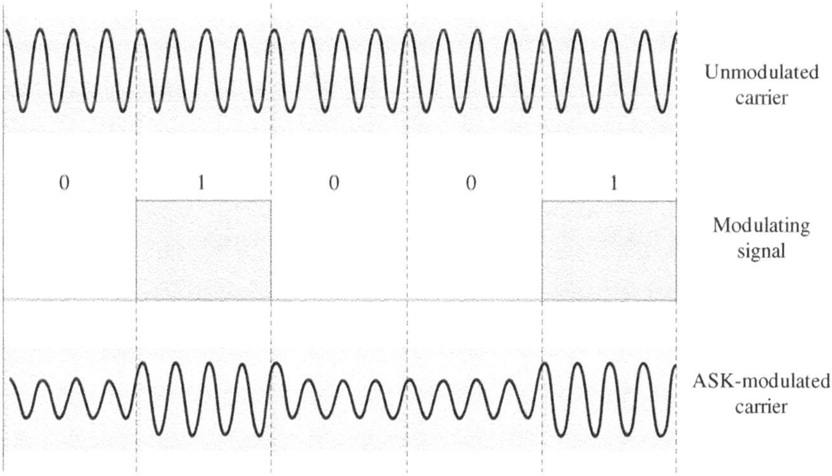

Figure 2.30 LED-based ASK Scheme.

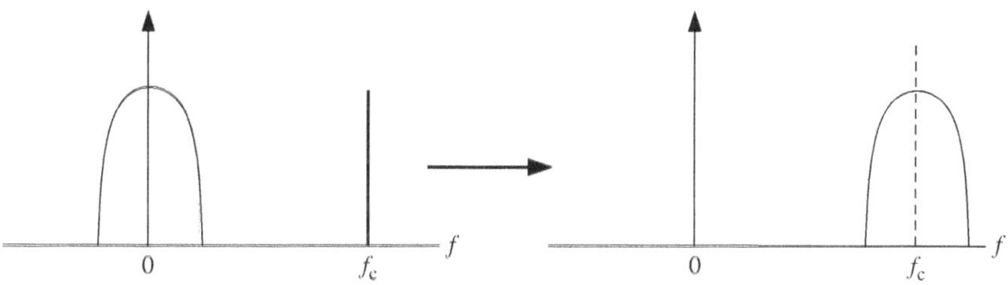

Figure 2.31 Spectral Structure of ASK.

a carrier to indicate a binary 0. When light-emitting diode (LED) transmitters are used, there is a slightly different behavior. LED transmitters normally have a fixed "bias" current that causes the device to emit a low light level. In this case, the low-amplitude signal represents a binary 0 while the higher-amplitude signal represents a binary one. This is illustrated in Figure 2.30.

ASK is said to be *spectrally efficient* because the bandwidth occupied by the signal in the ASK-modulated carrier is the same as the bandwidth of the original signal. This is illustrated in Figure 2.31.

2.11.2 Frequency-Shift Keying

FSK describes the modulation of a carrier using two frequencies: one frequency is for a binary 1 and another frequency is for a binary 0. The resultant modulated signal may be regarded as the sum of two amplitude-modulated signals of different carrier frequencies. That is,

$$s(t) = m_1(t)\sin(2\pi f_1 t + \varphi) + m_2(t)\sin(2\pi f_2 t + \varphi)$$

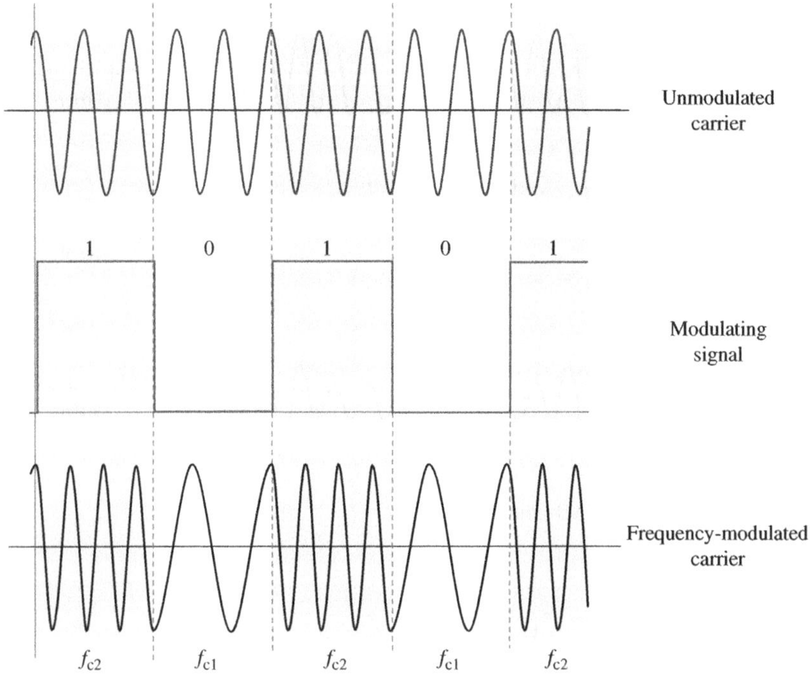

Figure 2.32 FSK Scheme.

If we set the phase to zero, then FSK implies the following:

For a binary 1, we set $m_1(t) = 1,\ m_2(t) = 0$ and obtain $s(t) = \sin(2\pi f_1 t)$.
For a binary 0, we set $m_1(t) = 0,\ m_2(t) = 1$ and obtain $s(t) = \sin(2\pi f_2 t)$.

The FSK scheme is illustrated in Figure 2.32.

In FSK, the spectrum of the modulated signal appears as two separate ASK signals. Thus, FSK may be thought of as the sum of two separate ASK waveforms, one of which represents the occurrence of binary 1s and the other the binary 0s. The two carriers must be sufficiently widely spaced to avoid overlap of the individual ASK spectra. This restriction tends to limit the use of FSK to lower speed modulation when compared to PSK. In comparing the bandwidth required for FSK, it can be seen that it is at least twice that of ASK. Thus, FSK is not as spectrally efficient as ASK. However, unlike ASK, it is a constant-envelope scheme, which makes it immune to noise. The spectral structure is illustrated in Figure 2.33.

2.11.3 Phase-Shift Keying

PSK describes the modulation technique that alters the phase of the carrier. In the most basic case, which is called *binary phase shift keying* (BPSK), the

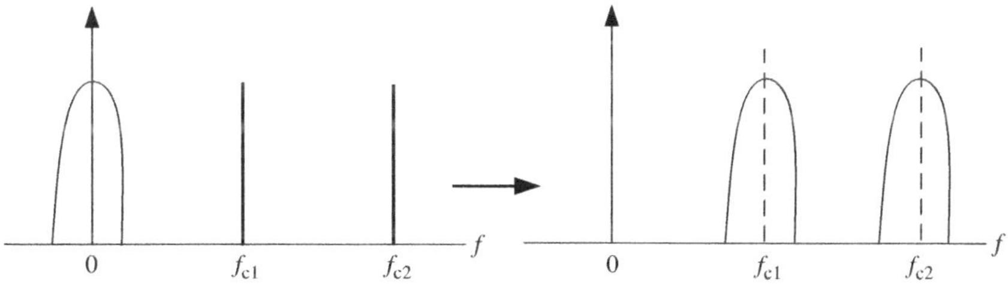

Figure 2.33 Spectral Structure of FSK.

phase of a constant amplitude carrier signal is switched between two values that correspond to binary 1 and binary 0, respectively. Usually the two phases are separated by 180°. Thus, a BPSK signal may be represented mathematically by the following:

Binary 1 corresponds to $s(t) = \sin(2\pi ft)$.
Binary 0 corresponds to $s(t) = \sin(2\pi ft + \pi) = -\sin(2\pi ft)$.

The PSK scheme is illustrated in Figure 2.34.

BPSK enjoys the advantages of FSK in the sense that it is a constant-envelope signal. However, unlike FSK it enjoys the reduced bandwidth of ASK, as shown in Figure 2.35. Thus, PSK is spectrally efficient like ASK, and like FSK it performs better than ASK in the presence of noise. On the other hand, unlike ASK and FSK, PSK does not lend itself to noncoherent detection. Coherent detection means that the receiver uses the carrier phase

Figure 2.34 PSK Scheme.

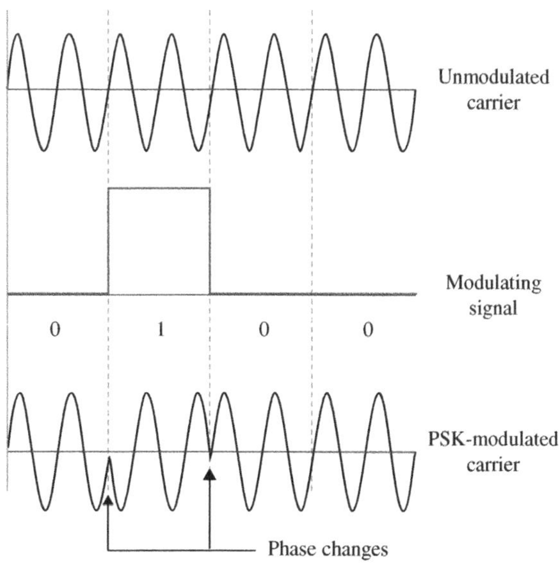

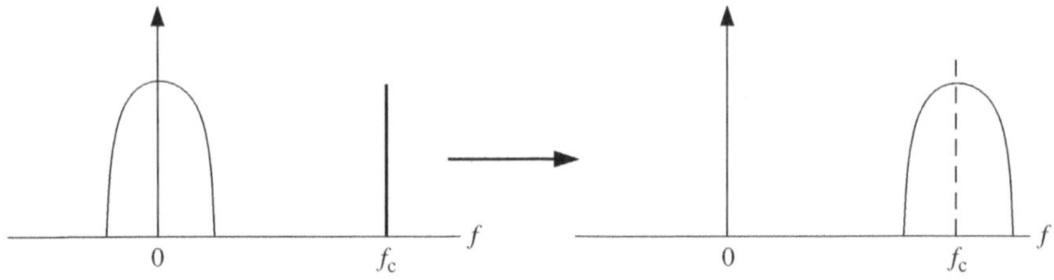

Figure 2.35 Spectral Structure of PSK.

to detect the signal. In other words, the transmitter and receiver must be perfectly synchronized. This requires a more complex circuitry to achieve than we use for ASK and FSK.

2.12 Media Sharing Schemes

In Chapter 1, we discussed three schemes for sharing media among the users to increase the utilization of a medium that no one user can use efficiently. They are as follows:

- Frequency division multiplexing (FDM)
- Time division multiplexing (TDM)
- Spread spectrum (SS).

In this section, we describe in greater detail how FDM and TDM can be implemented.

2.12.1 Frequency Division Multiplexing

The basic idea of FDM is that if a signal to be transmitted occupies a narrow band of frequencies, then other signals can be transmitted at the same time in other parts of the medium's frequency spectrum as long as their frequencies do not overlap. This is achieved by using each signal to modulate a unique carrier whose frequency is chosen so that the spectra of the different modulated signals do not overlap. In this way, the bandwidth of the medium is divided up into a number of separate frequency bands (or *channels*) each of which accommodates one signal, hence the term "frequency division." Figure 2.36 shows the FDM process at the transmitting end.

At the receiving end, the arriving signals are presented to a filter bank that provides band-pass filtering at the different carrier frequencies that are used at the transmitter. Then, each signal is demodulated using the same carrier that is used for that channel at the transmitter. Figure 2.37 illustrates the process at the receiver.

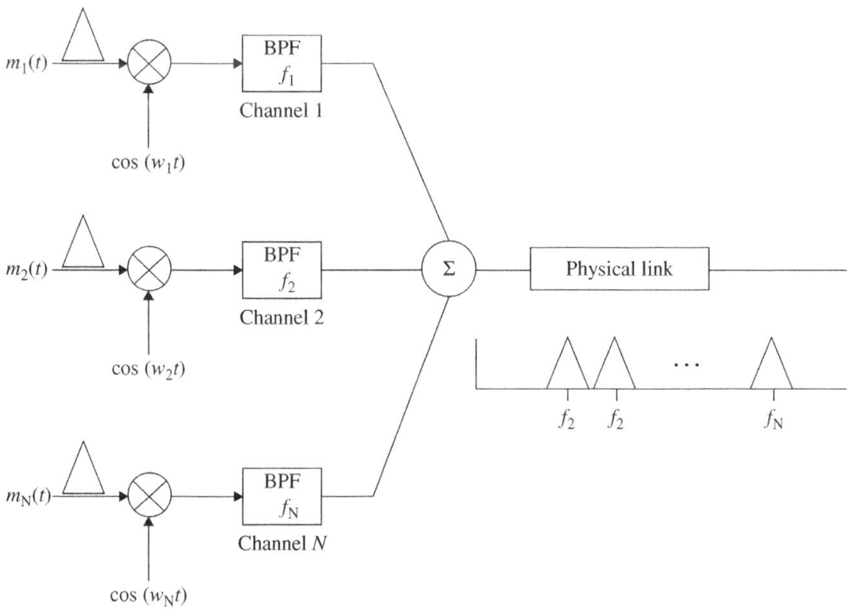

Figure 2.36 FDM Generation at the Transmitter End.

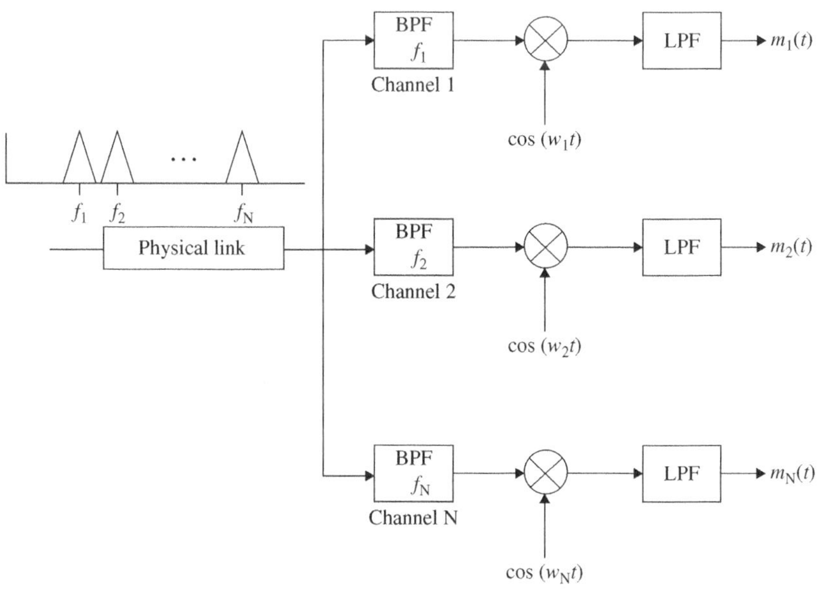

Figure 2.37 Signal Extraction at the Receiver.

FDM is the mechanism that permits us to have different TV and radio stations. Also, because of FDM a cable TV system can transmit a few dozen TV programs simultaneously on the same cable without interfering with each other.

2.12.1.1 Wavelength Division Multiplexing

Wavelength division multiplexing (WDM) is a variation of FDM used in optical fiber channels. Here, different wavelengths are applied to different fibers that are connected to a *combiner*. The output of the combiner is a single fiber that carries signals at the different wavelengths to a distant location, where a splitter is used to separate the different wavelengths. This is illustrated in Figure 2.38.

2.12.2 Time Division Multiplexing

TDM is the process of switching a number of signal sources (or channels) in strict rotation one at a time to a single output. Specifically, in TDM, the transmission time is divided into time slots of equal duration, and the sources are polled in a fixed order to transmit in the slots allocated to them. For example, in a four-source system, source 1 transmits in slots 1, 5, 9, and so on; source 2 transmits in slots 2, 6, 10, and so on; source 3 transmits in slots 3, 7, 11, and so on; and source 4 transmits in slots 4, 8, 12, and so on. At the destination, the receivers extract the data in the slots into which their communicating parties inserted their data. The data from one complete round of transmissions constitutes a frame. Since a particular slot in each frame is assigned to a particular source, the frame length is fixed. The TDM process is illustrated in Figure 2.39.

2.12.2.1 Synchronous Versus Asynchronous TDM

The TDM scheme we described earlier is called *synchronous TDM* because the time slots are allocated to the sources and each destination knows precisely the slots from which it can extract information. One of the problems with the scheme is that because the slots are "reserved" for the sources, the slots assigned to an idle source are unused and therefore wasted. To solve this problem, an alternative scheme called *asynchronous TDM* (or *statistical multiplexing*) is

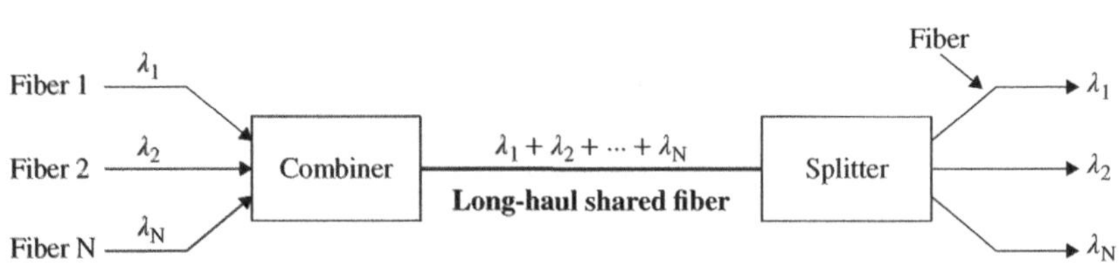

Figure 2.38 Wavelength Division Multiplexing.

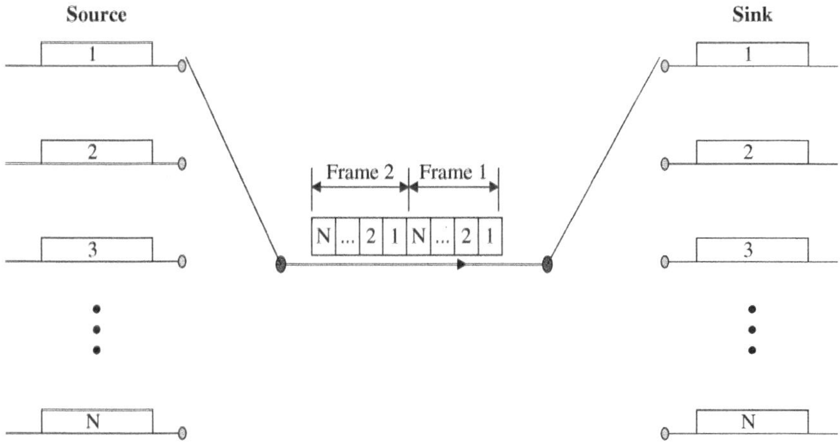

Figure 2.39 The TDM Process.

used in which slots are allocated on demand. As in synchronous TDM, the sources are polled in a fixed order, but only active sources that have data to transmit in any round are allocated slots in the frame for that round. Thus, the frame length is no longer fixed but varies from one round to another depending on the number of active users in a round. In this way, the available capacity is used more efficiently than in the synchronous TDM. Also, since the sources and sinks are no longer synchronized with respect to the time slots, each data must now include the address of the intended recipient sink. Asynchronous TDM is illustrated in Figure 2.40, where the first frame contains data from sources 1, 2, and 5 that were the stations that were active in that round, and the second frame has data from sources 1, 2, 4, and N, which were the only active sources in that round.

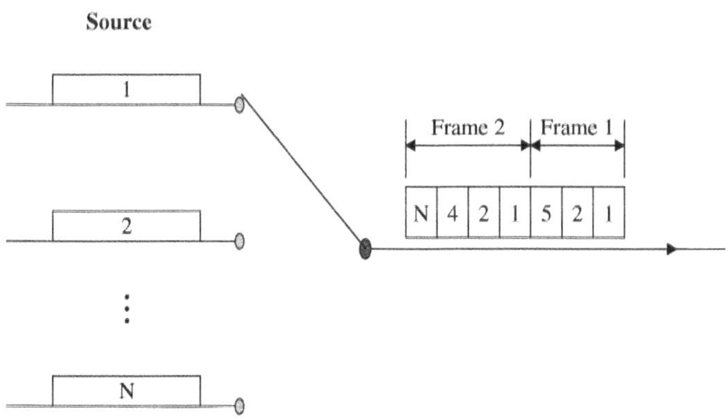

Figure 2.40 Asynchronous TDM Process.

2.13 Modems

The term "modem" is a contraction of the words "**mo**dulator–**dem**odulator." A modem connects a computer to the Internet, and the quality of the modem can affect the speed of the Internet connection. Modems came into existence in the 1960s as a way to allow terminals to connect to computers over the phone lines. The sending modem modulates the data into a signal that is compatible with the phone line, and the receiving modem demodulates the signal back into digital data.

The same operational model still holds today. That is, a modem modulates outgoing digital signals from a digital device like a computer to analog signals for an applicable transmission medium and demodulates the incoming analog signal and converts it to a digital signal for the digital device. An example of a modem is a DSL modem that is used to connect a home network or a single computer to the Internet via the telephone network.

There is one standard interface called RC-232 for connecting external modems to computers, which means that any external modem can be attached to any computer that has an RS-232 port.

How fast a modem can transmit and receive data is measured by its bit rate. Modems that operate at slow data rates are measured in *baud* rate. Baud rate measures the number of symbols that are transmitted per second. If a symbol is the bit, then the baud rate and the bit rate become identical.

There are different types of modems. These include the following:

- *Onboard modem*, which is a modem that is built onto the computer motherboard. Such a modem cannot be removed but can be disabled.
- *Internal modem*, which is a modem that connects to a peripheral component interconnect (PCI) slot inside a newer desktop computer or industry standard architecture (ISA) slot in an older computer.
- *External modem*, which is a modem within a box that connects to the computer externally, usually through a serial port or a universal serial bus (USB) port.

2.14 Transmission Media

The purpose of the physical layer is to transport a raw bitstream from one node to another. These nodes can be interconnected by twisted pair, coaxial cable, optical fiber, or air. Each type of medium has its strengths and weaknesses with respect to bandwidth, delay, cost and ease of installation, and maintenance. Media like twisted pair, coaxial cable, and optical fiber are called *guided media* because they enable the transfer of information between two or more points that are connected by an electrical conductor. Wireless media are called *unguided media* because they do not require physical links between two or more devices. Alternatively, we can say that guided media essentially provide a confined transmission "pipe" for a signal between two points while unguided media do not guide signals to any specific direction. In unguided media, the information signal propagates through free space in the form of an electromagnetic (EM) wave.

2.14.1 Twisted Pair

The twisted pair is one of the oldest and most common transmission media. It consists of two insulated copper wires that are typically about 1 mm thick and twisted together in a helical form.

Twisting is done because two parallel wires constitute an antenna. When the wires are twisted, the waves from different twists cancel out essentially allowing the wire to radiate less.

Twisted pair can run several kilometers without amplification, but longer distances require amplifiers. It can be used for both digital and analog transmission and the capacity depends on the thickness of the wire and the distance between the two interconnected nodes. Usually up to several Mbps can be achieved for a few kilometers.

It comes in several varieties, but the more common ones used in computer networks are the following unshielded twisted pairs (UTP):

- Category 3 UTP that consists of two insulated wires gently twisted together and were more prominent prior to 1988. It is capable of handling signals with a bandwidth of 16 MHz.
- Category 5 UTP that is similar to Category 3 UTP but with more twists per centimeter resulting in less cross talk and better-quality signal over longer distances, making it more suitable for high-speed applications. It is capable of handling signals with bandwidth of 100 MHz.
- New categories 6 and 7 UTP are becoming popular and can handle signals with bandwidth of 250 and 600 MHz, respectively.

Shielded twisted pair is very bulky and expensive, and was popularly used in IBM installations.

2.14.2 Coaxial Cable

The coaxial cable (coax) is used primarily for TV hookup at home. It is also used in the hybrid fiber coax (HFC) networks where modems communicate with the headend in a cable network. To facilitate two-way communication, the cable bandwidth is split into two sets of channels: upstream channels that carry information from the modem to the headend, and downstream channels that carry information from the headend to the modems. A guard band separates the two sets of channels. This is illustrated in Figure 2.41.

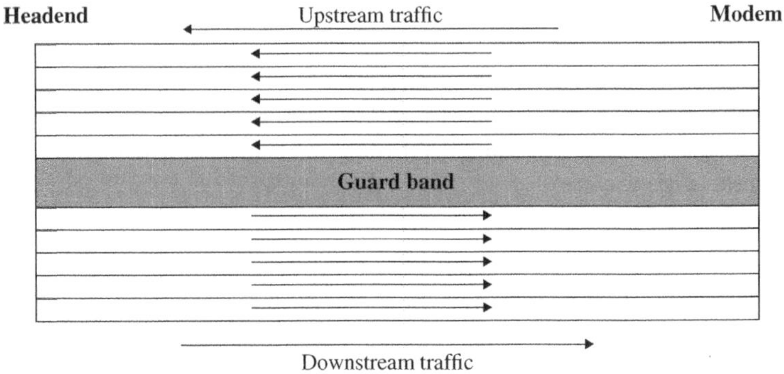

Figure 2.41 Partitioning of the Coaxial Cable.

Cable can be split in one of the three ways:

- *Mid split,* where bandwidths above and below the guard band are fairly equal; the guard band occupies from 108 to 168 MHz. The upstream channels occupy from 5 to 108 MHz, while downstream channels occupy from 168 MHz to the upper frequency limit (up to about 750 MHz).
- *High split,* where more bandwidth is allocated to the upstream channels than to the downstream channels and the guard band occupies from 174 to 234 MHz. The upstream channels are located from 5 to 174 MHz, while downstream channels are located from 234 MHz to the upper frequency limit.
- *Subsplit,* where more bandwidth is allocated to the downstream channels than to the upstream channels and the guard band occupies from 42 to 54 MHz. The upstream channels are located from 5 to 42 MHz, while downstream channels are located from 54 MHz to the upper frequency limit.

2.14.3 Optical Fiber

We know that light is both a wave and a particle. As a wave, light behaves like radio waves and so is subject to reflection, refraction, diffraction, interference, polarization, and fading. Also, as a wave, light is characterized by frequency (and wavelength), phase, and propagation speed. As a particle, light can move and exert pressure. The smallest quantity of monochromatic light is called a *photon,* which has energy associated with it that depends on the frequency of the light.

Some materials allow all light energy to propagate through them; such materials are defined to be *optically transparent.* Materials that are not optically transparent are said to be *opaque.* Some materials are *semitransparent* in the sense that they pass a portion of light energy through them and absorb the remainder. Semitransparent materials are also said to be *translucent* and are used to make optical attenuators.

Optical fiber refers to the medium and the technology associated with the transmission of information as light pulses along a glass or plastic wire or fiber. It can carry more information than conventional copper wire and is in general not subject to EM interference and the need to retransmit signals. Transmission in optical fiber wire requires *repeaters* at periodic intervals.

The glass fiber requires more protection than copper wire because it is more fragile than traditional copper wire. A typical single optical fiber consists of a strand of ultrapure silica mixed with special elements called *dopants* that are added to adjust the refractive index of the silica and thus the light-propagation characteristics. The optical cable consists of a strand of fiber, many miles long and containing several layers as shown in Figure 2.42. It also consists of the following:

- Innermost layer is the silica *core*, which carries most of the light.
- The core is surrounded by another layer of silica with a different mix of dopants, called *cladding*; typically, the cladding has a diameter of about 125 μm and bends the light and confines it to the core.
- The cladding is covered with a *primary buffer coating* that absorbs mechanical stress during handling of the cable; it provides the first layer of mechanical protection.
- The primary buffer coating is covered by a *secondary buffer coating*, which protects the relatively fragile primary coating and the underlying fiber.
- The final layer is a *plastic* material that covers these layers.

Fiber cable used in long-haul communications consists of a bundle of up to 432 optical fibers.

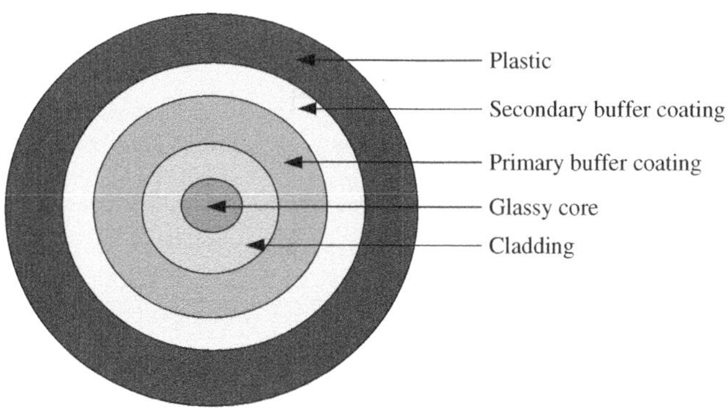

Figure 2.42 Structure of the Optical Fiber.

2.14.3.1 Fiber Modes

As stated earlier, the cladding has a typical diameter of 125 μm, but the core comes in two different dimensions depending on the intended fiber application. According to the ITU-T G.652 Recommendation:

- Fiber with a core diameter of about 50 μm is known as *multimode* fiber.
- Fiber with a core diameter of about 8.6–9.5 μm is known as *single-mode* fiber.

Multimode and single-mode fibers have different manufacturing processes, different refractive index profiles, different dimensions, and, therefore, different transmission characteristics. Thus, they are used for different applications in optical transmission. The structural differences between the two types of fiber are shown in Figure 2.43.

A mode of light is a distribution of the EM field that satisfies boundary conditions for a waveguide, such as an optical fiber. A mode can be visualized as the path of a single ray of light in the fiber.

- *Single-mode fiber* allows only one pathway, or mode, of light to travel within the fiber. It is used in applications where low signal loss and high data rates are required, such as in long spans between two system or network devices where repeater/amplifier spacing needs to be maximized.
- *Multimode fiber* has more paths available for rays of light to propagate because the core is larger. Thus, multimode fiber allows more than one mode of light. It is suited for shorter distance applications that require less than 500 m between the transmitter and receiver because the modes tend to disperse over longer lengths (this is called *modal dispersion*). There are two types of multimode fiber:
  - Step-index fiber, which is characterized by an abrupt change in refractive index.
  - Graded-index fiber, which is by a continuous and smooth change in refractive index. It can support a higher bit rate over a greater distance than step-index fiber and is more expensive than step-index fiber.

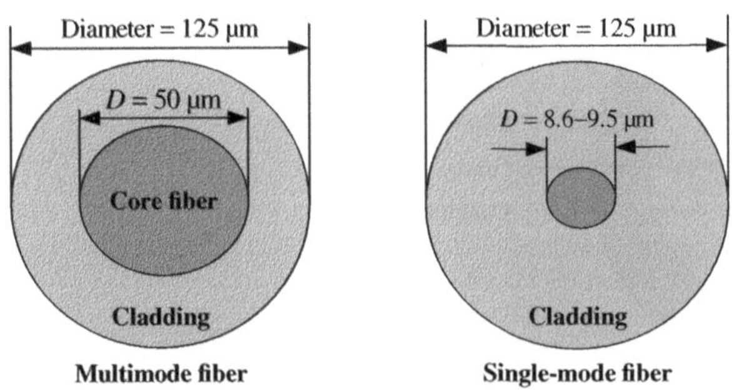

Figure 2.43 Structural difference between multimode and single-mode fiber.

2.14.4 Wireless Medium

As stated earlier, wireless media are unguided media that permit the transfer of information between two or more points that are not connected by an electrical conductor. Wireless communication is based on radio waves whose distances can be as short as a few meters or as far as thousands of kilometers for satellite communications.

The radio frequency (RF) spectrum is a part of the EM spectrum that is the range of all types of EM radiation. EM radiation can be expressed in terms of energy (measured in electron volts), wavelength (measured in meters), or frequency (measured in cycles/second, or Hertz). Generally, the shorter the wavelength, the higher the frequency, and the higher is the energy. Conversely, the longer the wavelength, the lower the frequency, and the lower is the energy. Figure 2.44 shows the EM spectrum that consists of the following:

- Radio
- Microwave
- Infrared
- Visible light
- Ultraviolet
- X-ray
- Gamma ray.

The RF spectrum is broken down into different bands that have their own characteristics. The classification is shown in Table 2.2.

Radio communication involves the transmission of radio waves at the source by an antenna and the reception of these waves by another antenna. As stated

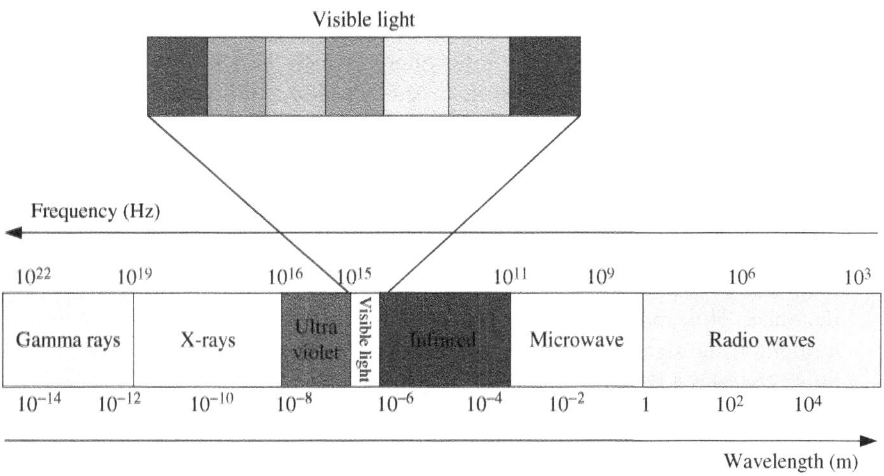

Figure 2.44 The Electromagnetic Spectrum.

Table 2.2 The Radio Frequency Spectrum.

No.	Band name	Frequency range	Wavelength range
1	Very low frequency (VLF)	3–30 kHz	100–10 km
2	Low frequency (LF)	30–300 kHz	10–1 km
3	Medium frequency (MF)	0.3–3 MHz	1 km–100 m
4	High frequency (HF)	3–30 MHz	100–10 m
5	Very high frequency (VHF)	30–300 MHz	10–1 m
6	Ultrahigh frequency (UHF)	0.3–3 GHz	1 m–10 cm
7	Super high frequency (SHF)	3–30 GHz	10–1 cm
8	Extra high frequency (EHF)	30–300 GHz	1 cm–1 mm
9	Tremendously high frequency (THF)	0.3–3 THz	1 mm–0.1 mm

earlier, the dimensions of these antennas are on the order of the wavelength of the radio signal. This simple fact has a great impact on wireless communication. For example, the very-low-frequency (VLF) and low-frequency (LF) bands have wavelengths that are in the kilometer range (30 kHz has a wavelength of 10 km or 6.2 miles) and very gigantic antennas need to be used.

Similarly, the medium-frequency (MF) and HF bands are used by commercial AM broadcasting stations. Signal radiation in these frequency ranges have the important property of being reflected by the ionosphere. The ionosphere is a layer of electrically charged particles at the top of the earth's atmosphere. The layer is caused by the strong solar radiation entering the upper atmosphere. When a radio wave in the MF or HF range hits this layer, it is reflected back to earth. The disadvantage of this type of propagation is that it depends on the characteristics of the ionosphere, which varies widely, especially during daylight hours. As a result of this variation, the waves are reflected differently and take different paths over a period of time. This causes the signal at the receiver to vary in strength, which in turn causes the output to fade in and out.

Also, signal radiation in the very-high-frequency (VHF) and ultrahigh-frequency (UHF) bands are not reflected by the ionosphere. For this reason, communication in these bands tends to be line of sight and over a short distance. This means that the transmitter and receiver must be within a straight visual sighting line from each other. Buildings and uneven terrain may affect the signal transmission in these bands.

The frequency band from 30 to 300 GHz (i.e., the EHF) is called the *millimeter wave* band, which has short wavelengths that range from 10 to 1 mm. The band has high atmospheric attenuation and the radio waves are absorbed by gases in the atmosphere, which reduces the range and strength of the waves. It is also subject to *rain fade*, which results from the fact that rain and humidity affect the performance of the band by reducing the signal strength. (Similar to the microwave oven that is used for cooking, the signal heats up the rain and loses its energy or strength in the process. Thus, the rain causes its strength to fade or to be attenuated.) Also, because of its short range of about a kilometer, a millimeter wave travels by line of sight and can be blocked by physical objects such as buildings and trees.

2.15 Channel Impairments

As a signal propagates along a communication path from its source to its destination, it is subject to different types of impairments. These impairments degrade the quality of the signal by decreasing its amplitude or increasing its potential to interfere with other signals. The different types of impairments include the following:

(a) Attenuation
(b) Noise
(c) Distortion.

2.15.1 Attenuation

The strength of a signal decreases as it travels along a transmission medium. The amount of attenuation depends on the medium, but in general it increases with distance. There is a need to increase the strength of a signal, so it can be detected at the receiver. This is accomplished by amplifying the signal at the receiver. Note that the signal can only be amplified if its level at the receiver is above the amplifier's *sensitivity threshold*, which is the smallest detectable signal level of the amplifier. If the distance between the transmitter and the receiver is great, the signal may need to be amplified several times at intermediate points along the path to prevent it from falling below the sensitivity threshold at the receiver.

2.15.2 Noise

Noise is usually defined as an unwanted signal that is superimposed on a desired signal. Figure 2.45 illustrates how noise affects a signal along a communication channel.

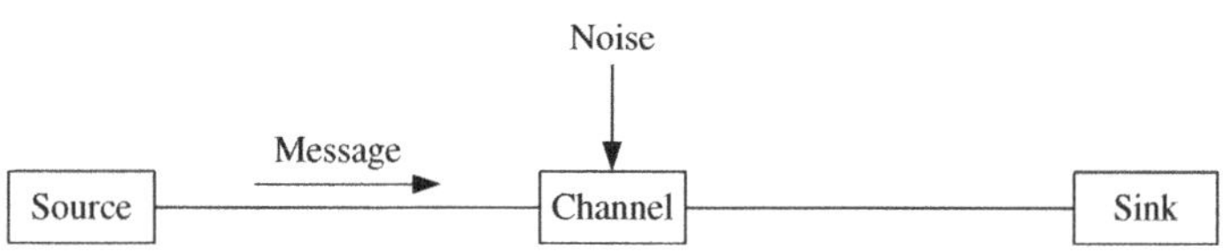

Figure 2.45 Model of a Practical Communication System.

There are different types of noise, which include the following:

(a) Atmospheric noise
(b) Man-made noise
(c) Extraterrestrial noise
(d) Thermal noise
(e) Shot noise.

Atmospheric noise is the noise that is caused by such natural atmospheric phenomena as lightning discharge in thunderstorms and other electrical disturbances that occur in nature. It is usually an impulsive phenomenon, which means that it has a short duration but is intensive. This means that in a digital communication system, an atmospheric noise will affect a group of bits that are clustered together. Thus, it leads to burst errors.

Man-made noise is an EM noise that is caused by human activities that are associated with the use of electrical equipment. Thus, activities such as car ignition, turning on a lawn mower, aircrafts, and using a blow dryer can generate EM waves that can interfere with a signal. High-voltage wires and fluorescent lamps also produce this type of noise. Generally, the amplitude of the noise varies with location, being greater in urban areas than in rural areas.

Extraterrestrial noise is the noise that comes from outside the earth and includes solar noise and cosmic noise. Solar noise is the noise that originates from the sun. Under normal conditions, there is constant radiation from the sun due to its high temperature. Electrical disturbances such as corona discharges, as well as sunspots can produce additional noise. Cosmic noise is generated by distant stars. These stars are too far away to individually affect terrestrial communications systems. However, their large number leads to appreciable collective effects.

Thermal noise occurs in electrical conductors and is caused by the thermal agitation of the charges in the material. It is present even when no voltage is applied. The amplitude of the noise is given by:

$$P_{\mathrm{TH}} = k_{\mathrm{B}}TB$$

where k_{B} is the Boltzmann's constant (i.e., $k_{\mathrm{B}} = 1.38064852 \times 10^{-23}$ J/K), T is the temperature in K and B is the bandwidth of the receiver. Thermal noise is sometimes called *Johnson–Nyquist noise, Johnson noise,* or *Nyquist noise.* As the temperature increases, the agitation of the charge carriers increases and so does the level of noise. Thus, to reduce the noise the associated equipment must be operated at very low temperatures.

Shot noise arises from the time-dependent fluctuations in electrical current. This is caused by the discrete nature of electron charges. Specifically, current is not a continuous flow but the sum of discrete pulses in time, each corresponding to the transfer of an electron through the conductor. Shot noise is particularly noticeable in semiconductor devices. Unlike thermal noise, shot noise cannot be eliminated by lowering the temperature.

2.15.2.1 Concept of Decibel

The *bel* (named after Alexander Graham Bell) is a logarithmic unit of the ratio of a given power P to a reference power, P_{REF}. Thus, it is expressed as follows:

$$\text{Power ratio in bels} = \log_{10}\left(\frac{P}{P_{REF}}\right)$$

The bel is usually too large a number for many applications. Thus, it is customary to use the *decibel* (dB), which is one-tenth of a bel. Thus, the power ratio in decibels is defined by:

$$\text{Power ratio in decibels} = 10\ \log_{10}\left(\frac{P}{P_{REF}}\right)$$

Decibels are sometimes expressed as voltage ratios. Since power is directly proportional to the square of voltage, we have that

$$\text{Voltage ratio in decibels} = 10\ \log_{10}\left(\frac{V^2/R}{V_{REF}^2/R}\right) = 10\ \log_{10}\left(\frac{V}{V_{REF}}\right)^2$$

$$= 20\ \log_{10}\left(\frac{V}{V_{REF}}\right)$$

where R is the resistance and V is the voltage. Sometimes the reference power is 1 mW, and the power ratio is expressed in *dBm*. Thus, a power of 1 W relative to 1 mW is equal to:

$$10\ \log_{10}\left(\frac{1\text{ W}}{10^{-3}\text{ W}}\right) = 10\ \log_{10}(10^3) = 30\ \log_{10}(10) = 30\text{ dBm}$$

Power gains and losses are usually expressed in decibels. For example, consider an amplifier that boosts 1 mW power to 2 W. The power gain of the amplifier is:

$$10\log_{10}\left(\frac{2\text{ W}}{10^{-3}\text{ W}}\right) = 10\log_{10}(2\times 10^3) = 10\log_{10}(2) + 10\log_{10}(10^3) = 33\text{ dB}$$

Similarly, a signal whose strength goes from 2 W to 1 mW has a gain of:

$$10\ \log_{10}\left(\frac{10^{-3}\text{ W}}{2\text{ W}}\right) = 10\ \log_{10}(2^{-1}\times 10^{-3})$$

$$= 10\ \log_{10}(2^{-1}) + 10\ \log_{10}(10^{-3}) = -33\text{ dB}$$

The negative sign indicates a loss; thus, the signal has suffered a loss of 33 dB. Just as dBm refers to a reference power of 1 mW, *dBW* is used to describe a reference power of 1 W.

Note that decibels are measured using a logarithmic scale. For this reason, decibels cannot be added arithmetically. For example, consider two sources that are next to each other and each produces 80 dB. The total power in decibels can be computed as follows. Let D denote the power in decibels; that is,

$$D = 10\log_{10}\left(\frac{P}{P_{\text{REF}}}\right) = \log_{10}\left(\frac{P}{P_{\text{REF}}}\right)^{10} \Rightarrow \left(\frac{P}{P_{\text{REF}}}\right)^{10} = 10^{D}$$

$$\Rightarrow \frac{P}{P_{\text{REF}}} = 10^{D/10}$$

From this, we obtain:

$$P = 10^{D/10} P_{\text{REF}}$$

Let P_1 denote the power of the first source and let P_2 denote the power of the second source. Similarly, let D_1 be the power of the first source in decibels and let D_2 denote the power of the second source in decibels. Then, we have that the combined noise power of the two sources is given by:

$$P_1 + P_2 = 10^{D_1/10} P_{\text{REF}} + 10^{D_2/10} P_{\text{REF}} \Rightarrow \frac{P_1 + P_2}{P_{\text{REF}}} = 10^{D_1/10} + 10^{D_2/10}$$

Since we assume that the two sources have the same power, we obtain:

$$\frac{P_1 + P_2}{P_{\text{REF}}} = 2 \times 10^{D_1/10}$$

Thus, the combined power in decibels is:

$$D = 10\log_{10}\left(\frac{P_1 + P_2}{P_{\text{REF}}}\right) = 10\log_{10}(2 \times 10^{D_1/10})$$

$$= 10\log_{10}(2) + 10\log_{10}(10^{D_1/10})$$

$$= 10\log_{10}(2) + \frac{10D_1}{10}\log_{10}(10) = 10\log_{10}(2) + D_1\log_{10}(10) = 3 + D_1$$

$$= D_1 + 3$$

This means that the combined power in decibels is $80 + 3 = 83$ dB and not 160 dB. In the same way, the combined power of n identical sources each of which generates D decibels power is $D + 10\log_{10}(n)$. For the case of $n = 10$, we obtain $D + 10$ dB.

2.15.2.2 Signal-to-Noise Ratio

Signal-to-noise (S/N) ratio is a measure of signal strength relative to background noise. It is usually written as S/N or SNR and measured in decibels (dB). If the incoming signal strength is V_S volts and the noise level is V_N volts, then the SNR is given by:

$$\begin{aligned}\mathrm{SNR} &= 10\log_{10}\left(\frac{P_\mathrm{S}}{P_\mathrm{N}}\right) = 10\log_{10}\left(\frac{V_\mathrm{S}^2/R}{V_\mathrm{N}^2/R}\right) = 10\log_{10}\left(\frac{V_\mathrm{S}}{V_\mathrm{N}}\right)^2 \\ &= 20\log_{10}\left(\frac{V_\mathrm{S}}{V_\mathrm{N}}\right)\ \mathrm{dB}\end{aligned}$$

A high value of SNR indicates a better quality of service compared to a low SNR. This is because a high SNR implies that the amplitude of the signal is much greater than that of the noise.

SNR is used to determine two other performance measures, which are *noise factor* and *noise figure*. The noise factor of a system, such as a receiver, is a measure of the degradation introduced by the system. It is defined as the ratio of the SNR at the input to the system to the SNR at the output of the system. That is, the noise factor, F, of a system is defined by:

$$F = \frac{\mathrm{SNR_{in}}}{\mathrm{SNR_{out}}}$$

where $\mathrm{SNR_{in}}$ is the SNR at the input of the system and $\mathrm{SNR_{out}}$ is the SNR at the output of the system. As stated earlier, it is a measure of the degradation of SNR due to the noise added by the system. Generally, the SNR at the output will always be smaller than the SNR at the input due to the fact that a system such as a receiver always adds to the system noise.

The noise figure, NF, is the noise factor expressed in decibels. That is,

$$\begin{aligned}\mathrm{NF} &= 10\log_{10}(F) = 10\log_{10}\left(\frac{\mathrm{SNR_{in}}}{\mathrm{SNR_{out}}}\right) \\ &= 10\log_{10}(\mathrm{SNR_{in}}) - 10\log_{10}(\mathrm{SNR_{out}})\end{aligned}$$

Thus, the noise figure is the difference between the input SNR in decibels and the output SNR in decibels. Finally, we define a measure of a system's sensitivity called *noise floor*. The noise floor of a system is the minimum detectable input signal level for a given output SNR.

2.15.3 Distortion

Distortion refers to the change or alteration of an object. Thus, with respect to data transmission, distortion means that the signal changes its form or shape. Delay distortion is a phenomenon that is peculiar to guided transmission media. The velocity of propagation of a signal through a guided medium varies with frequency. Every complex signal consists of many frequency components each of which travels at a different velocity. For a band-limited signal, the propagation velocity tends to be highest near the center frequency and lower toward the two edges of the band. Thus, various frequency components of a signal will arrive at the receiver at different times. This variation in the arrival

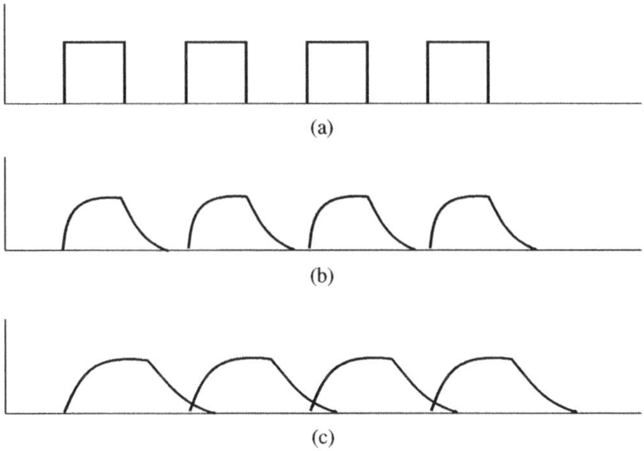

Figure 2.46 Illustration of Delay Distortion and ISI. (a) Transmitted Signal; (b) Received Signal without ISI; and (c) Received Signal with ISI.

of the different components of the signal leads to the distortion of the signal, an effect that is referred to as *delay distortion*.

Delay distortion is particularly critical for digital data. Consider that a sequence of bits is being transmitted, as in Figure 2.46. The received signal is a distorted version of the transmitted signal and each received bit tends to spill over into the position of the next bit. In Figure 2.46(b), the distortion is not enough to allow one received bit to spill over the next bit. However, in Figure 2.46(c), the received signal does spill over into the next bit position giving rise to *intersymbol interference* (ISI).

Another type of distortion is *attenuation distortion*. As stated earlier, attenuation is the reduction in the strength of a signal as it travels along a transmission medium. This reduction varies with frequency. Thus, because a typical signal has many frequency components each of which is attenuated to a different degree, even without delay distortion the received signal will be a distorted version of the original signal since the components do not recombine proportionately due to their differential attenuation. In general, attenuation level tends to increase with increase in the frequency.

2.15.4 Equalization

If the *frequency response* of a channel is approximately constant, all the frequency components of a signal would be attenuated at the same level and this would prevent the occurrence of attenuation distortion and delay distortion. The frequency response of a channel is ratio of the magnitude and phase of an output signal to those of the input signal. Equalization is a technique that is used to compensate for attenuation distortion and delay distortion. Any device that performs this function is called an *equalizer*.

An equalizer is fundamentally a filter that has the property that the product of its frequency response and that of the channel has a constant amplitude and linear phase over the range of frequencies of interest. Equalization is a postprocessing activity that is performed at the receiver. Its goal is to introduce the opposite effect of what the channel introduced in a signal. Specifically, let $H_C(w)$ denote the frequency response of the channel, and let $H_E(w)$ denote the frequency response of the equalizer. If a is a constant, then the composite frequency response of a channel that has an equalizer at the receiver is given by:

$$H(w) = H_C(w)H_E(w) = a$$

The frequency response of the equalizer compensates for the channel distortion and is related to the frequency response of the channel by:

$$H_E(w) = \frac{a}{H_C(w)} = \frac{a}{|H_C(w)|}e^{-j\theta_C(w)}$$

where $|H_E(w)| = a/|H_C(w)|$ is the amplitude of the frequency response of the equalizer and $\theta_E(w) = -\theta_C(w)$ is its phase response. In this way, the equalizer can be seen as an inverse channel. Figure 2.47 is a block diagram of a system with an equalizer and Figure 2.48 shows the combined frequency response of a channel and an equalizer that is approximately constant over the range of frequency of interest.

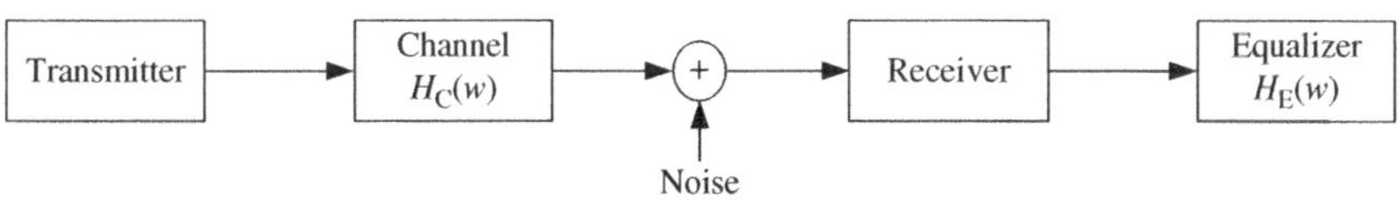

Figure 2.47 Block Diagram of a Communication System with an Equalizer.

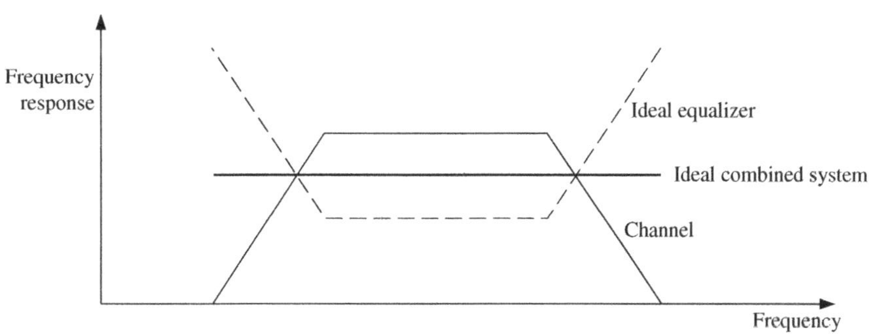

Figure 2.48 Example of Frequency Response of an Equalized Channel.

There are two main types of equalizers: *linear* and *nonlinear equalizers.* Nonlinear equalizers include *decision-feedback equalizers* and *maximum likelihood sequence detection equalizers.* A detailed discussion of these equalizers is outside the scope of our study and thus is not presented here.

2.16 Summary

This chapter has discussed different physical layer issues. These include signal classification, Fourier analysis including Fourier series and Fourier transform, modulation and demodulation, sampling theorem, analog-to-digital conversion, channel sharing schemes, guided and unguided transmission media, and channel impairments. The data link layer that is discussed in Chapter 3 uses these services to forward frames.

Exercises

1 Find the Fourier series of the periodic function shown in Figure 2.49, which is defined by

$$x(t) = \begin{cases} 1 & 0 \le t \le 0.5 \\ -1 & -0.5 \le t \le 0 \end{cases}$$

$$x(t+1) = x(t) \Rightarrow \text{period} \ \ T = 1$$

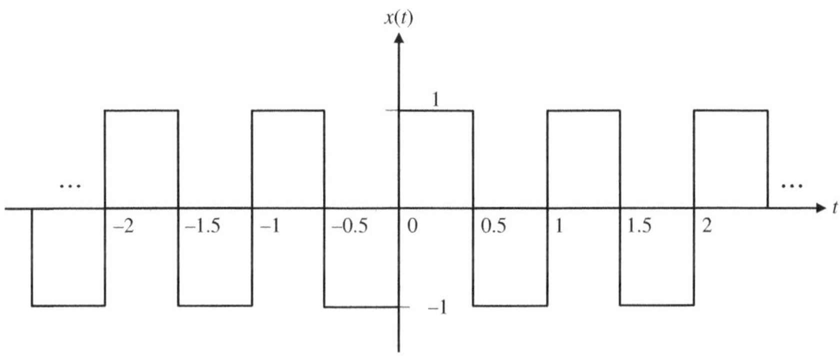

Figure 2.49 Figure for Problem 1.

2 Find the Fourier series of the periodic function shown in Figure 2.50, which is defined by

$$y(t) = \begin{cases} 1 & -1 \le t \le 1 \\ -1 & 1 \le t \le 3 \end{cases}$$

$$y(t+4) = y(t) \Rightarrow \text{period} \ \ T = 4$$

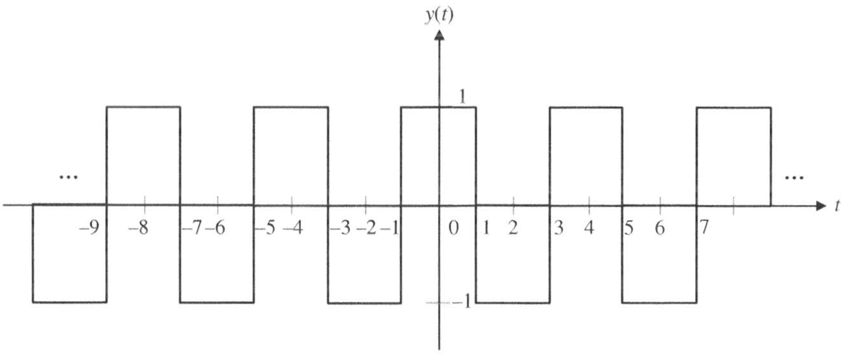

Figure 2.50 Figure for Problem 2.

3 Let $y(t)$ be a periodic function with period 2π such that

$$y(t) = \begin{cases} \pi + t & -\pi < t < 0 \\ \pi - t & 0 < t < \pi \end{cases}$$

(a) Sketch a graph of $y(t)$ in the interval $-3\pi < t < 3\pi$
(b) Obtain the Fourier series of $y(t)$

4 Find the Fourier transform of the rectangular function shown in Figure 2.51, which is defined by

$$x(t) = \begin{cases} 1 & -1 \leq t \leq 1 \\ 0 & \text{otherwise} \end{cases}$$

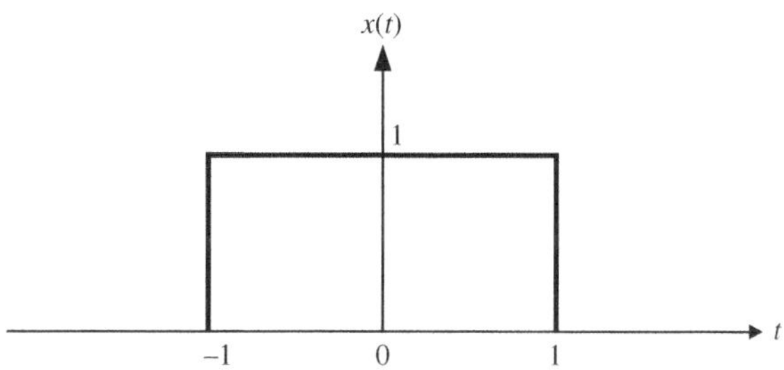

Figure 2.51 Rectangular Function.

5 Find the Fourier series of the periodic triangular wave function shown in Figure 2.52, which is defined by

$$x(t) = |t| = \begin{cases} t & 0 \leq t \leq 1 \\ -t & -1 \leq t \leq 0 \end{cases}$$

$$x(t+2) = x(t) \Rightarrow \text{period } \ T = 2$$

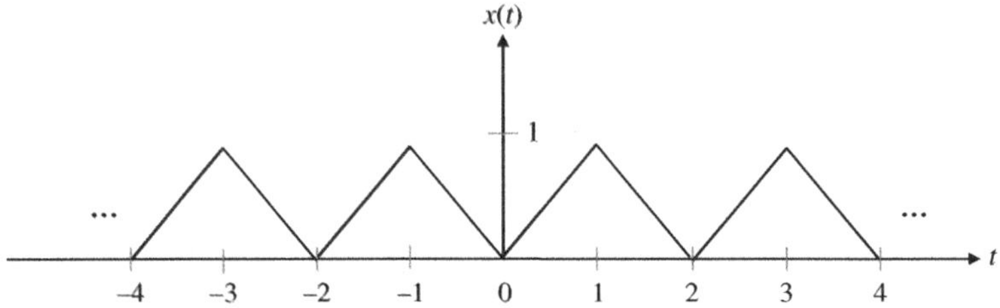

Figure 2.52 Periodic Triangular Wave Function.

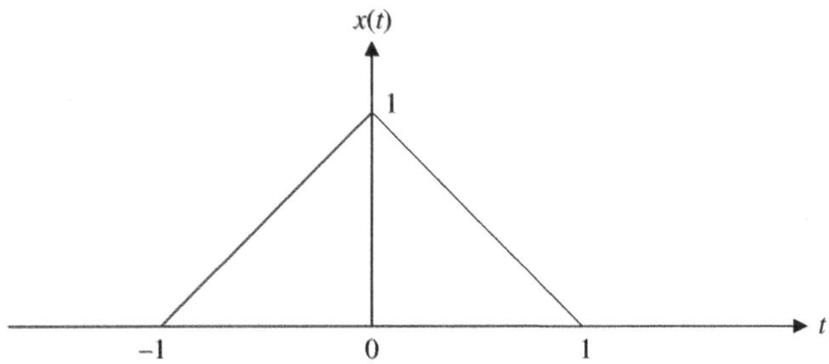

Figure 2.53 Triangular Function.

6 Find the Fourier transform of the triangular function shown in Figure 2.53, which is defined by

$$x(t) = \begin{cases} 1-t & 0 \leq t \leq 1 \\ 1+t & -1 \leq t \leq 0 \end{cases}$$

7 Find the Fourier transform of the following function:

$$x(t) = e^{-a|t|} = \begin{cases} e^{-at} & t \geq 0 \\ e^{at} & t < 0 \end{cases} \quad a > 0$$

8 Let $X(w)$ be the Fourier transform of $x(t)$. Using the basic definition of Fourier transform, find the Fourier transform of the time-shifted signal $x(t-3)$, which is the original signal that has been delayed by three time units.

9 Let $X(w)$ be the Fourier transform of $x(t)$. Using the basic definition of Fourier transform, find the Fourier transform of the signal $x(3t)$, which is the original signal that has been expanded three times in time.

Consider the following bitstream: $b = 11010110$.

(a) Sketch the non-return-to-zero (NRZ) line code of the bitstream.

(b) Sketch the Manchester line code for the bitstream.
(c) Sketch the return-to-zero (RZ) line code for the bitstream.

10 A modulating signal used to amplitude-modulate a carrier wave is given by $m(t) = 6\cos(2\pi f_m t)$. If the modulation index is 0.4 and the frequency of the carrier is 10 MHz, write down the exact expression for the carrier signal, $c(t)$.

11 Consider an amplitude modulating system that works by multiplying a modulating signal $m(t) = 6\cos(2000\pi t)$ and a carrier wave $c(t) = 10\cos(2{,}000{,}000\pi t)$. Thus the modulated signal, which is the output of the system, is $s(t) = m(t)c(t)$. Give an exact expression of $s(t)$ showing all the frequency components present.

12 The bandwidth of a modulating signal is 4 kHz. If it is used to amplitude-modulate a carrier wave, what is the bandwidth of modulated signal?

13 When does overmodulation occur in amplitude modulation, and what is the result of overmodulation?

14 A modulating signal used to amplitude-modulate a carrier wave is given by $m(t) = 3\cos(2\pi f_m t + \varphi)$. If the modulation index is 0.8 and the frequency of the carrier is 10 MHz, write down the exact expression for the carrier signal, $c(t)$, assuming that the phase shift $\varphi = 0$.

15 Consider a system with a power rating of 40 dBm. What is the power of the system in milliwatts?

16 Consider a system of four sources, each of which generates a power of 40 dB. What is the total power of the system in decibels?

3

Data Link Layer Protocols

3.1 Introduction

As discussed in Chapter 1, the data link layer is responsible for organizing data in frames and for detecting errors that occur in a frame. It is also responsible for hop-by-hop (or link-by-link) flow control. In this chapter, we examine the structure of a frame, the data link layer flow control schemes, and two of the protocols defined for this layer, namely the high-level data link control (HDLC) and the point-to-point protocol (PPP).

3.2 Framing

The data link layer breaks up the bitstream it receives from the network layer into discrete *frames* and computes the checksum for each frame. In the past, many schemes were used to create frames, especially when character-oriented transmission was still used. Today, the common method is the bit-oriented scheme in which a frame consists of the following:

- A leading *flag*
- A header
- The data to be transmitted (or payload)
- The checksum
- A trailing flag.

The format of a frame is shown in Figure 3.1. The exact structure of the header depends on the type of data link control (DLC) protocol in use. The checksum is used for error detection in the frame at the receiver.

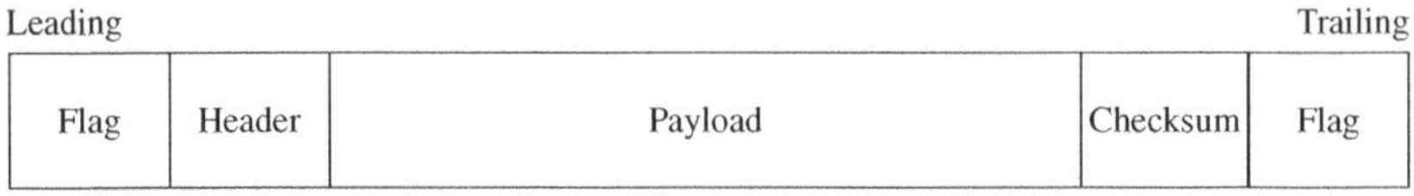

Figure 3.1 Format of Data Link Layer Frame.

3.3 Bit Stuffing

The flag is a special bit pattern 01111110, which is 1 byte (or octet) long. A process called *bit stuffing* is used to preserve the uniqueness of the flag in a frame. Thus, bit stuffing ensures that the flag's bit pattern never appears anywhere else in a frame. This is carried out in the following way. When the transmitter encounters five consecutive 1s following a 0 in the data, it automatically stuffs (i.e., inserts) a 0 into the outgoing bit stream. At the receiving end, when the receiver sees five consecutive 1s followed by a 0 bit, it automatically destuffs (or deletes) the 0 bit.

For example, consider the bit stream 011011101111100110 that is to be transmitted. We have the following:

(a) Data to be transmitted: 011011101111100110
(b) Data actually transmitted: 0110111011111**0**00110
(c) Data received after destuffing: 011011101111100110.

The 0 in bold is the stuffed bit that is inserted because there are five 1s after a 0 before the stuffed bit. At the receiving end, the receiver strips the zero and recovers the original bitstream.

As another example, assume that we want to send the bitstream:

$$m = 01101110111111111100110$$

What is actually sent is $m_1 =$ 0110111011111**0**11111**0**100110. In this case two bits are stuffed, as shown in bold. The first bit is stuffed after the five 1s following a 0. The second bit is stuffed as a result of the first stuffed bit. Introducing the first stuffed bit restarts the counter, which means that if there are five consecutive 1s after that 0, then another bit needs to be inserted. At the destination, the two bits in bold are discarded.

3.4 Flow Control

Flow control is a technique used to ensure that a sender transmits data at a rate that the receiver can accept. Thus, it is used to ensure that the transmitter does not overwhelm the receiver. Flow control can be exercised at the data link layer on a hop-by-hop basis and at the transport layer on an end-to-end basis. In this chapter, we consider flow control protocols that are used in the data link layer.

These include the *stop-and-wait* protocol and the *sliding window flow control* protocol.

3.4.1 The Stop-and-Wait Protocol

In the stop-and-wait protocol, the sender sends one frame and then waits for an acknowledgment (or ACK) from the receiver before sending another frame. Since the sender cannot send a new frame until the receiver has issued an ACK for the previous frame, it is obvious that the protocol ensures that the sender is transmitting at a rate that the receiver can handle. Figure 3.2 is an illustration of the protocol.

One of the drawbacks of the scheme is its poor channel utilization. A single frame travels from the source to the destination, and the associated acknowledgment travels from the destination to the source. This round trip delay, which is the time for a frame to go from the source to the destination and for the ACK to be received at the source, can be appreciable, particularly if the distance between the source and the destination is long.

3.4.2 The Sliding Window Flow Control

In the sliding window flow control, each outgoing frame contains a sequence number that ranges from zero to a predefined maximum number. The maximum number is usually $2^n - 1$, which means that the sequence numbers are represented by an n-bit field in the frame. The essence of the window scheme is that at any time instant, the sender maintains a set of sequence numbers corresponding to the frames that it is allowed to send. These frames are said to fall within the *sending window*. Similarly, the receiver maintains a *receiving window*, which corresponds to the set of frames it is permitted to receive.

The sequence numbers within the sending window represent the frames that have been sent or can be sent but are yet unacknowledged. The sequence numbers within the receiving window represent the frames that the receiver may accept; any frame falling outside the window is silently discarded. Note that the sending and receiving windows need not have the same lower and upper limits. Also, the window size need not be equal to the maximum number of frames that the sequence number field can support. Thus, even though the maximum number of frames that an n-bit sequence number can support is 2^n, the window size need not be 2^n; however, it cannot exceed 2^n. For example, consider a system

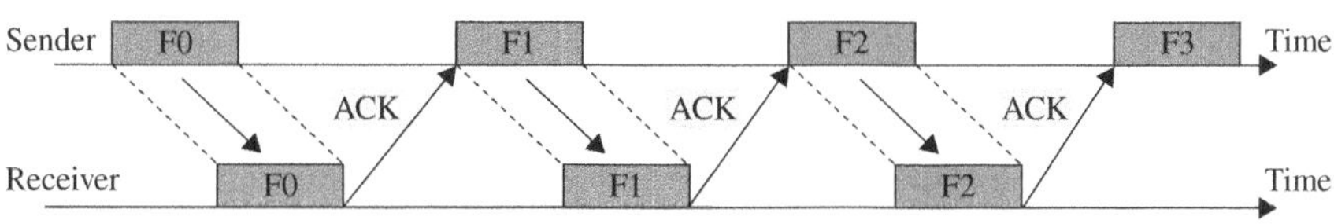

Figure 3.2 Example of Stop-and-Wait Protocol.

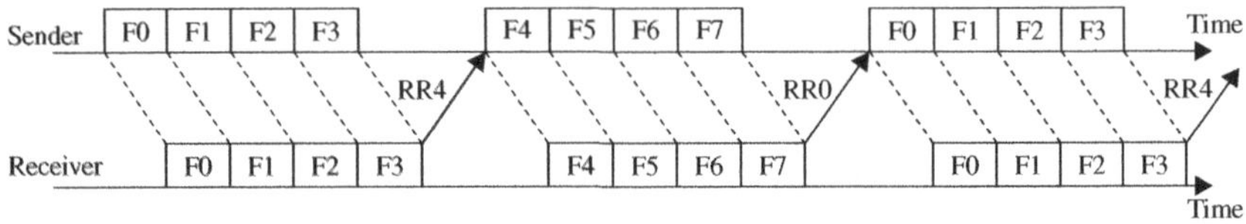

Figure 3.3 Example of the Window Flow Control.

with $n = 3$ and window size $W = 4$, which is shown in Figure 3.3 where the frames are numbered modulo-8 (since $2^3 = 8$). The sender usually does group acknowledgment with the RR (or receive ready) message, where RRk means "I have received frames up to and including frame $k - 1$; I am ready to receive frame k."

The sliding window flow control is a major improvement over the stop-and-wait protocol in terms of the channel utilization. The channel utilization increases as the window size increases.

3.5 Error Detection

Every practical communication channel is prone to noise that can cause errors to occur in the received information. Two techniques that are used to control transmission errors are error correction coding and error detection coding. Error correction codes add enough redundant bits, called *error correction bits,* to the information bit stream to enable the receiver to determine which bits have been corrupted by noise. Error detecting codes are concerned with detecting that errors have occurred without knowing precisely which bits are in error. Thus, while they also add redundant bits to the data bitstream that is to be transmitted, the overhead in terms of the number of redundant bits they use for this function is much smaller than that of error correcting codes.

In this chapter, we consider only error detection that usually calls for a retransmission of the information. We consider two error detection schemes:

- Parity checking
- Cyclic redundancy checking (CRC).

3.5.1 Parity Checking

A simple method of error detection is by adding redundant bits called *parity bits* to each character. This method is called parity checking and is commonly used for ASCII characters where seven bits are used for actual character encoding and the eighth bit is for parity. The value of the parity bit (i.e., the eighth bit) is chosen to make the number of 1s an even number (for even parity) or an odd number (for odd parity).

As an example of even parity, assume that the bit sequence 1001101 is to be transmitted; the message actually transmitted is 10011010 – the number of 1s is 4 (even), so the bit 0 is appended after the original sequence. The problem with the scheme is that if, for example in even parity, an even number of errors occur, the parity check scheme will fail to detect the fact the frame should be discarded. For example, if we transmitted 10011010 and received 10011000, the parity check will fail because the second to the last bit was corrupted, which is fine. But if the bit sequence 11011000 is received, the parity check will pass, even though the second bit and second to the last bit were corrupted in transit.

3.5.2 Two-Dimensional Parity

In the two-dimensional parity check, blocks of data are organized as a two-dimensional array. Specifically, each row of the array is a data block that is to be transmitted. A parity bit is appended to each row based on if even or odd parity is used. The parity bit for each column is similarly calculated. This is illustrated in Figure 3.4, which has four blocks of data and even parity is used in both row and column parities.

As shown in Figure 3.5(a), when exactly one error occurs the scheme can detect the location by the failure of the horizontal and parity checks associated with that bit. Then it can flip the bit thereby correcting the error. It can detect certain configurations of errors, particularly if they do not occur on the same

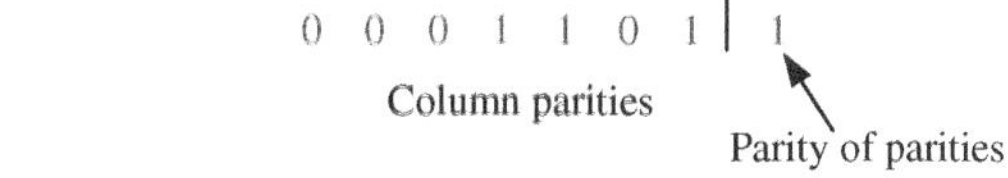

Figure 3.4 Example of Two-Dimensional Parity Check.

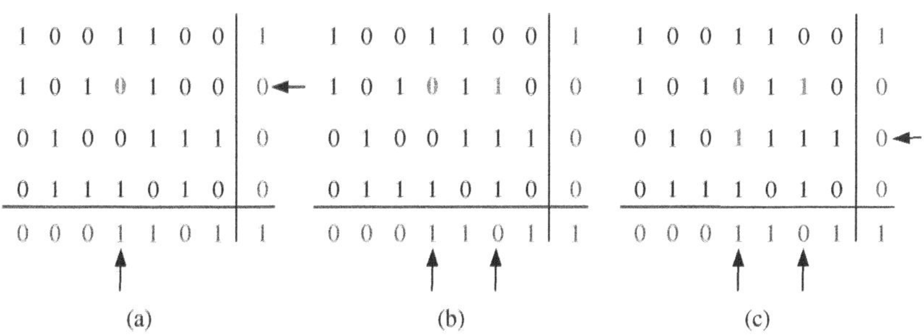

Figure 3.5 Some Possible Error Configurations. (a) 1 Error; (b) 2 Errors; and (c) 3 Errors.

row or column which it cannot detect. However, it cannot correct such errors as Figure 3.5(b) illustrates. Similarly, it can detect three errors, as illustrated in Figure 3.5(c). Finally, it cannot detect all the four error configurations.

3.5.3 Cyclic Redundancy Checking

A more sophisticated error detection scheme used by almost all communication networks is the CRC. CRC detects the presence of transmission errors by adding a few bits called CRC bits or *checksum* bits or *frame check sequence* (FCS) bits to each frame.

Let m be a string of $n+1$ bits, where n is usually a large number. m is the message to be transmitted and is represented by the binary string $m = b_n, b_{n-1}, \dots, b_2, b_1, b_0$, where $b_i = 0$ or 1. The message is sometimes represented as the polynomial $M(x)$ in which the bits of the message are considered to be the coefficients of the polynomial; that is, for the above message m we have the following polynomial of order n (which implies that the highest power of x is n and thus the number of bits in the bit stream m is $n+1$):

$$M(x) = b_n x^n + b_{n-1} x^{n-1} + \cdots + b_2 x^2 + b_1 x + b_0$$

For example, the bit stream $m = 101101$ is represented as the polynomial $M(x) = x^5 + x^3 + x^2 + 1$, which is a polynomial of order 5.

Before it is transmitted, the message polynomial is known only to the transmitter. Another polynomial $G(x)$ called the *generator polynomial* of order k is known to both the transmitter and the receiver. Typical values of k are 12, 16, and 32.

The sender and receiver use an agreed-upon *generator polynomial* $G(x)$. The high and low order bits of $G(x)$ are 1. To compute a *checksum* for a frame with n bits (whose polynomial is $M(x)$), the frame must be longer than the generator polynomial.

CRC operates as follows:

(a) $M(x)$ is multiplied by x^k, which is equivalent to shifting the bit stream k bits to the left (recall that $k+1$ is the number of bits in g).
(b) The new polynomial $x^k M(x)$ is divided by $G(x)$ to generate a remainder $R(x)$ whose degree is at most k.
(c) The division is not done in the usual manner; in the subtraction portion of division, the remainder is obtained through an XOR operation rather than by traditional subtraction.
(d) The quotient is discarded, and the message is concatenated with the remainder, and the concatenated bit stream is transmitted.
(e) Upon receiving the bit stream, the receiver divides it by $G(x)$. If a nonzero remainder is found, the message is discarded because it is in error; otherwise, it is accepted.

The reason is the following: Let

$$\frac{x^k M(x)}{G(x)} = Q(x) + \frac{R(x)}{G(x)}$$

where $Q(x)$ is the quotient. Then, using modulo-2 arithmetic,

$$\frac{x^k M(x) + R(x)}{G(x)} = Q(x)$$

That is, $x^k M(x) + R(x)$ is divisible by $G(x)$, where in modulo-2 arithmetic, addition and subtraction operations produce identical results. Thus, if we discard $Q(x)$ and add the remainder bitstream r to the shifted version of the message (equivalently, if we append the remainder bitstream to the original message bitstream), the cascaded sequence obtained is divisible by g.

Note: In the remainder of this discussion, we use the upper case for polynomials and the lower case for bitstreams. For example, $G(x)$ is the generator polynomial and g is the associated bitstream.

Example 3.1 Assume that $M(x) = x^7 + x^5 + x^4 + x^2 + x + 1$, which means that $m = 10110111$. Let $G(x) = x^5 + x^4 + x + 1$, which means that $g = 110011$ and $k = 5$. Let

$$A(x) = x^k M(x) = x^5 M(x) = x^{12} + x^{10} + x^9 + x^7 + x^6 + x^5$$

This corresponds to the bitstream $a = 1011011100000$, which is the original bitstream m shifted five places to the left. Next we divide a by g to generate the remainder, which is the FCS:

```
                 11010111 = the quotient that is discarded
         ______________________________
  110011 | 1 0 1 1 0 1 1 1 0 0 0 0 0
           1 1 0 0 1 1
           -----------
             1 1 1 1 0 1
             1 1 0 0 1 1
             -----------
                 1 1 1 0 1 0
                 1 1 0 0 1 1
                 -----------
                     1 0 0 1 0 0
                     1 1 0 0 1 1
                     -----------
                       1 0 1 1 1 0
                       1 1 0 0 1 1
                       -----------
                         1 1 1 0 1 0
                         1 1 0 0 1 1
                         -----------
                           0 1 0 0 1 = r (or FCS)
```

The FCS is appended to the message to generate the bitstream b, which is then transmitted, where $b = 1011011101001$. At the receiver, b is divided by g, and if the remainder of this division is zero, the last $k = 5$ bits of b are discarded to generate the message bitstream m:

```
                  11010111 = the quotient that is discarded
         ________________________________
  110011 | 1 0 1 1 0 1 1 1 0 1 0 0 1
           1 1 0 0 1 1
           -----------
             1 1 1 1 0 1
             1 1 0 0 1 1
             -----------
                 1 1 1 0 1 0
                 1 1 0 0 1 1
                 -----------
                     1 0 0 1 1 0
                     1 1 0 0 1 1
                     -----------
                       1 0 1 0 1 0
                       1 1 0 0 1 1
                       -----------
                         1 1 0 0 1 1
                         1 1 0 0 1 1
                         -----------
                           0 0 0 0 0  No error
```

Different generator polynomials are used in CRC computations for different applications. These include the following:

(a) CRC-8: $x^8 + x^2 + x + 1$
(b) CRC-10: $x^{10} + x^9 + x^5 + x^4 + x + 1$
(c) CRC-12: $x^{12} + x^{11} + x^3 + x^2 + x + 1$
(d) CRC-16: $x^{16} + x^{15} + x^2 + 1$
(e) CRC-CCITT: $x^{16} + x^{12} + x^5 + 1$
(f) CRC-32: $x^{32} + x^{26} + x^{23} + x^{22} + x^{16} + x^{12} + x^{11} + x^{10} + x^8 + x^7 + x^5 + x^4 + x^2 + x + 1$

3.6 Error Control Protocols

We now consider mechanisms that deal with errors that occur in the transmission of a frame. These mechanisms deal with *error correction by retransmission*; that is, when an error occurs, the receiver will detect it and inform the source, and the source will have to retransmit the frame either on account of an explicit negative acknowledgment (NAK) from the receiver or after a timeout period with no positive ACK. Three error control schemes based on the automatic repeat request (ARQ) are used depending on how retransmission is done when errors occur:

(a) Stop-and-wait ARQ
(b) Go-back-*N* ARQ
(c) Selective repeat ARQ.

The stop-and-wait ARQ is based on the stop-and-wait flow control protocol, while the other two are based on the sliding window flow control protocol.

3.6.1 Stop-and-Wait ARQ

As stated earlier, the stop-and-wait ARQ is used to deal with errors that occur when the stop-and-wait flow control protocol is used. Errors can occur in four ways:

(a) The frame was corrupted in transit when going from source to sink.
(b) The frame was OK, but the ACK was corrupted in transit.
(c) The frame was lost in transit.
(d) The ACK was lost in transit.

Under the stop-and-wait ARQ, the source sends a frame and waits for a response, which can be an ACK or a NAK that can also be in the form of a timeout. When a NAK is received, the source resends the frame and keeps resending it until it receives an ACK after the frame has been correctly received at the destination. Some protocols permit a fixed maximum number of retransmissions after which the link is declared unusable. Figure 3.6 is an illustration of the scheme. In the figure, frame 1 is received in error and is retransmitted after the sender received the NAK.

3.6.2 Go-Back-*N* ARQ

The go-back-*N* ARQ deals with errors that occur when the sliding window protocol is used. When a NAK is received for a particular frame, the source resends that frame and all the frames that have been transmitted since that frame was sent as well as any new frames, provided the total number of frames being sent does not exceed *N*. Consider the example in Figure 3.7 where $N = 4$, which is essentially the window size. Frame F4 is received in error, but the sender got this information after it had already sent frames F5 and F6. Thus, these frames will be ignored by the receiver and the sender resends frames F4, F5, F6, and additionally frame F7 since its window size is 4.

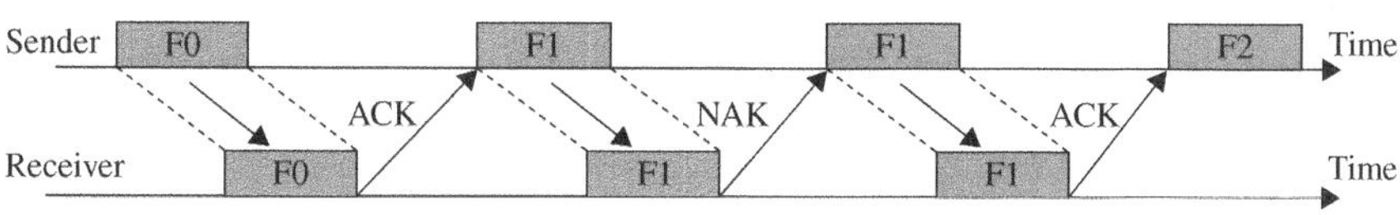

Figure 3.6 Illustration of the Stop-and-Wait ARQ.

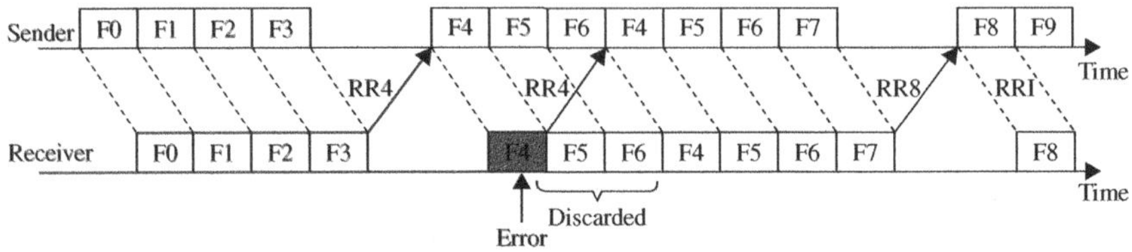

Figure 3.7 Example of Go-Back-4 ARQ.

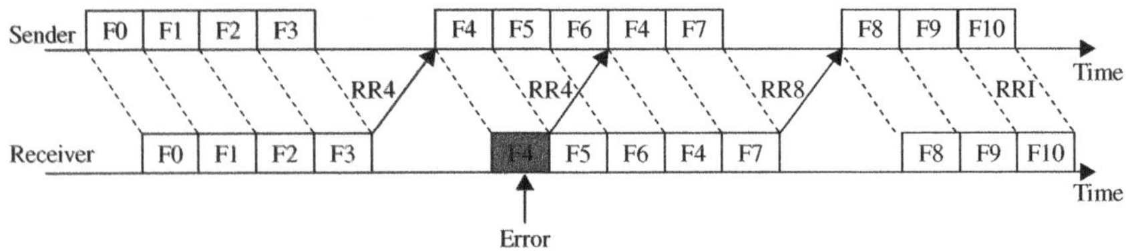

Figure 3.8 Illustration of the Selective Repeat ARQ.

3.6.3 Selective Repeat ARQ

One of the problems with the go-back-N scheme is that once a frame is found to be in error, all subsequent frames, including those that were not in error, are retransmitted. When the round-trip delay is high, as in satellite communication, the throughput of the channel becomes low. The selective repeat ARQ addresses this problem. Under this scheme, only the frame in error is retransmitted. The drawback is that the receiver must provide enough buffer to store the frames that were transmitted after the erroneous frame until the frame has been retransmitted. The reason for this is that the destination must resequence the frames and deliver them in the same order that they appear at the source. Thus, until the errored frame has been retransmitted and correctly received, the frames that are not errored must be stored in the buffer at the receiver.

Consider the operation of the scheme in the previous example assuming $W = 4$. This is illustrated in Figure 3.8 where unlike the go-back-4 scheme, only frame F4 is retransmitted instead of F4, F5, and F6 as in the go-back-4 scheme.

3.7 Data Link Control Protocols

DLC protocols supervise the retransmission of corrupted frames and regulate the transfer of frames. Thus, the job of DLC protocols is to ensure that data passed up to the next layer has been received *exactly as transmitted* (i.e., *error free, without loss, and in the correct order*), and also practice *flow control,* which ensures that data is transmitted only as fast as the receiver can receive it. As stated earlier, frames have a specific format and carry sequence numbers that enable the receiver to acknowledge them either individually or in groups. Different DLC protocols have been defined and they include the following:

(a) HDLC protocol
(b) PPP
(c) Frame relay.

The most commonly used DLC protocols are based on the same principles, namely, as follows:

(a) They are all bit-oriented and use bit stuffing for data transparency.
(b) They have the same frame structure where each frame has a leading and a trailing flag; there is an address field, a control field, and a data field.

They differ only in minor ways.

3.7.1 High-level Data Link Control

HDLC is a DLC protocol defined by the ISO for use in both point-to-point (P2P) and point-to-multipoint (or multidrop) data links. It supports full-duplex operation and is widely used in computer networks. HDLC operates in three modes:

(a) Normal response mode (NRM)
(b) Asynchronous response mode (ARM)
(c) Asynchronous balanced mode (ABM).

NRM is an unbalanced or asymmetric mode in which one end of the line is the master and the other end is the slave. It is used mainly in terminal-based networks where the slaves (or secondary stations) can transmit only when polled by the master (or primary). Thus, the primary station initiates transfers to the secondary station; the secondary station can only transmit a response when, and only when, it is instructed to do so by the primary station. In other words, the secondary station must receive explicit permission from the primary station to transfer a response. After receiving permission from the primary station, the secondary station initiates its transmission. This transmission from the secondary station to the primary station may be much more than just an acknowledgment of a frame. After each transmission, the secondary station must wait once again for explicit permission from the primary station to transfer anything. A *command frame* is a frame sent from the primary to the secondary, and a *response frame* is a frame sent from the secondary to the primary.

ARM is a scheme in which the secondary station does not have to wait to receive explicit permission from the primary station to transfer any frames. That is, the primary station does not need to initiate transfers by a secondary station. The frames may be more than just acknowledgment frames; they may contain data or control information regarding the status of the secondary station. This mode can reduce overhead on the link as no frames need to be transferred in order to give the secondary station permission to initiate a transfer. However, some limitations do exist. Due to the fact that this mode is asynchronous, the secondary station must wait until it detects an idle channel before it can transfer any frames. This is when the ARM link is operating at half duplex. If the ARM link is operating at full duplex, the secondary station can transmit at any time. In this mode, the primary station still retains responsibility for error recovery, link setup, and link disconnection.

ABM is used mainly on P2P links and each station has equal status and performs both primary and secondary functions. This mode uses combined stations. There is no need for permission on the part of any station in this mode because combined stations do not require any sort of instructions to perform any task on the link.

The NRM is used most frequently in multipoint lines, where the primary station controls the link. The ARM is better for point-to-point links as it reduces overhead. The ABM is not used widely today. The "asynchronous" in both ARM and ABM does not refer to the format of the data on the link; it refers to the fact that any given station can transfer frames without explicit permission or instruction from any other station.

3.7.1.1 HDLC Frame Format

The HDLC frame format is shown in Figure 3.9.

The frame consists of the following fields:

- *Flag*: Every frame on the link must begin and end with a flag sequence field containing the sequence 01111110.
- *Address*: The address field identifies the primary or secondary station involved in the frame transmission or reception. Each station on the link has a unique address. In an unbalanced configuration, the address field in both command and response frames refers to the secondary station. In a balanced configuration, the command frame contains the destination station's address and the response frame has the sending station's address.
- *Control*: HDLC uses the control field to determine how to control the communication process. This field contains the commands, responses, and sequence numbers used to maintain the data flow accountability of the link,

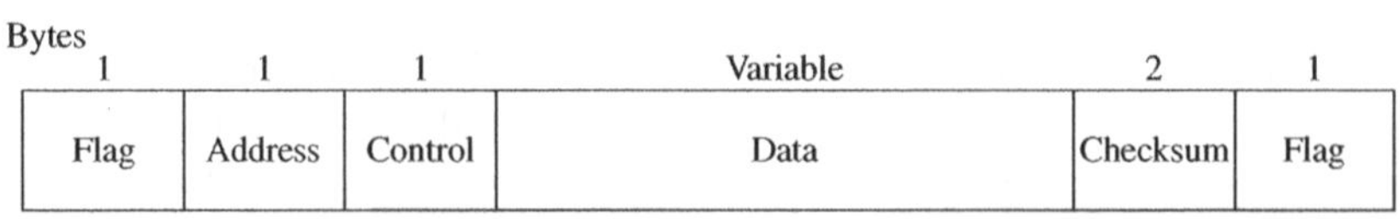

Figure 3.9 HDLC Frame Format.

defines the functions of the frame, and initiates the logic to control the movement of traffic between sending and receiving stations.

- *Information/data*: This field is not in all HDLC frames; it is only present when the information format is being used in the control field. The information field contains the actual data the sender is transmitting to the receiver.
- *FCS*: This field contains a 16-bit CRC that is used for error detection.

3.7.1.2 Control Field Format

The control field has three formats:

(a) *Information format*, which is used to indicate that the frame is used to transmit end-user data between two devices.
(b) *Supervisory format*, which is used to indicate that the control field is performing control functions such as acknowledgment of frames, requests for retransmission, and requests for temporary suspension of frames being transmitted. Its use depends on the operational mode being used.
(c) *Unnumbered format*, which is used to indicate that a frame is being used for control purposes. It is used to perform link initialization, link disconnection, and other link control functions.

These formats are shown in Figure 3.10.

- *Seq* is the sequence number of the current frame.
- *Next* is the piggybacked acknowledgment and the value is the sequence number of the next frame expected by the station transmitting this frame.
- *Poll/final bit* (P/F) is called the poll/final (or P/F) bit, which is ignored when it is set to 0 but is used to provide dialogue between the primary and the secondary stations. The primary station uses $P=1$ to acquire a status response from the secondary station; the P bit signifies a poll. The secondary station responds to the P bit by transmitting a data or status frame to the primary

Bits					
	1	3		1	3
(a) Information format	0	**Seq**		**P/F**	**Next**
	1	1	2	1	3
(b) Supervisory format	1	0	**Type**	**P/F**	**Next**
	1	1	2	1	3
(c) Unnumbered format	1	1	**Type**	**P/F**	**Modifier**

Figure 3.10 Control Filed Formats.

station with the *P*/*F* bit set to $F = 1$. The *F* bit can also be used to signal the end of a multiframe transmission from the secondary station under NRM.

In supervisory frame, different *types* are defined:

(a) Type 0 is *RR*, used by the primary or secondary station to indicate that it is ready to receive an information frame and/or acknowledge previously received frames.
(b) Type 1 is *reject* (REJ), which denotes negative acknowledgment due to transmission error and thus is used to request the retransmission of frames.
(c) Type 2 is *receive not ready* (RNR), which is used to indicate that the primary or secondary station is not ready to receive any information frames or acknowledgments.
(d) Type 3 is *selective reject* (SREJ), which is used by a station to request retransmission of specific frames.

The unnumbered format frames have five bits (two from type and three from modifier) that provide up to 32 additional commands and 32 additional response functions. These include the following:

- *Set normal response mode* (SNRM) places the secondary station into NRM. NRM does not allow the secondary station to send any unsolicited frames. Hence the primary station has control of the link.
- *Set asynchronous response mode* (SARM) allows a secondary station to transmit frames without a poll from the primary station.
- *Set asynchronous balanced mode* (SABM) sets the operational mode of the link to ABM.
- *Disconnect* (*DISC*) places the secondary station into a disconnected mode.
- *Set normal response mode extended* (SNRME) increases the size of the control field to two octets instead of one in NRM. This is used for extended sequencing. The same applies for *SARME* and *SABME*.
- *Set initialization mode* (SIM) is used to cause the secondary station to initiate a station-specific procedure(s) to initialize its data link level control functions.
- *Unnumbered information* (UI) is used to send information to a secondary station.
- *Unnumbered acknowledgment* (UA) is used by the secondary station to acknowledge the receipt and acceptance of an *SNRM, SARM, SABM, SNRME, SARME, SABME, RSET, SIM,* or *DISC* commands.

3.7.2 Point-to-Point Protocol

PPP was originally designed as an encapsulation protocol for transporting IP traffic over P2P links. It also established a standard for the assignment and management of IP addresses, asynchronous (start/stop) and bit-oriented synchronous encapsulation, network protocol multiplexing, link configuration, link quality testing, error detection, and option negotiation for such capabilities as network layer address negotiation and data-compression negotiation. PPP supports these functions by providing an extensible link control protocol (LCP) and a family of network control protocols (NCPs) to negotiate optional configuration parameters and facilities. In addition to IP, PPP supports other network layer protocols, such as the Novell's internetwork packet exchange (IPX) and DECnet.

3.7.2.1 PPP Components

PPP provides a method for transmitting datagrams over serial P2P links and contains three main components:

(a) A method for encapsulating datagrams over serial links. PPP uses the HDLC protocol as a basis for encapsulating datagrams over P2P links.
(b) An extensible LCP to establish, configure, and test the data link connection.
(c) A family of NCPs for establishing and configuring different network layer protocols; PPP is designed to allow the simultaneous use of multiple network layer protocols.

To establish communications over a P2P link, the originating PPP first sends LCP frames to configure and (optionally) test the data link. After the link has been established and optional facilities have been negotiated as needed by the LCP, the originating PPP sends NCP frames to choose and configure one or more network layer protocols. When each of the chosen network layer protocols has been configured, packets from each network layer protocol can be sent over the link. The link will remain configured for communications until explicit LCP or NCP frames close the link, or until some external event occurs (e.g., an inactivity timer expires or a user intervenes). PPP uses the principles, terminology, and frame structure of the HDLC procedures.

3.7.2.2 PPP Frame Format

The format of PPP is shown in Figure 3.11.

The fields are as follows:

- *Flag*: A single byte consisting of the binary sequence 01111110 that indicates the beginning or end of a frame.

Bytes

1	1	1	2	Variable	1 or 2	1
Flag	Address	Control	Protocol	Data	Checksum	Flag

Figure 3.11 PPP Frame Format.

- *Address*: A single byte that contains the binary sequence 11111111, the standard broadcast address; PPP does not assign individual station addresses.
- *Control*: A single byte that contains the binary sequence 00000011, (hexadecimal 0x03), which is the unnumbered information (UI) command in HDLC with the *P*/*F* bit set to zero. Frames with other control field values are silently discarded.
- *Protocol*: Two bytes that identify the protocol encapsulated in the information field of the frame and content is written in hexadecimal format. The protocol field values in the "cxxx" range identify datagrams as belonging to the LCP or associated protocols. Values in the "8xxx" range identify datagrams belonging to the family of NCPs. Finally, values in the "0xxx" range identify the network protocol of specific datagrams.
- *Data* consists of zero or more bytes that contain the datagram for the protocol specified in the protocol field. The end of the information field is found by locating the closing flag sequence and allowing two bytes for the FCS field. The default maximum length of the information field is 1500 bytes. By prior agreement, consenting PPP implementations can use other values for the maximum information field length.
- FCS is normally 16 bits (2 bytes), but by prior agreement, consenting PPP implementations can use a 32-bit (4-byte) FCS for improved error detection.

The LCP can negotiate modifications to the standard PPP frame structure, but modified frames will always be clearly distinguishable from standard frames.

3.7.2.3 PPP Link Control

The PPP LCP provides a method of establishing, configuring, maintaining, and terminating the P2P connection. LCP goes through four distinct phases. First, link establishment and configuration negotiation occur. Before any network layer datagrams (e.g., IP) can be exchanged, LCP first must open the connection and negotiate configuration parameters. This phase is complete when a configuration-acknowledgment frame has been both sent and received. This is followed by link quality determination. LCP allows an optional link quality determination phase following the link-establishment and configuration-negotiation phase. In this phase, the link is tested to determine whether the link quality is sufficient to bring up network layer protocols. (This phase is optional.) LCP can delay transmission of network layer protocol information until this phase is complete. At this point, network layer protocol configuration negotiation occurs. After LCP has finished the link quality determination phase, network layer protocols can be configured separately by the appropriate NCP and can be brought up and taken down at any time. If LCP closes the link, it informs the network layer protocols so that they can take appropriate action. Finally, link termination occurs; LCP can terminate the link at any time. This usually is done at the request of a user but can happen because of a physical event, such as the loss of carrier or the expiration of an idle-period timer.

Three classes of LCP frames exist:

(a) Link-establishment frames are used to establish and configure a link.
(b) Link-termination frames are used to terminate a link, and
(c) Link-maintenance frames are used to manage and debug a link.

These frames are used to accomplish the work of each of the LCP phases.

3.8 Summary

The data link layer is responsible for organizing data in frames and for detecting errors that occur in a frame. It is also responsible for hop-by-hop (or link-by-link) flow control. In this chapter, we have discussed different data link layer flow control schemes as well as error control schemes. We have also discussed the HDLC protocol that is the basis on which many other data link layer protocols were defined and the PPP.

Exercises

1 Assume that the following bitstream is to be transmitted as the payload of a system that uses the HDLC protocol:

011111001101111101111101110111110.

What bitstream is actually transmitted, given that bit stuffing is practiced?

2 Consider the following bit-stuffed bitstream that is received at the destination:

1100111110110011101111101111110011110

What is the bitstream that was presented to the source where the bit stuffing took place? (Note that this is asking for the original bitstream before bit stuffing was practiced at the source.)

3 Assume that the following block of messages is to be transmitted together:

0011011, 1100110, 1010110, 0101110, 1110001

Construct a two-dimensional parity check matrix for the block of messages using even parity.

4 Consider a sender–receiver arrangement for the following go-back-3 ARQ system:

- Frames 0, 1, 2, and 3 are sent, and the four frames are acknowledged.
- Frames 4, 5, 6, and 7 are sent, and NAK is received for frame 5.

(a) Draw the diagram for this sequence of transmissions.

(b) Assuming the sender has frames 8 and 9 to send after receiving the NAK for frame 5, show the next frame transmission sequence after frame 5 is retransmitted.

5 Consider a system with the generator bitstream $g = 1011$. We are interested in transmitting the data (or message) $m = 1001001$.

(a) Write down the generator polynomial $G(x)$.

(b) What is the order of the generator polynomial?

(c) Write down the message polynomial $M(x)$.

(d) Perform the cyclic redundancy check (CRC) computation on the message and indicate the bitstream that is transmitted.

(e) Assume that the bitstream 1001001011 is received at the destination. Show why it should or should not be accepted as being error-free.

6 A data link layer protocol uses the cyclic redundancy check (CRC) error detection technique with the generator polynomial $G(x) = x^5 + x^4 + x^2 + 1$.

(a) If the codeword 1010001101 is to be transmitted in this system, what codeword is actually sent?

(b) If the codeword 101000101001111 is received in this system, should it be accepted or rejected?

7 What is the difference between the normal response mode of operation of the high-level data link control (HDLC) protocol and the asynchronous response mode of operation of the HDLC protocol?

4

Multiple Access Schemes

4.1 Introduction

Multiple access schemes are used whenever there is a need for a number of independent users to share a resource that may be scarce or expensive. For example, in a communication environment, users may need to communicate with one another over a shared medium. The term "multiple access" means that many devices can connect to and share the same medium. One feature of such a shared medium is that when any user transmits their information, many of the users that are connected to the medium can receive the transmission. The role of a multiple access scheme is to coordinate the transmissions from the users to avoid a "collision" in the medium and define steps that can be taken to resolve collisions when they occur. A collision occurs when two or more users transmit their information in the medium in such a manner that at least one bit from one user overlaps with another bit from another user. In this case, the transmissions are unusable, and the information needs to be retransmitted.

A multiple access scheme must be able to handle several users without mutual interference. Also, it should be able to maximize the spectrum efficiency. In this chapter, we consider some of the popular multiple access schemes. These schemes can be divided into three categories:

(a) *Orthogonal access schemes*, in which transmissions are perfectly scheduled to avoid collision of two or more user packets on the channel.
(b) *Controlled access schemes*, in which users are using the same channel but are allowed to transmit in a round-robin manner; that is, exactly one user is "invited" to use the channel at a time.
(c) *Random access schemes*, in which transmitters transmit their packets either immediately or almost immediately upon their arrival hoping that no interference from other transmitters will be encountered. Thus, random access schemes are used in media where collisions can occur.

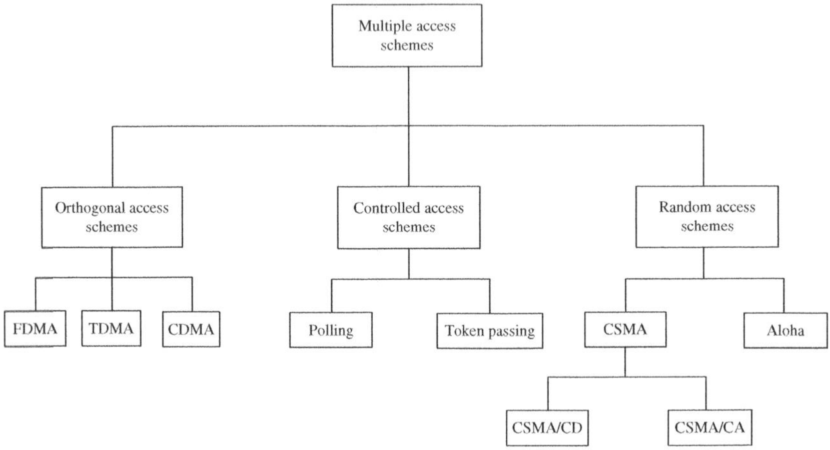

Figure 4.1 Classification of Multiple Access Schemes.

The classification is illustrated in Figure 4.1. Orthogonal access schemes include frequency division multiple access (FDMA), time division multiple access (TDMA), and code division multiple access (CDMA). Similarly, controlled access schemes include polling schemes and token-passing schemes. Finally, random access schemes include the Aloha system, carrier sense multiple access (CSMA), carrier sense multiple access with collision detection (CSMA/CD), and carrier sense multiple access with collision avoidance (CSMA/CA).

Multiple access schemes are used in a medium access control sublayer that lies between the data link layer and the physical layer. Before we discuss these schemes, we first review the basic fixed channel access schemes that are the foundation of the orthogonal access schemes.

4.2 Multiplexing Schemes Revisited

As discussed in Chapter 2, multiplexing is the transmission of more than one signal on the same physical link, staggered either in time or frequency. Most multiplexing schemes are based on the concept of dividing a communication link into independent communication channels. Three methods used to divide a communication link into independent channels are as follows:

- Frequency-division multiplexing (FDM), which is used for analog systems
- Time-division multiplexing (TDM), which is used for digital systems
- Code-division multiplexing (CDM), which is a form of spread spectrum.

4.2.1 FDM

The basic idea of FDM is that if a signal to be transmitted occupies a narrow band of frequencies, then other signals can be transmitted at the same time provided that their frequencies do not overlap. This is achieved by using each signal to modulate a unique carrier whose frequency is chosen so that the frequency spectra of the different modulated signals do not overlap. In this way, the bandwidth of the physical link is divided up into a number of separate frequency bands (or channels), each of which accommodates one signal, hence the term "frequency division."

4.2.2 TDM

TDM is the process of switching a number of signal sources (or channels) in strict rotation one at a time to a single output. Specifically, in TDM, the transmission time is divided into time slots of equal duration, and the sources are polled in a fixed order to transmit in the slots allocated to them. For example, in a four-source system, source 1 transmits in slots 1, 5, 9, and so on; source 2 transmits in slots 2, 6, 10, and so on; source 3 transmits in slots 3, 7, 11, and so on; and source 4 transmits in slots 4, 8, 12, and so on. At the destination, the sinks extract the data in the slots into which their communicating parties (i.e., the sources) inserted their data. All the data from the sources in each complete round of transmission constitutes a frame; thus, a particular slot in each frame is assigned to a particular source.

4.2.3 CDM

CDM is a scheme in which each user is assigned a unique code that can be used any time over the entire operating frequency spectrum. Thus, we may characterize FDM as a scheme where a user is allocated a portion of the frequency spectrum to be used any time, TDM as a scheme in which a user is allowed to use the entire operating spectrum only during an allocated time interval, and CDM as a scheme in which a user can use the entire operating frequency spectrum any time but with an allocated code. The code allocated to one user is uncorrelated to the codes allocated to the other users. This means that even though all users are transmitting at the same time over the same frequency band, their transmissions do not interfere with one another. Thus, they are assigned orthogonal codes.

4.3 Orthogonal Access Schemes

Orthogonal multiple access schemes refer to techniques that allow two or more users to share radio frequency spectrum in a manner that avoids collision. Thus, users' transmissions are perfectly scheduled by some mechanism to prevent their packets from colliding on the channel. There are three primary orthogonal multiple access schemes, which are based on the multiplexing schemes discussed earlier. These are as follows:

(a) FDMA
(b) TDMA
(c) CDMA.

4.3.1 FDMA

In FDMA, the system uses FDM to divide the medium into different channels so that when a station requests a channel, one is assigned to it for the duration of the session, if it is available. When the station is done with its transmission, the channel will be put back to the pool of available channels that can be assigned on demand to another station. If the same station requests a channel later, the channel that is assigned to it may be the same one used before or another channel depending on what is available at the time the request is received. Thus, there is no guarantee that the station will be assigned the same channel whenever it makes a request for a channel.

Note that there is a difference between FDM and FDMA. First, FDM is a physical layer multiplexing technique, while FDMA is a data link layer access method. Thus, FDM is used to create the channels that can be accessed on demand by the multiple access scheme called FDMA. Also, in FDM, the user uses the same fixed channel all the time, while in FDMA, the user is allocated any available channel whenever he requests a channel to use.

4.3.2 TDMA

In TDMA, similar to FDMA, the system uses TDM to create time slots that constitute the channels on the medium. Any station that requests a channel will be assigned a time slot that it can use for the duration of its session. At the end of the session, the station releases the channel that can be made available to another station that will need it later. If the same station later requests a channel, there is no guarantee that the channel it used previously will be assigned to it.

TDM is different from TDMA in the sense that TDM is a multiplexing scheme that can be used to generate channels that can be used on a TDMA basis. In TDMA, there is no guarantee that a station will be assigned the same channel whenever it has something to transmit.

In some situations, a hybrid FDMA and TDMA is used. FDM is used to create different frequency bands, each of which is slotted and can be used in a TDMA manner. Thus, a channel is defined by both the FDM channel and the TDM time slot. This is used in cellular wireless networks.

4.3.3 CDMA

CDMA is a form of spread spectrum communication and the main principle of spread spectrum communication is that a spreading code is used to spread a narrowband signal over a bandwidth that is much larger than the signal's original bandwidth. Because of this much larger bandwidth, the signal looks like noise in the channel. The spreading is done by combining the data signal with a code that is independent of the transmitted data message. Figure 4.2 is an illustration of a given narrow band signal that is spread over a much larger frequency band. To recover the original signal, the received signal is despread with the same code used to spread it at the source.

In CDMA, the spreading code is a Walsh function that is derived from the 64×64 Walsh–Hadamard matrix. The Walsh–Hadamard matrices are characterized as follows:

(a) There is single row of 0s.
(b) In each of the other rows, there is an equal number of 0s and 1s.
(c) The matrices have N rows (and N columns) such that $N = 2^n$, where n is an integer.
(d) They can be recursively constructed as follows:

$$H_1 = [0], \ H_2 = \begin{bmatrix} 0 & 0 \\ 0 & 1 \end{bmatrix}, \ H_{2N} = \begin{bmatrix} H_N & H_N \\ H_N & \overline{H}_N \end{bmatrix}$$

where the elements of $\overline{H}_N$ are the binary complements of the corresponding elements of H_N. The rows of the matrix constitute the spreading codes. Thus, as a multiple access scheme, when a station requests a channel, it will be assigned one of the rows, if any is available, with which it can use to spread its signal on the transmission medium. As discussed earlier, even though different stations are using the medium at the same time, the spreading codes are such that there is minimal interference among the signals on the medium.

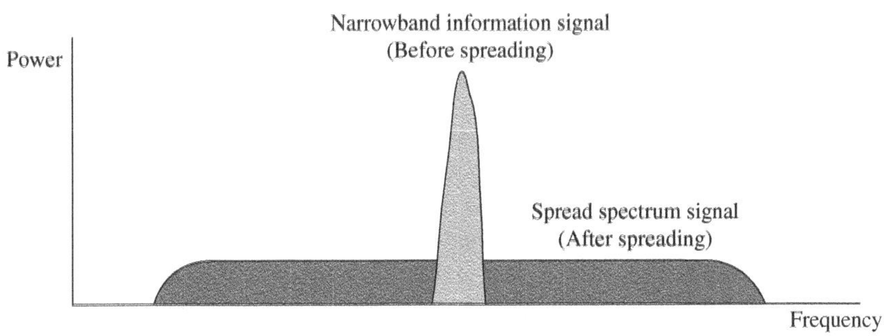

Figure 4.2 The Pre- and Postspreading Behavior of a Signal.

4.4 Controlled Access Schemes

As discussed earlier, in controlled access schemes, users are using the same channel but are allowed to transmit in a round-robin manner. The right to transmit can be gained in one of the two ways: *centralized polling*, which is also called *roll call*, or *distributed polling*, which is also called *token passing*.

4.4.1 Centralized Polling

In the centralized polling, a controller sequentially invites each station to transmit by explicitly polling the station, which is why it is also called a roll call. When a station is polled, it will transmit its frame, if it has one to transmit; otherwise, it does nothing. After the station has completed its transmission, or at the expiration of the timeout if the station has nothing to transmit, the controller polls the next station in the polling order. No station can transmit until it is polled, which means that frame collision on the medium is not possible.

4.4.2 Token Passing

Token passing is a peer-to-peer system used in a network such as a ring network in which stations are addressed in a fixed order. In this network, a special frame called the token circulates in the network. A station can only transmit when it captures the token. After transmitting its frame, the station releases the token and the next station in the polling order will receive it. In the case of the token ring network, the next station is the station that is downstream from the current station. The station will be the first to capture the token and transmit its frame, if it has any; otherwise, it allows the token to move to the next station downstream from it.

4.4.3 Service Policies

Regardless of whether centralized or distributed polling is used, there are three broad service policies in polling systems: *exhaustive service*, *gated service*, and *limited service*. In exhaustive service, when a station is polled it will transmit all the frames in its buffer, including the frames that arrive while it is still transmitting. Thus, the station will relinquish control of the channel only when there is no frame left in its buffer.

In gated service, when a station is polled it will transmit only those frames that were already in its buffer before it was polled. Frames that arrive while the station is busy transmitting will be transmitted the next time the station is polled. Thus, one way to view this policy is that at the instant a station is polled a gate is closed such that frames that arrive when the gate is closed will not be transmitted in the current round; they will be transmitted in the next round.

In limited service, a limit, say k, is placed on the number of frames that a station can transmit each time it is polled. Thus, the station continues to transmit until all frames in its buffer have been transmitted or a maximum of k frames have been transmitted, whichever comes first. A special case of the limited service polling scheme is the single-service polling system in which $k = 1$.

4.5 Random Access Schemes

The orthogonal multiple access schemes and the controlled access schemes discussed earlier are also called *conflict-free schemes* because every scheduled transmission is guaranteed to succeed. With random access (or contention-based) schemes the success of any transmission is not guaranteed because two or more sources may be transmitting at the same time, which results in the collision of their packets. For this reason, packets may have to be transmitted and retransmitted a number of times until they are successfully transmitted. Thus, the main concern of the random access schemes is transmission scheduling to minimize the probability of packet collision. The random access schemes to be discussed in the chapter include the following:

(a) The Aloha system
(b) CSMA
(c) CSMA/CD
(d) CSMA/CA.

4.5.1 Aloha System

The Aloha system was developed by Norman Abramson at the University of Hawaii in the 1970s. The goal was to develop an inexpensive method of data communication among the different University of Hawaii campuses scattered across the Pacific Ocean. The original scheme is called the *pure Aloha*, and its basic idea is to permit a user to transmit a packet as soon as the packet is generated, hoping that it does not suffer interference from other users' packets. If two or more sources transmit and their packets overlap in time, even partially, interference results and all involved transmissions are deemed unsuccessful and such packets will need to be retransmitted.

When a collision occurs, each source involved in the collision independently chooses a random backoff time after which it retransmits the packet. The randomness is required to ensure that the same set of packets does not continue to collide indefinitely. If the retransmitted packet suffers another collision, the random backoff scheme is used again until the packet is successfully transmitted or the user aborts the transmission. (Some systems define the maximum number of collisions allowed before packet retransmission is discontinued.)

The Aloha scheme is well suited for bursty traffic since a source does not hang onto the shared channel when it has nothing to transmit. It is a completely

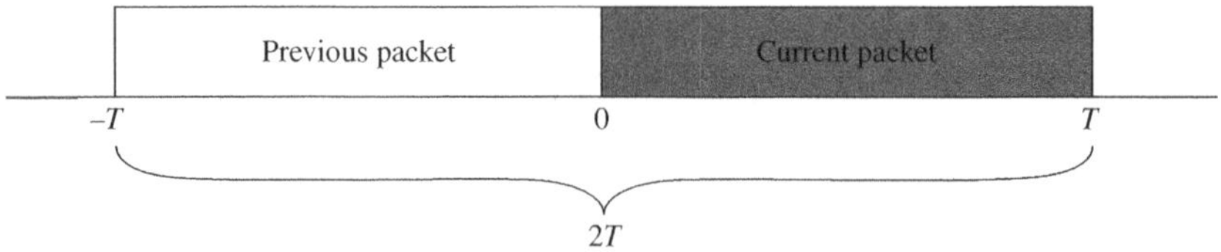

Figure 4.3 Window of Vulnerability of an Aloha Packet.

distributed scheme that permits every source to operate independently of the others. The major drawback is that the network performance significantly deteriorates at medium to high traffic levels due to the excessive number of collisions.

Let T seconds be the time to transmit a packet. Since partial overlap causes a packet to suffer a collision, a packet will be successfully transmitted if there is no other packet that arrived T seconds earlier and no other packet arrives during the time it takes to transmit this packet. Thus, the *window of vulnerability* of a packet is $2T$, as shown in Figure 4.3.

4.5.2 Slotted Aloha

Recall that in the pure Aloha even a very tiny overlap in time between two packets causes both packets to be considered unusable. The slotted Aloha was developed to address this drawback using slotted channels. Thus, packet transmission can only commence at the beginning of a slot and two or more packets either completely overlap or do not overlap at all. This scheme has been shown to perform better than the pure Aloha scheme. In this case, the window of vulnerability of a packet is T.

4.5.3 CSMA

CSMA is an evolutionary extension of the Aloha system. The Aloha scheme was developed for satellite-based systems where the satellite is on the order of miles up in the sky, which means that sensing to see if there is an ongoing transmission does not make sense because if we hear the transmission on the return channel, then the packet has already cleared the uplink.

CSMA was developed for packet radio communication where the signal is reflected in the ionosphere that is much lower than the satellite location. This means that if we are receiving a packet on the downlink, there is a very high probability that part of it is still using the uplink. Thus, the scheme operates as follows. When a station generates a frame for transmission, it first senses (or listens to) the channel to see if a transmission is currently taking place. If the channel is idle, the station commences its transmission. If the transmission is successful, the station is done. If it encounters a collision, the station will wait a random time and try again using the same transmission policy we described.

If the channel is busy when a station generates a frame to be transmitted, the station follows one of the following policies:

(a) Nonpersistent CSMA
(b) p-persistent CSMA
(c) 1-persistent CSMA.

In nonpersistent CSMA, the station backs off a random time and returns to try again, following the steps discussed earlier. In the p-persistent CSMA, the station will keep sensing the channel until the channel becomes idle. When it becomes idle, the station will transmit the frame immediately with probability of p, or with probability of $1 - p$ it waits one time unit to transmit the frame, sensing first to ensure that the channel is idle before transmitting. The special case of $p = 1$ is called the 1-persistent CSMA. Thus, in the 1-persistent CSMA, when the channel becomes idle the station transmits immediately.

4.5.4 CSMA/CD

As discussed earlier, CSMA was developed for packet radio communication where the signal is reflected in the ionosphere that is much lower than the satellite location. This means that if we are receiving a packet on the downlink, there is a very high probability that part of it is still using the uplink. When the transmission medium is on the order of kilometers long, then we can detect a collision after transmitting only a few bits of a packet. This is the idea behind the CSMA with collision detection that is used in local area networks, such as the Ethernet.

CSMA/CD operates as follows. When a station generates a frame to be transmitted, it first listens to make sure that no other source is transmitting. If it senses no ongoing transmission, it will commence transmitting the frame; otherwise, it takes one of the following actions:

(a) *Nonpersistent CSMA/CD*: It defers the transmission and tries again after a random time.
(b) *1-persistent CSMA/CD*: It waits until the end of the current transmission and starts its transmission when the current transmission ends.

While transmitting a frame, the station continues to monitor the channel to receive its own transmission and compares what it receives with what it transmitted. If the two are the same after an end-to-end propagation delay, it knows that it has sole access to the channel and will continue the transmission and monitoring the channel until it has completed the transmission. If what is received is different from what was transmitted, it aborts the transmission since its frame has suffered a collision with at least one other frame

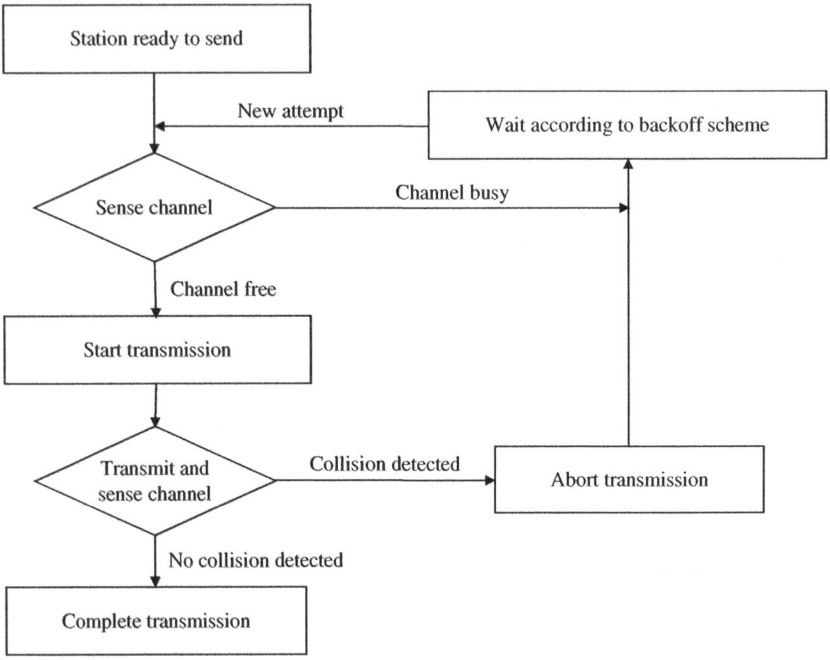

Figure 4.4 Summary of the Nonpersistent CSMA/CD Operation.

from another station. If a transmission results in a collision, each of the affected stations independently waits for a random time and reschedules its frame for transmission following the steps listed above. The operation of the nonpersistent CSMA/CD is summarized in the flow diagram shown in Figure 4.4, while that for the 1-persistent CSMA/CD is shown in Figure 4.5.

4.5.4.1 Why Listen While Transmitting in CSMA/CD

CSMA/CD was developed when local area networks were based on the line-drop architecture shown in Figure 4.6. Consider a cable that is 1 km long with a capacity of 1 Mb/s. Assume that the speed of signal propagation in the cable is 2/3 the speed of light; that is, $v = 2 \times 10^8$ m/s. Thus, the time it takes a bit to travel from one end of the cable to the other end is

$$t = \frac{\text{Distance}}{\text{Velocity}} = \frac{10^3}{2 \times 10^8} = 0.5 \times 10^{-5} = 5 \times 10^{-6} \text{ s} = 5 \text{ μs}$$

The number of bits that can be transmitted in 5 μs at the transmission rate of 1 Mb/s is

$$(10^6 \text{ bits/s}) \times (5 \times 10^{-6} \text{ s}) = 5 \text{ bits}$$

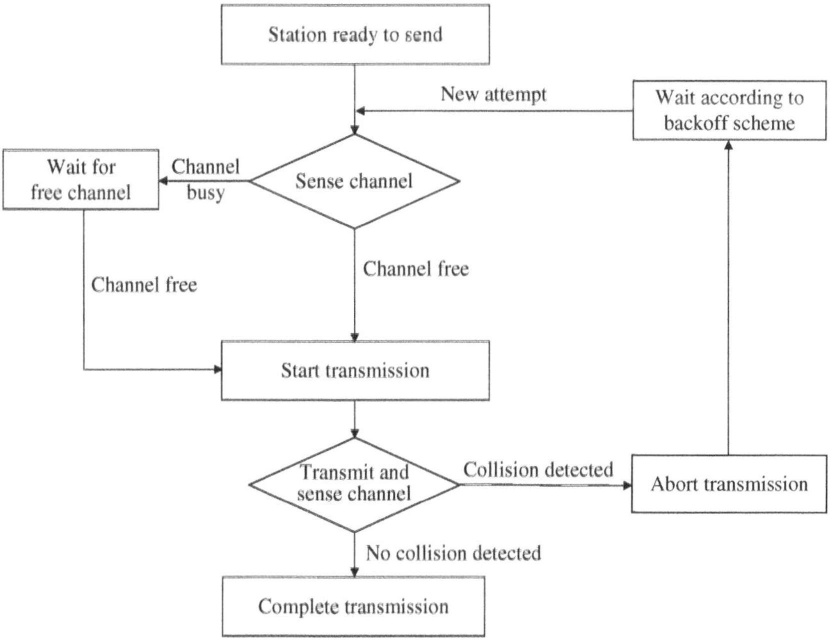

Figure 4.5 Summary of the 1-Persistent CSMA/CD Operation.

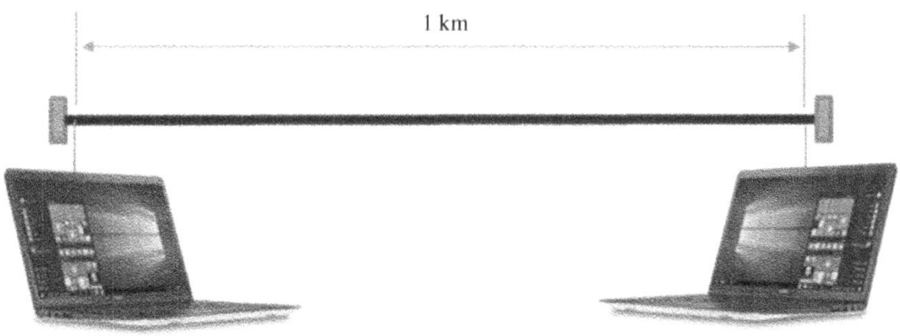

Figure 4.6 Illustration of the Rationale for Active Sensing in CSMA/CD.

Thus, after a node at one end of the cable has transmitted at most 5 bits, it will be able to determine if it has captured the channel. This is because after this time every other node that is connected to the cable will have known that a transmission from this node is taking place and will thus defer to the node, if such a node has not already started transmitting its packet. For a packet of 1500 bytes, the overhead of 5 bits required to decide whether to stop or continue transmission is very small. Thus, 5 bit times constitute the window of vulnerability.

4.5.5 CSMA/CA

CSMA/CA was developed for local area networks. Unlike wired local area networks where a station can transmit and receive at the same time, wireless local area networks are half-duplex systems where a wireless device can transmit or listen but not do both at the same time. Thus, while wired technologies have the ability to detect a collision as it is taking place, wireless LANs cannot detect an ongoing collision. This means that they have to make every effort to avoid a collision, which is the basis for CSMA/CA.

CSMA/CA avoids the collisions by using basic techniques defined in the IEEE 802.11 standard. Specifically, when a station is transmitting a frame, it must indicate how long it will take to complete the transmission, which is a parameter called *network allocation vector* (NAV) that every other station on the network notes. NAV is in the header of the frame being transmitted and is fundamentally a counter that needs to count down to zero before a station can transmit. It is also regarded as a *virtual carrier sensing* mechanism that is combined with the physical sensing mechanism to implement collision avoidance.

When a station generates a frame and finds the value of NAV to be zero, it senses the channel. If the channel is idle, it does not commence transmission immediately. It waits for a period of time called *distributed interframe space* (DIFS). The rationale for this is because if the channel is sensed to be idle, it may be possible that some distant station may have already started transmitting and the signal of that distant station has not yet reached other stations. Thus, by waiting for the duration defined as DIFS time, such a signal transmitted by another station will be received. If after this DIFS time the channel is still idle, the station can commence transmission of its frame. If either mechanism (i.e., NAV and physical sensing) indicates that the medium is in use during the interval DIFS, the station will choose a backoff interval and come back to try again.

4.6 Summary

This chapter has been concerned with methods of accessing a channel that is available to multiple users. These include the following:

1. Orthogonal access schemes, such as FDMA, TDMA, and CDMA, that prevent collisions and once a source accesses the channel it is guaranteed a collision-free transmission.
2. Controlled access schemes, such as polling systems and token-passing systems, that use a round-robin scheduling scheme to grant access to the medium. They are different from orthogonal systems because in orthogonal access schemes multiple transmissions can be taking place concurrently, while in controlled access schemes only one user can use the medium at a time.

3. Random access schemes, such as the Aloha scheme, CSMA, CSMA/CD, and CSMA/CA, where there is no guarantee that collision will not occur when a source starts its transmission.

Exercises

1 What is the difference between pure Aloha and slotted Aloha?

2 What is the difference between the carrier sense multiple access (CSMA) and the carrier sense multiple access with collision detection (CSMA/CD)?

3 Consider a cable that is 2 km long with a capacity of 10 Mb/s and used in a CSMA/CD manner. Assume that the speed of signal propagation in the cable is $v = 2 \times 10^8$ m/s.

a. How many bit times will it take a user at one end of the cable to capture the channel?

b. If the channel capacity is 10 kb/s, how many times will it take a user at one end of the cable to capture the channel?

4 What is the difference between an orthogonal multiple access scheme and a controlled access scheme?

5 What is the difference between a roll-call polling scheme and a token-passing scheme?

6 What is network access vector (NAV)?

7 Why is NAV referred to as a virtual carrier sensing mechanism?

8 What is the difference between the gated-service polling system and the exhaustive-service polling system?

5

Local Area Networks

5.1 Introduction

In this chapter, we discuss different flavors of local area networks (LANs). The specific topics to be covered include the following:

(a) Ethernet LANs
(b) Gigabit Ethernet
(c) Wireless LANs
(d) Token ring LANs.

5.2 Ethernet

The Ethernet is the most widely used LAN technology. The data rates of the most popular versions of Ethernet are as follows:

- 10 Mbps
- 100 Mbps, which is usually referred to as the Fast Ethernet
- 1000 Mbps (or 1 Gbps), which is the Gigabit Ethernet.

An Ethernet LAN may use coaxial cable, special grades of twisted pair wiring, or fiber optic cable. "Bus" and "Star" wiring configurations are supported. Ethernet devices compete for access to the network using the carrier sense multiple access with collision detection (CSMA/CD) protocol. Ethernet LANs are defined in the IEEE 802.3 standards. The IEEE 802.3 LAN network architecture follows the hierarchical model, as shown in Figure 5.1. The MAC client sublayer is called the logical link control (LLC) when an end system, such as PC, is transmitting data. It is called the bridge entity when it is used for LAN-to-LAN interconnection.

Fundamentals of Data Communication Networks, First Edition. Oliver C. Ibe.

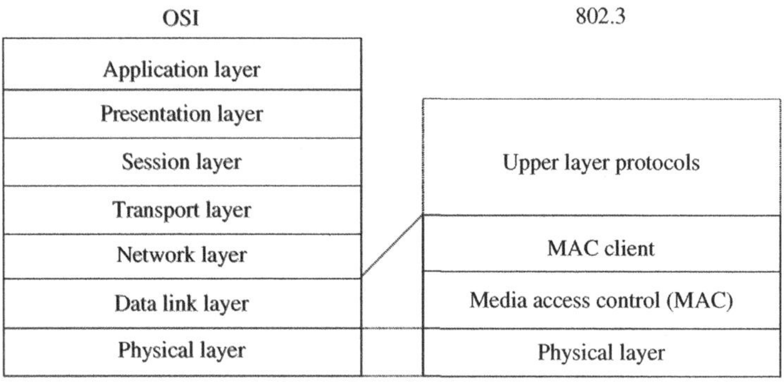

Figure 5.1 IEEE 802.3 Protocol Architecture.

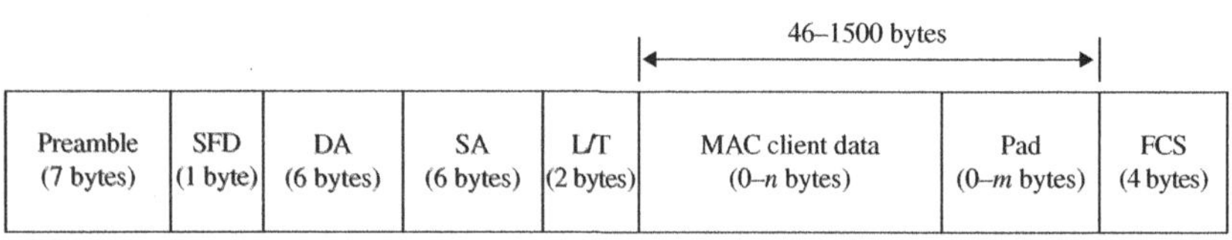

Figure 5.2 Ethernet Frame Format.

5.2.1 Ethernet Frame Structure

Figure 5.2 illustrates the format of an Ethernet frame, which is similar to the IEEE 802.3 LAN frame. In the remainder of this chapter, we use the term "Ethernet" for both systems.

The fields are as follows:

- The *preamble* consists of seven bytes all of the form 10101010 and is used by the receiver to allow it to establish bit synchronization.
- The *start frame delimiter* (SFD) is a single byte, 10101011, which is a special data structure that indicates the start of a frame.
- The *destination MAC address* (DA) is 48 bits long and is the MAC address of the destination device.
- The *source MAC address* (SA) is also 48 bits long and is the MAC address of the source device that is generating the frame.
- The *length/type* field (L/T) is used to distinguish between an IEEE 802.3 frame and an Ethernet frame. IEEE 802.3 is the international standard, while Ethernet is the original network that was developed at the Xerox Lab. This field is fundamentally the only place where the two differ. L is used in 802.3 frame to indicate the number of octets of data in the frame's payload, while T is used in the Ethernet to indicate the type of payload carried in the frame. While today's payload is mainly IP, the Ethernet was designed at a time when there were other types of payload, such as DECnet and IPX.

- The *MAC client data* is the payload or user data that is being transmitted.
- The *frame check sequence* (FCS, or Checksum) field is a 4-byte field that uses a CRC-32 polynomial code.

The minimum length of the MAC client data field is 46 bytes and maximum is 1500 bytes. If it is less than 46 bytes, it is padded up to 46 bytes using the Pad field. Similarly, the maximum frame length was originally set to 1518 bytes. However, with the introduction of the virtual local area network (VLAN) that has a VLAN header of 4 bytes, the current maximum length is 1522 bytes.

5.2.2 IEEE 802.3 LAN Types

The Ethernet has undergone a great deal of evolutionary change with respect to its data rate. The different incarnations are labeled by either "Base" to indicate that it is a baseband communication-based scheme or "Broad" to indicate that it is a broadband communication-based scheme. Another label that is associated with each incarnation is "T" (including T2, T4, and TX) for twisted pair cable, "F" (including FX) for fiber optic cable, "LX" for long wavelength laser transmitters over fiber optic cable, "SX" for short wavelength laser transmitters over fiber optic cable, and "CX" for gigabit signaling over copper cable. Finally, the number that precedes "Base" and "Broad" indicates the data rate in Mbits/s. For example, 10Base-T implies a 10-Mbits/s system that uses baseband signaling over twisted pair. Also, some standards are of the form "kX," where k is C, F, L, S, or T, and the X indicates full-duplex operation. The following are some of the past and current Ethernet versions:

- 1Base5: Original IEEE 802.3 LAN at 1 Mb/s data transfer rate
- 10Base2: IEEE 802.3 shorthand term for 10 Mb/s Ethernet based on Manchester signal encoding over thin coaxial cable; also called "Thinnet" or "Cheapernet"
- 10Base5: IEEE 802.3 shorthand term for 10 Mb/s Ethernet based on Manchester signal encoding over thick coaxial cable; also called "Thicknet"
- 10Base-F: IEEE 802.3 shorthand term for 10 Mb/s Ethernet based on Manchester signal encoding over fiber optic cable
- 10Base-T: IEEE 802.3 shorthand term for 10 Mb/s Ethernet based on Manchester signal encoding over category 3 or better twisted pair cable
- 10Broad36: IEEE 802.3 shorthand term for 10 Mb/s Ethernet on broadband cable
- 100Base-FX: IEEE 802.3 shorthand term for 100 Mb/s Fast Ethernet based on 4B/5B signal encoding over fiber optic cable
- 100Base-T: IEEE 802.3 shorthand term for entire 100 Mb/s Fast Ethernet system
- 100Base-T2: IEEE 802.3 shorthand term for 100 Mb/s Fast Ethernet based on PAM5x5 signal encoding and using two pairs of category 3 twisted pair cable

- 100Base-T4: IEEE 802.3 shorthand term for 100 Mb/s Fast Ethernet based on 8B6T signal encoding and using four pairs of category 3 twisted pair cable
- 100Base-TX: IEEE 802.3 shorthand term for 100 Mb/s Fast Ethernet based on 4B/5B signal encoding and using two pairs of category 5 twisted pair cable
- 100Base-X: IEEE 802.3 shorthand term for any 100 Mb/s Fast Ethernet system based on 4B/5B signal encoding; includes 100Base-TX and 100Base-FX
- 1000Base-CX: IEEE 802.3 shorthand term for 1000 Mb/s Gigabit Ethernet based on 8B/10B signaling over copper cable
- 1000Base-LX: IEEE 802.3 shorthand term for 1000 Mb/s Gigabit Ethernet based on 8B/10B signaling using long wavelength laser transmitters over fiber optic cable
- 1000Base-SX: IEEE 802.3 shorthand term for 1000 Mb/s Gigabit Ethernet based on 8B/10B signaling using short wavelength laser transmitters over fiber optic cable
- 1000Base-T: IEEE 802.3 shorthand term for 1000 Mb/s Gigabit Ethernet over Cat 5e unshielded twisted pair (UTP) cable
- 1000Base-TX: IEEE 802.3 shorthand term for 1000 Mb/s Gigabit Ethernet over Cat 6 UTP
- 10GBase-T: shorthand term for10 Gbps Ethernet over Cat 6a/Cat 7 UTP
- 10GBase-LX: shorthand term for 10 Gbps Ethernet based on single-mode fiber
- 10GBase-SX: shorthand term for 10 Gbps Ethernet based on multimode fiber
- 40GBase-CR4: shorthand term for 40 Gbps Ethernet based on copper cable assembly
- 40GBase-SR4: shorthand term for 40 Gbps Ethernet based on parallel multimode fiber
- 40GBase-LR4: shorthand term for 40 Gbps Ethernet based on parallel single-mode fiber
- 100GBase-CR10: shorthand term for 100 Gbps Ethernet based on copper cable assembly
- 100GBase-SR10: shorthand term for 100 Gbps Ethernet based on parallel multimode fiber
- 100GBase-LR10: shorthand term for 100 Gbps Ethernet based on parallel single-mode fiber
- 100GBase-ER4: shorthand term for 100 Gbps Ethernet based on parallel single-mode fiber.

5.2.3 Ethernet Topologies

Ethernet is a contention-based broadcast technology that uses baseband signaling. In the early days of its development, the Ethernet consisted of a single bus to which all computers (or nodes) were connected. It is a contention-based system in the sense that every node must compete with every other node for

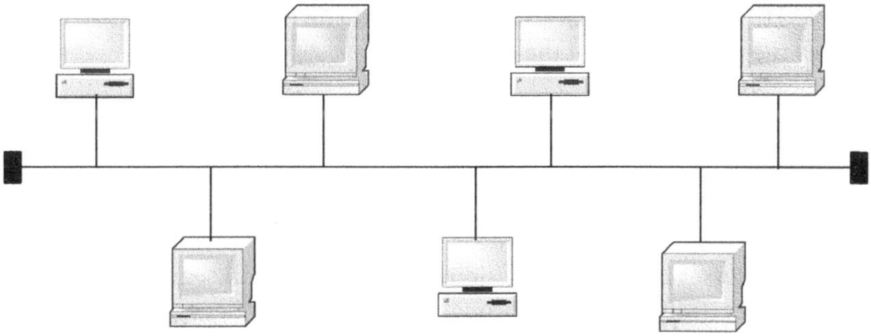

Figure 5.3 Ethernet Physical Bus Topology.

access to the network. At any point in time only one node can use the bus; that is, a node can transmit only when no other node is transmitting. It is a passive system in which no one node controls the network. The original topology is shown in Figure 5.3.

The topology later became the star topology in which all nodes are wired directly to a central hub. The advantage of a star topology is that it is easy to isolate a problem node. Unfortunately, the hub becomes a single point of failure because its failure brings the network down. The physical structure of the topology is shown in Figure 5.4.

The role of the hub is to enable packet flow from one node to all the other nodes connected to the hub. Each node attaches to the hub via a port. Note that while the hub is physically a star topology, it is logically a bus. Thus, a packet transmitted by each node that is connected to the hub is seen by every other node that is connected to the hub. Also, as the number of nodes that are

Figure 5.4 The Star Topology.

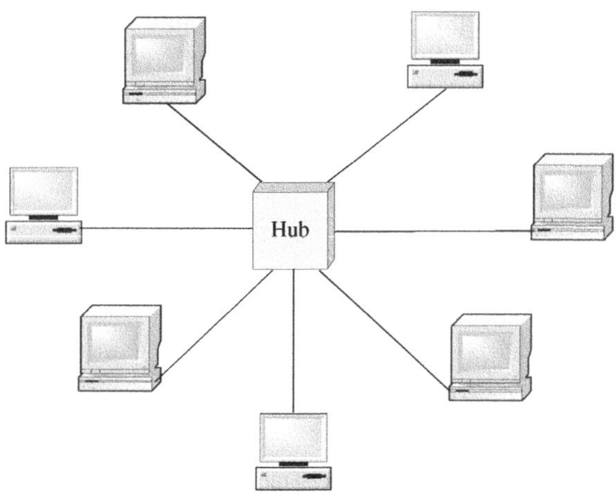

connected to the hub increases, the collision rate on the "bus" increases, which degrades the performance of the network.

5.2.4 LAN Switching

A *bridge* is a network element that is used to connect separate networks together. Bridges connect different networks types (such as Ethernet and Fast Ethernet) or networks of the same type. Ethernet bridges map the Ethernet addresses of the nodes residing on each network segment and allow only necessary traffic to pass through the bridge. Bridges are also called "store-and-forward" devices because they look at the whole Ethernet packet before making filtering or forwarding decisions. Specifically, when a packet is received, the bridge determines the destination and source LAN segments. If the LAN segments are the same, the packet is dropped (or "filtered"). If the segments are different, then the packet is "forwarded" to the correct segment. In addition, bridges do not forward bad or misaligned packets.

The Ethernet switch was developed in an attempt to improve the performance of the hub. While the switch physically looks like a hub, it operates in a different way. As in the hub each node is attached to the switch via a port. However, unlike the hub no collision can occur in the switch. The switch is an intelligent device that maintains a record of the MAC address of each node connected to a port. When it receives a packet, it reads the destination address of the packet and forwards the packet to the port to which the node that has the MAC address is connected. This not only improves the performance of the network but also ensures that no other node can "see" a packet that is destined for another node. Also, unlike a hub in which only one node can transmit at a time, a switch permits parallel transmissions to take place as long as no two packets are destined for the same port.

Figure 5.5 shows the difference between a hub and a switch. The node attached to port C is sending a frame to the node attached to port A. In Figure 5.5(a), the hub forwards the frame out of all ports except the port through which it received the frame. In Figure 5.5(b), the switch forwards the frame only to port A where the recipient of the frame is located. Thus, in a hub the node attached to port B receives the frame that is destined for the

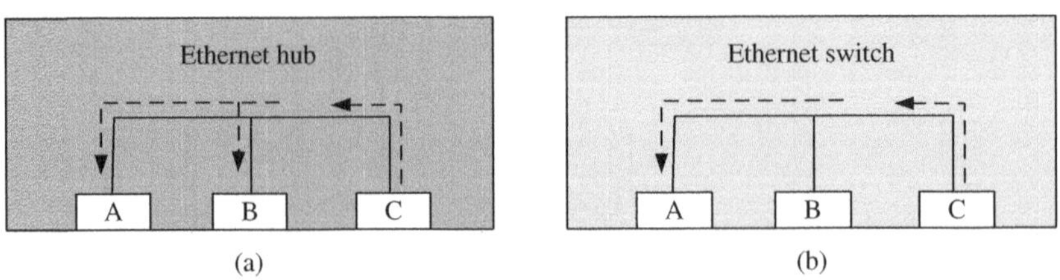

Figure 5.5 Difference Between Frame Handling in a Hub and a Switch.

node attached to port A; it will drop the frame because the destination MAC address on the frame is not its own MAC address. This is not the case in a switch; only the node attached to port A will receive the frame.

An Ethernet switch behaves like a learning bridge in the sense that it maintains a forwarding table. When it is first turned on the table is empty and when it receives a frame it will enter that MAC address of the sender against the port through which it received the frame and then it retransmits the frame to all ports, except the one on which the frame was received. When and if it receives an ACK to the frame, it will record the MAC address of the sender against the port through which it came. Thus, for a few minutes after the switch is turned on, it will behave like a hub until it has learned the location of all the nodes and thus fully populated the forwarding table when it begins to send frames to their output ports. At any time if a frame is destined for a node whose MAC address is not on the forwarding table, the switch will behave like a hub when handling the frame.

LAN switches are an expansion of the concept in Ethernet bridging. A LAN switch can link many LANs together. LAN switching allows many users to communicate in parallel through the use of virtual circuits and dedicated network segments in a collision-free environment. A LAN switch is also used to partition a large LAN into multiple *collision domains* thereby reducing collisions and improving network performance. The switch allows the different collision domains to belong to one *broadcast domain*.

A collision domain is a part of a network where packet collisions can occur. A collision can occur in one LAN segment or in a hub environment. Thus, all the ports on a hub are in the same collision domain. Similarly, a broadcast domain is a domain in which a broadcast is forwarded. A broadcast domain contains all devices that can reach each other at the data link layer using a broadcast. Thus, all ports on a hub or a switch are in the same broadcast domain because when a broadcast frame is received in one port, it is forwarded to all ports. Note that all ports on a router are in the different broadcast domains and routers do not forward broadcasts from one broadcast domain to another. Figure 5.6 illustrates the concepts of collision domain and broadcast domain.

5.2.5 Classification of Ethernet Switching

There are different ways to classify Ethernet switches. These include the following:

(a) Port capacity configuration
(b) Frame forwarding method
(c) Highest layer used for forwarding.

With respect to port capacity configuration, an Ethernet switch can be classified as an *asymmetric* or a *symmetric* switch. An asymmetric switch is one that

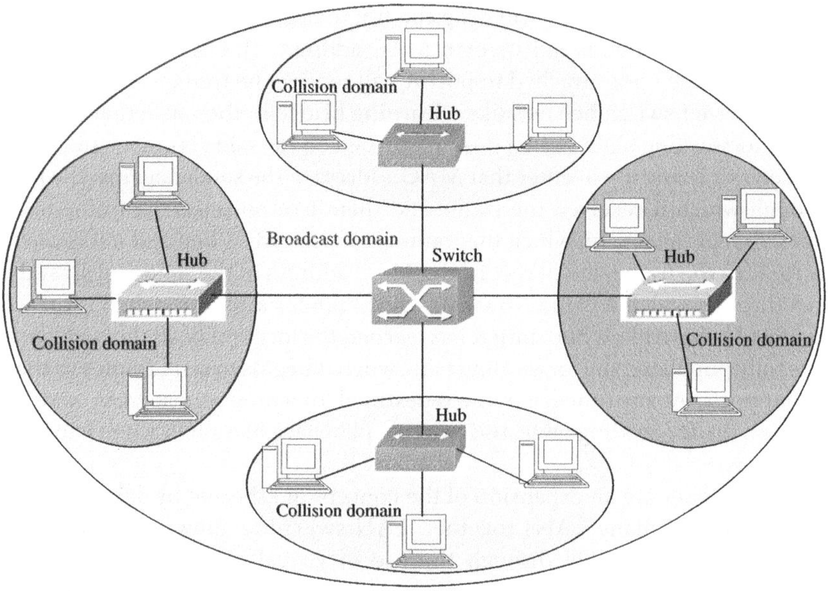

Figure 5.6 Illustration of Collision and Broadcast Domains.

provides switched connections between ports with the different data rates, such as some 10 Mbps and one or more 100 Mbps ports. A symmetric switch provides switched connections between ports with the same data rate, such as all 10 Mbps or all 100 Mbps ports.

5.2.6 Frame Forwarding Methods

There are three modes in which switches can operate. These are as follows:

(a) Store-and-forward switching
(b) Cut-through switching
(c) Fragment-free switching.

5.2.6.1 Store-and-Forward Switching

With *store-and-forward switching*, the switch does not begin transmitting the outgoing frame until it has received the entire incoming frame, buffered it, and has checked to make sure it contains no errors. If the frame is error-free, the switch will transmit it on the outgoing port to which the destination node is connected. If errors are found, the switch simply discards the frame. Store-and-forward switching prevents an errored frame from consuming network capacity thereby ensuring high reliability. However, it generates high latency and thus results in degradation of network performance. Because the entire frame must be stored in the switch before it is forwarded on the outgoing port, store-and-forward switching can be used for both symmetric and asymmetric switches.

5.2.6.2 Cut-Through Switching

With cut-through switching, the switch begins to forward the incoming frame on the proper outgoing circuit as soon as it has read the destination MAC address in the frame, which means the first 6 bytes. In other words, the switch begins transmitting before it has received the entire frame. The advantage of this is low latency through the switch and results in a very fast network. The disadvantage is that the switch begins transmitting before it has read and processed the frame check sequence at the end of the frame; if the frame contains an error, it will use up network resources only to be discarded at the destination. However, because of the high reliability of most current networks, the corruption of frames is not nearly as much of an issue as it used to be. This makes cut-through switching a reasonable choice. Cut-through switching can only be used when the input port has the same data rate as the output port. That is, it cannot be used in an asymmetric switch.

5.2.6.3 Fragment-Free Switching

With fragment-free switching, the switch will read the first 64 bytes of the frame and check them for errors. If these first 64 bytes are error-free, then the switch presumes that the rest of the frame is error-free and will begin transmitting the frame. It has been determined that errors are very likely to occur within the first 64 bytes of a frame. If no error is found in first 64 bytes it is highly unlikely to occur after that range. These 64 bytes are thus known as the "collision window." Fragment-free switching is a compromise between cut-through and store-and-forward switching because it has higher latency and better error control than cut-through switching, but lower latency and worse error control than store-and-forward switching. Fragment-free switching is sometimes called *runtless* switching. (An Ethernet *runt* frame is a frame that is smaller than 64 bytes.)

Figure 5.7 shows the ranges of the Ethernet frame where the forwarding decisions are made for the three switching schemes.

5.2.7 Highest Layer used for Forwarding

Based on which layer of the OSI Protocol Reference Model where forwarding decisions are made, there are three types of Ethernet switches:

(a) Layer 2 switches

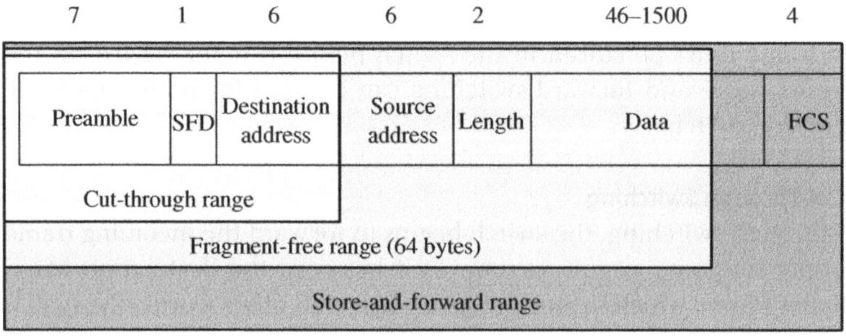

Figure 5.7 Frame Ranges where Forwarding Decisions are Made.

(b) Layer 3 switches
(c) Layer 4 switches.

5.2.7.1 Layer 2 Switching

A Layer 2 LAN switch is operationally similar to a multiport bridge but has a much higher capacity and supports many new features, such as full-duplex operation. A Layer 2 LAN switch performs switching and filtering based on the OSI data link layer (Layer 2) MAC address. As with bridges, it is completely transparent to network protocols and user applications.

Layer 2 switches use hardware to forward network traffic at Layer 2, based on the physical MAC address of each network device. They behave like learning bridges by learning, filtering, and forwarding packets, thereby reducing the amount of data that will be forwarded to the routers. Specifically, when a switch does not know the location of the destination of the packet it has received, it forwards copies of the packet to all its output ports except the port through which the packet was received. It keeps handling these packets this way until it has learned the location of the destination at which time it stops the flooding operation and will forward the packet to only the port through which the destination can be reached.

5.2.7.2 Layer 3 Switching

A Layer 3 LAN switch makes switching and filtering decisions based on Layer 2 (MAC) address and Layer 3 (IP) address. This type of switch dynamically decides whether to switch (Layer 2) or route (Layer 3) incoming traffic: It switches within a workgroup and routes between different workgroups.

Similar to routers, Layer 3 switches calculate routes based on IP addresses. Also, they have the ability to connect to WANs and are intelligent devices that can further segment network traffic to reduce congestion by calculating routes between various network links. Unlike traditional routers, which implement routing through software and a general purpose processor that tends to have a slower data throughput, switches perform these operations at full wire speeds.

5.2.7.3 Layer 4 Switching

Layer 4 switches can operate up to the transport, thus making switching decisions based on information held in the transport layer. As we discussed in Chapter 1, the transport layer takes care of things such as flow control and reliable and accurate delivery of the data to the destination. Thus, Layer 4 switch has the ability to not only examine the MAC and IP addresses but also control the traffic based on the Layer 4 information of the OSI model. (Layer 4 contains information that is used to identify the different applications.) With this information, a Layer 4 switch can perform the following functions:

- *Packet filtering*: Layer 4 switches have the ability to implement a variety of services that take advantage of packet filtering. For example, routers are often used as a network firewall, filtering packets, and providing security features by either allowing or blocking certain connections. A Layer 4 switch can offer this same service but implement it by means of hardware, thus offering the same service with a much higher speed of data throughput.
- *Prioritization*: Layer 4 switches can also use Layer 4 information to prioritize traffic flow. As we will see in a later chapter, applications are distinguished by their Layer 4 label called *port address*. Because it can see both the port address and IP address of a data packet, a Layer 4 switch can give priority to data intended for mission critical applications.
- *Load balancing*: Another service that can be made possible by a Layer 4 switch is load balancing, which is implemented to more efficiently control the amount of information that a particular server, among a group of servers supporting the same application, may receive.

5.3 Virtual LANs

VLANs divide a single existing physical network into multiple logical networks each of which forms its own broadcast domain. Thus, fundamentally a VLAN is a broadcast domain. The only difference between a traditional broadcast domain and one defined by a VLAN is that the boundaries of a traditional broadcast domain consist of a router while a router is not required to delineate a broadcast domain defined by a VLAN. Communication between two different VLANs is only possible through a router that is connected to both VLANs. The implementation of VLANs requires "managed" LAN switches that provide the ability to make ports members of different VLANs.

5.3.1 Advantages of VLANs

The primary advantages of VLANs are in the following areas:

(a) Cost reduction

(b) Performance improvement
(c) Ease of management
(d) Enhanced network security
(e) Flexible user location
(f) Better use of network resources.

With respect to cost reduction, prior to the development of VLANs the method used to separate traffic from different departments was by putting the different departments into different broadcast domains. This in turn means buying as many switches as there are departments with the hope that there would be as many ports on the switch as there are devices to be connected in the department. With VLAN, this requirement of a separate switch for each department is no longer an issue as members can easily be grouped into VLANs using a fewer number of switches thereby reducing cost.

With respect to performance improvement, VLANs offer functions that could only be provided by routers. Since they are implemented with switches, they provide these functions at a faster rate than routers can. Also, VLANs create broadcast domains that confine traffic to a small section of the network, and this prevents network congestion and performance degradation. Thus, VLANs provide an effective control of broadcast traffic without using routers. This frees up bandwidth by limiting broadcast traffic to devices within the VLAN.

With respect to ease of management, because VLANs are created and modified by network administrators via software, the need to recable networks for user additions, moves, and changes is virtually eliminated. Thus, VLANs simplify moves, additions, and changes in a network as this can now be done quickly and conveniently from the management console rather than the wiring closet.

With respect to network security, VLANs create virtual boundaries that can only be crossed through a router, which enables standard, router-based security measures to be used to restrict access to each VLAN as required. This also explains why there is increased performance: Routers prevent broadcasts from reaching nodes that are not in a segment from where a broadcast was generated.

With respect to flexible user location, network administrators can create VLANs for specific groups regardless of the geographic location. Also, VLANs allow greater mobility because a user can continue to be a member of a VLAN regardless of their location.

With respect to better use of network resources, with VLAN a server can be a member of multiple VLANs thereby reducing the need to route traffic to and from the server.

5.3.2 Types of VLANs

Three types of VLAN are usually defined, depending on switching criteria and the level at which the VLAN is implemented. These are as follows:

1. Port-based VLAN
2. MAC address-based VLAN
3. Protocol-based VLAN.

The protocol-based VLAN is further classified into two subgroups, namely,

(i) IP address-based VLAN
(ii) Network layer protocol-based VLAN.

A switch can support only one of these types of VLANs. Each type of VLAN corresponds to one of the three lower layers of the OSI reference model. The port-based VLAN corresponds to VLAN at the physical layer (PHY); the MAC address-based VLAN corresponds to VLAN at the data-link layer, and the protocol-based VLAN corresponds to VLAN at the network layer. Thus, an alternative method of classification is as follows:

1. Port-based (or Layer 1) VLAN
2. MAC address-based (or Layer 2) VLAN
3. Protocol-based (or Layer 3) VLAN.
   (a) IP address-based VLAN
   (b) Network layer protocol-based VLAN.

5.3.2.1 Port-Based VLAN

In the port-based VLAN, a port is regarded as a member of a VLAN. Thus, a device inherits the VLAN membership of the port to which it is connected to the network. Port-based VLANs are usually implemented in static environments where users are not mobile or it does not matter the subnet to which users are connected.

It is logical to connect devices that are physically close to a switch to the ports of the switch. Thus, port-based VLAN is based on physical location; that is, the devices that become members of the VLAN are close to the ports that comprise the VLAN. Figure 5.8 is an illustration of a port-based VLAN system. VLAN 1 and VLAN 2 have three ports assigned to them. Any communication between a device in VLAN 1 and another device in VLAN 2 must pass through the router.

If the VLAN membership spans more than one switch, then a link is used to connect one port in one switch to another port in the next switch and both ports will be part of the VLAN. When a frame enters the network it is assigned a tag that identifies the VLAN to which the port belongs. Figure 5.9 illustrates a port-based VLAN system that includes two switches. Because there are two VLANs associated with each switch, two ports in each switch are connected with two ports in the other switch. One pair of interconnected ports belongs

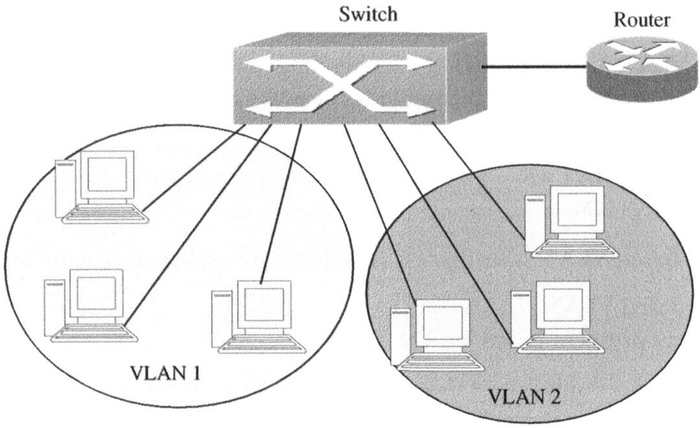

Figure 5.8 Example of a Port-Based VLAN System.

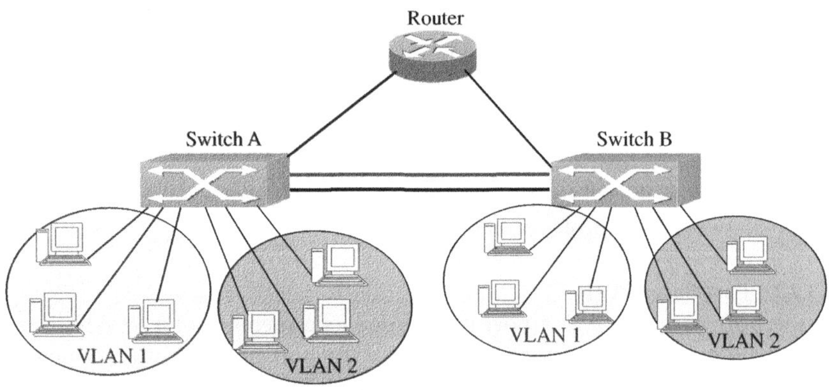

Figure 5.9 A Port-Based VLAN System that Spans Two Switches.

to one of the VLANs and the other pair of interconnected ports belongs to the other VLAN.

The major drawback of the scheme is that when a user moves to another port, the network administrator must reconfigure the VLAN membership, which can be time consuming, particularly when moves are frequent.

5.3.2.2 MAC Address-Based VLAN

As stated earlier, one problem with port-based VLANs is that if a device moves to a new port that is not in the same VLAN as the previous port, the device will be in a new VLAN. The MAC address-based VLAN avoids this problem by using the MAC addresses of the devices to define VLANs. This type of VLAN is much more flexible than the port-based VLAN because the network is independent of the location of the device. When a device is connected to the network via a switch, the switch queries a VLAN server to determine the VLAN to which the device belongs. When the server responds to the query the switch port will be set to the proper VLAN. This is particularly useful in dynamic organizations where users are constantly changing their locations; that is, it is the recommended option for mobile users.

MAC-based VLANs have the advantage over port-based VLANs in that as they are based on the hardwired address of the network device, a user can move the device to anywhere in the network without requiring the intervention of the network administration staff. This greatly reduces the load on the network administration staff.

One of the drawbacks of the MAC address-based VLAN solution is the initial requirement that each device must be individually placed in a specific VLAN. Thus, the configuration process can take substantially more time than is needed to configure a port-based VLAN. Another disadvantage of MAC address-based VLAN is that only one VLAN may be present on any port. Thus, they are most appropriate when each user on the network has a dedicated switch port. Also, a single MAC address cannot be a member of multiple VLANs and this makes it difficult to share server resources between more than one VLAN.

5.3.2.3 Protocol-Based VLANs

Protocol-based VLANs use the network layer address or network protocol type to define VLAN membership. They are particularly useful in environments with nonroutable legacy protocols that are used in special applications by a small set of users. These VLANs permit multiple VLANs to be configured in one port. They are efficient in reducing broadcast load on a network by only forwarding the traffic based on the protocol to the relevant ports. A Layer 3-based VLAN is constructed using information contained in the network layer header of packets. Thus, it is restricted to routers and Layer 3 LAN switches. There are different types of Layer 3 VLANs. We discuss two of them, which are as follows:

- The *IP address-based VLAN* links subnets according to the source IP address of the packets.
- The *network protocol-based VLAN* makes it possible to create a virtual network by protocol type (e.g., DECnet, TCP/IP, IPX, and AppleTalk), thereby grouping together all the machines using the same network layer protocol on the same network.

Since this VLAN is operating at Layer 3, it has some advantages. First, the Layer-3 VLAN permits moves to be accomplished without requiring a reconfiguration of the switch; thus, the cost and effort of supporting a Layer-3 VLAN can be less than that of other types of VLANs. Also, they support routing, which means they have built-in support for inter-VLAN communications.

One disadvantage of the scheme is that it requires a Layer 3 switch for its implementation, which makes it more expensive than the other schemes that are based on less expensive Layer 2 switches.

5.3.3 VLAN Tagging

VLAN tags are used to indicate VLAN membership within a frame going across the network. These tags are attached to the frame as it enters a switch port belonging to a VLAN and the tags are removed when the frame leaves a port belonging to the VLAN. The type of port within the VLAN will determine whether the VLAN tag is stripped from the frame or whether it remains attached to the frame. The two port types within a VLAN environment are known as access ports and trunk ports.

(a) *Access ports*: Access ports are used where a frame enters or exits the VLAN. When an access port receives a frame, the frame does not contain a VLAN tag. As the frame enters the access port, the VLAN tag is attached to the frame. While the frame is within the switch, it carries the VLAN tag that was attached when it entered through the access port. As the frame leaves the switch through the destination access port, the VLAN tag is removed. The transmitting device and the receiving device are not aware that the VLAN tag was ever attached.
(b) *Trunk ports*: In networks containing more than one switch, it becomes necessary to be able to send VLAN tagged frames from one switch to another. The difference between trunk ports and access ports is that trunk ports do not strip off the VLAN tag before sending the frame. With the VLAN tag preserved, the receiving switch will know the membership of the transmitted frame. This frame can then be sent out of the appropriate ports on the receiving switch.

Each VLAN tagged frame contains fields that denote its VLAN membership. There are two predominant formats for the VLAN tags: Cisco's inter-switch link (ISL) format and the standardized 802.1Q format. In this section, we discuss the IEEE 802.1Q tagging scheme.

The IEEE 802.1Q VLAN is a standard port-based VLAN that uses VLAN tagging, which is placed in the MAC header. It defines a 32-bit VLAN tag header field that is inserted in MAC frame after SA field and before the length field. The subfields in this header field are as follows:

- Tag protocol identifier (TPID) is a 16-bit field whose value is set to 0x8100 hex (33,024 decimal) to indicate frame is "tagged."
- Tag control information (TCI) is divided into three subfields:
  (a) Priority code point (PCP) a 3-bit field that contains IEEE 802.1p user priority value.

(b) Drop eligible indicator (DEI) (formerly called the canonical format indicator) bit is used to indicate frames that are eligible for dropping when congestion occurs.
(c) VLAN Identifier indicates VLAN (12 bits), allows up to 4093 VLANs numbered 2–4094 to be defined (VLAN IDs 0, 1, and 4095 are reserved).

The IEEE 802.1p standard specifies a mechanism for indicating frame priority in the 802.1Q VLAN standard. It supports up to eight traffic classes (priorities) labeled 0 through 7, with 7 the highest priority and 0 the lowest priority. In this way, it allows priority delivery of delay sensitive traffic. Network priority is signaled on a frame-by-frame basis. The IEEE 802.1Q VLAN header is shown in Figure 5.10 in comparison to the standard IEEE 802.3 header. Thus, it adds four extra octets to the standard LAN header.

5.3.4 Comments

A virtual LAN is a group of workstations, servers, and other network resources that behave as if they were connected to a single LAN segment even though they may not be. Thus, it is a group of devices in a network that are configured to be in the same broadcast domain thereby behaving as if they were all connected on the same physical LAN when they are actually spread throughout a network.

The port-based VLAN is the most widely used VLAN because of the relatively small administration overhead required to provision it. As stated earlier, the network administrator assigns each port of a switch to one VLAN. Once this

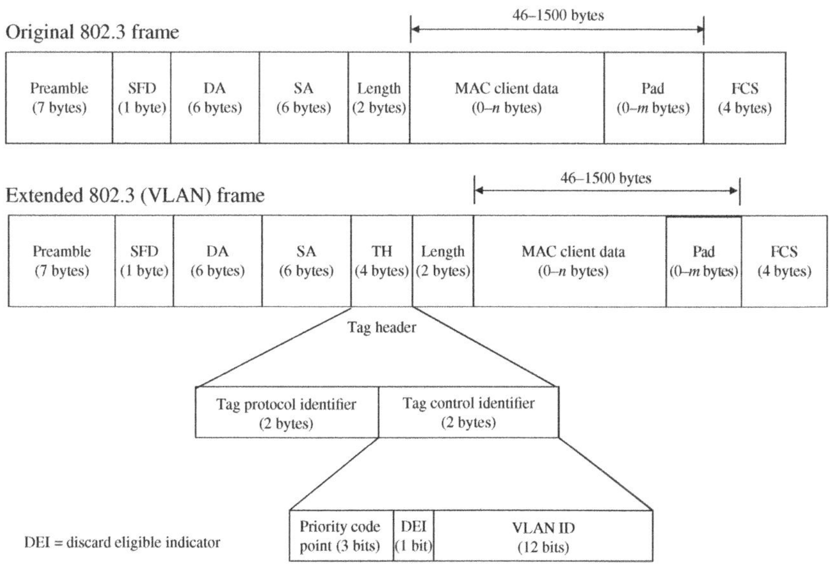

Figure 5.10 IEEE 802.1QVLAN Header.

is done, each device that connects to a port of the switch inherits the VLAN membership of the port. The main disadvantage of this method is that it does not allow user mobility. If a device is moved to a different port, the network administrator must reconfigure the VLAN.

Compared to port-based VLANs, the MAC-based VLANs provide true VLAN capabilities because membership in a MAC-based VLAN is not tied to a specific port. Instead, membership is based on the MAC address of the device. Thus, when a device is moved to another port, it still belongs to the same VLAN because its MAC address moves with it.

The protocol-based VLAN is not commonly used. However, when there are applications that use legacy network layer protocols, it makes sense to define a protocol-based VLAN.

5.4 Gigabit Ethernet

The Gigabit Ethernet is defined in the IEEE Std 802.3z standard for the 1000Base-X and the 1000Base-T is defined in the IEEE 802.3ab. The IEEE 802.3ae working group is developing the 10 Gbps Ethernet. One of the changes to the Ethernet CSMA/CD transmit specification was the addition of *frame bursting* for Gigabit operation. (Frame bursting is discussed as follows.) There are different flavors of Gigabit Ethernet, and they include the following:

- 1000Base SX, which uses short wavelength 850 nm lasers over multimode fiber and can span up to 525 m long depending on whether 62.5 or 50-μm fiber is used
- 1000Base LX, which uses long wavelength 1300 nm lasers over either multimode fiber which can span up to 550 m long or single-mode fiber which can be up to 3 km long
- 1000Base CX, which uses copper cable transmission over a maximum distance of 25 m, using a new type of shielded cable
- 1000Base-T uses standard Category 5 UTP cable for a distance of up to 100 m.

Although a half-duplex mode of Gigabit is defined, there are no half duplex devices being made. All Gigabit Ethernet is full-duplex transmission. Also, all fiber products are lasers, not LEDs, and so all the fiber-based equipment are much costlier than 10 or 100 Mbps multimode fiber-based products.

As stated earlier, there are now newer versions of the Gigabit Ethernet. These include the following:

- 40GBase-CR4: shorthand term for 40 Gbps Ethernet based on copper cable assembly
- 40GBase-SR4: shorthand term for 40 Gbps Ethernet based on parallel multimode fiber

- 40GBase-LR4: shorthand term for 40 Gbps Ethernet based on parallel single-mode fiber
- 100GBase-CR10: shorthand term for 100 Gbps Ethernet based on copper cable assembly
- 100GBase-SR10: shorthand term for 100 Gbps Ethernet based on parallel multimode fiber
- 100GBase-LR10: shorthand term for 100 Gbps Ethernet based on parallel single-mode fiber
- 100GBase-ER4: shorthand term for 100 Gbps Ethernet based on parallel single-mode fiber.

While their data rates are higher, they all operate in the same manner as discussed earlier.

5.4.1 Frame Bursting

Frame bursting is a feature that allows a MAC to send a short sequence (a burst) of frames without having to relinquish control of the medium. An interframe gap period is inserted between each frame in the burst. But instead of allowing the medium to go idle between frames, the transmitting station fills the interframe gaps with *extension bits* so that other stations on the network will see that the network is busy and will not attempt transmission until after the burst is complete. Extension bits are "nondata" symbols that maintain an active carrier and are readily distinguished from data bits by receiving stations.

The first frame of a burst is transmitted as normal and includes an "extension field" as required. Subsequent frames in the burst do not require an extension field, but each frame must be separated from the next one by an interframe gap (IFG). The IFG is a 96-bit time delay provided between frame transmissions to allow the network interfaces some recovery time between frames. However, during this IFG, the station must transmit the nondata symbols to prevent other stations from jumping in. A station can continue to send frames until a limit of 65,536 bits (or 8192 bytes) is reached; that is, the length of the burst is limited to 65,536 bits. Frame bursting is illustrated in Figure 5.11.

5.5 Wireless LANs

The IEEE 802.11 WLAN standard was released in 1997. The standard was designed to be similar to the other 802 LAN network standards. This is an example of an advantage of the hierarchical model because with respect to the wired LAN the only thing that has changed is the PHY: physically wired interface is now replaced by air interface, everything else remains the same.

The PHY is responsible for transmission over the air. In the original standard, three access schemes were specified: direct sequence spread spectrum (DSSS),

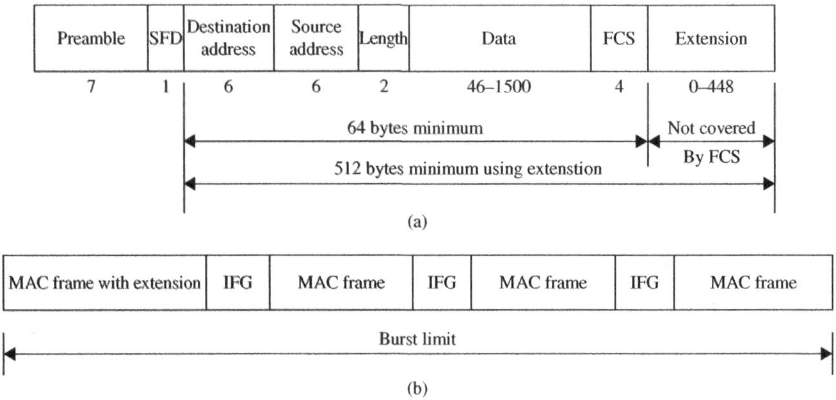

Figure 5.11 Example of Frame Bursting. (a) MAC Frame with Gigabit Carrier Extension and (b) Gigabit Frame Bursting.

frequency hopping spread spectrum (FHSS), and infrared. However, infrared is no longer supported in the later versions of 802.11 WLANs. Both DSSS and FHSS supported data rates of 1–2 Mbps. The first IEEE 802.11 WLAN was designed to operate in the so-called Industrial, Scientific, and Medical (ISM) band, which is the 2.4 GHz frequency band, with maximum allowable transmit power of 1000 mW (or 1 W) in North America and 100 mW in Europe.

The protocol architecture of the IEEE 802.11 WLAN is illustrated in Figure 5.12.

As stated earlier, the infrared WLAN is no longer available and FHSS has such a low throughput that it is rarely used these days. Each DSSS PHY channel allows for three noninterfering channels spaced 25 MHz apart in the 2.4 GHz frequency band. Equivalently, only three networks can operate in the same location. The channel spacing for DSSS WLAN networks in North America is illustrated in Figure 5.13.

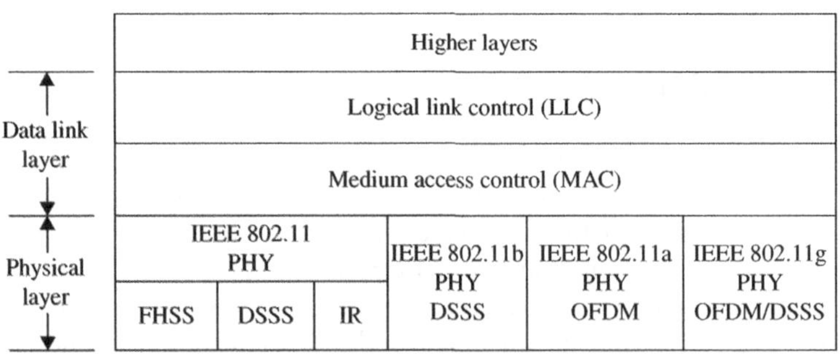

Figure 5.12 Protocol Architecture of the IEEE 802.11 WLAN.

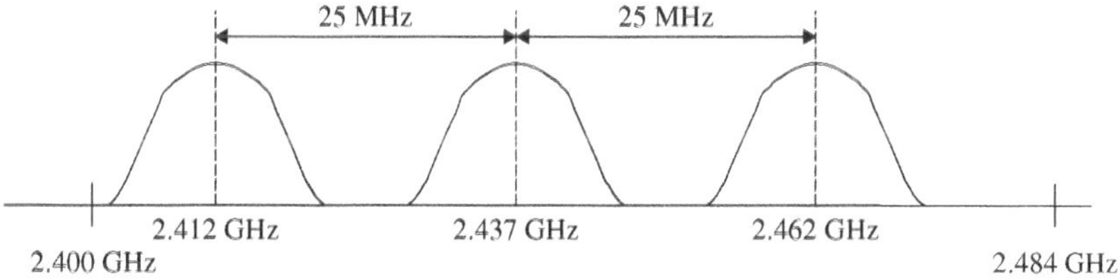

Figure 5.13 Minimum Channel Spacing for DSSS Networks in North America.

5.5.1 IEEE 802.11b WLAN

The 802.11b standard uses the 2.4 GHz band and operates in the DSSS mode only. It is generally referred to as *wireless fidelity* (Wi-Fi) because it incorporated the wired equivalent privacy (WEP) that has now been shown to have some weakness in providing security. It uses a coding scheme called *complementary code keying* (CCK) to attain a data rate of 11 Mbps. A second coding scheme called *packet binary convolutional code* (PBCC) was included as an option for providing 1, 2, and 5.5 Mbps data rates. Note that the 11 Mbps data rate represents the maximum attainable data rate; the actual data rate is usually less than that.

5.5.2 IEEE 802.11a WLAN

The 802.11a uses the so-called unlicensed national information infrastructure (U-NII) band, which is the 5 GHz band, to achieve data rates of 54 Mbps. It uses a multiplexing scheme called orthogonal frequency division multiplexing (OFDM). By operating in the 5 GHz spectrum and using a different modulation method, 802.11a is not interoperable with the 802.11b standard.

The standard stipulates that data rates of 6, 9, 12, 18, 24, 36, 48, and 54 Mbps may be supported, and support for 6, 12, and 24 Mbps is mandatory. The attraction of 802.11a is that it has more channels than the 802.11b that operates in the 2.4-GHz band. The 54 Mbps radio provides eight nonoverlapping channels compared to three nonoverlapping channels for the 11 Mbps radios. However, 5 GHz consumes more power and the range is restricted compared to the 2.4 GHz band.

5.5.3 IEEE 802.11g WLAN

The new standard 802.11g operates at the ISM (2.4 GHz) band delivering 54 Mbps. The standard uses the CCK and OFDM techniques with optional mode of PBCC. It is specified to be backward compatible with 802.11b standard. Using CCK ensures backward compatibility with the installed 802.11b base, while OFDM provides the speed required for today's high-bandwidth applications. Thus, the 802.11g client can select from the widest possible range of both OFDM data rates of 54, 48, 36, 24, 18, 12, 9, and 6 Mbps and CCK data rates of 11, 5.5, 2, and 1 Mbps.

5.5.4 Architecture of the IEEE 802.11 WLAN

The IEEE 802.11 wireless LANs operate in one of two modes:

- *Ad hoc mode,* which the 802.11 standard defines as the *independent basic service set* (IBSS) mode
- *Infrastructure mode,* which the 802.11 standard also calls the *basic service set* (BSS).

The ad hoc mode is designed such that only the hosts within the transmission range (or in the same cell) of each other can communicate. If any network node wishes to communicate with a host that is outside the cell, a member of the cell must operate as a gateway and a perform packet relay function.

In the infrastructure mode, each host sends all its packets to a central station called *access point* (AP), which acts as an Ethernet bridge that forwards the packet to the wired LAN. Thus, within the IEEE 802.11 environment, an ad hoc network is a communication framework in which stations communicate directly with each other, without the use of an AP.

5.5.5 Ad Hoc Mode Deployment

Figure 5.14 shows a simple ad hoc network with three nodes. The two outer nodes (i.e., laptops) are outside each other's transmitting range, so the middle node acts as a router. In general, each node acts as both a router that forwards traffic originated by other nodes and a host (or end system). Each node is able to dynamically discover and maintain routes to other nodes in the network.

5.5.6 Infrastructure Mode Deployment

Figure 5.15 is an example of the infrastructure mode of operation. The devices labeled APs are the access points through which the clients gain access into the wired LAN. As stated earlier, there is a finite number of WLANs that can be established in one location: three in the ISM band and eight in the UNI-I band.

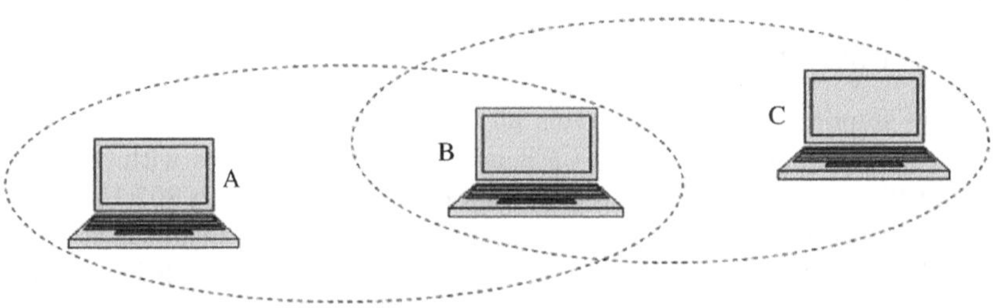

Figure 5.14 Ad Hoc Network Configuration.

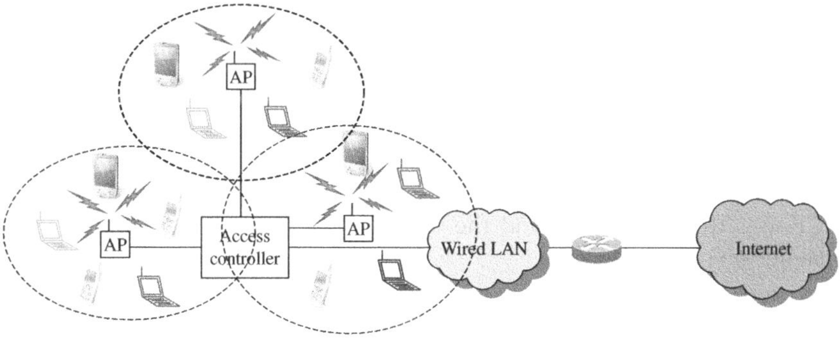

Figure 5.15 Infrastructure Mode Network Configuration.

5.5.7 IEEE 802.11 WLAN Timers

The IEEE 802.11 WLAN uses five timing intervals to provide both asynchronous and near-isochronous operations. These are as follows:

- Short interframe space (SIFS)
- Slot time
- Priority interframe space (PIFS)
- Distributed interframe space (DIFS)
- Extended interframe space (EIFS).

The shortest interval is the SIFS, followed by slot time. PIFS and DIFS are defined as follows:

PIFS = SIFS + 1 slot time
DIFS = SIFS + 2 slot times.

These timers are used in the WLAN operation, as explained in the following section.

5.5.8 IEEE 802.11 WLAN Operation

The minimal MAC frame exchange protocol consists of two frames that constitute an atomic unit, which means they cannot be interrupted by transmission from any other station:

- A frame sent from the source to the destination
- An ACK from the destination indicating that the frame was received correctly.

There is a retry counter associated with every MAC frame. The counter limits the number of times a single frame may be retransmitted. It starts at zero and is incremented after each unsuccessful retransmission attempt up to a predefined maximum and reset after the packet has been successfully transmitted.

The system uses two modes of operation: distributed coordination function (DCF) mode and point coordination function (PCF) mode. DCF uses timers to implement the basic access mechanism called carrier-sense multiple access with collision avoidance (CSMA/CA).

5.5.9 DCF Mechanism

DCF uses the CSMA/CA, and the operation of CSMA/CA is described in Chapter 4.

5.5.10 PCF Mechanism

PCF is an optional part of the standard, but every station is required to be able to respond to operation of PCF. It defines two types of operating intervals called *contention-free period* (CFP) and *contention period* (CP). CFP occurs frequently to provide near-isochronous service to the stations that need the service. It alternates with a CP where DCF rules operate and all stations may compete for access to the medium. The standard stipulates that the CP must be long enough to contain at least one maximum length frame and its ACK.

PCF uses a poll and response protocol to eliminate the possibility of contention for the medium during the CFP. A *point coordinator* (PC), which is located in AP, controls PCF. To participate in PCF, a station must register to be on the polling list, and the PC regularly polls the stations for traffic. This enables PCF to deliver near-isochronous service to stations on the polling list.

PCF uses PIFS < DIFS to seize and maintain control of medium. When the PC gains access to the medium, it sends a beacon at the beginning of the CFP that contains information on the maximum expected duration of the CFP. During CFP, the PC ensures that interval between frames on the medium is no longer than PIFS to prevent a station operating under DCF from gaining access to the medium. The PC will send a frame to a station and expect either an ACK or data frame in response to a CF-Poll within SIFS interval. If a response is not received before SIFS interval expires, the PC will transmit its next frame before the PIFS interval expires after the previous transmission, and so on until CFP is concluded.

As stated earlier, the PC sends a beacon at the beginning of each CFP, and the beacon contains the duration of the CFP. Every station that receives the beacon enters the information in its network allocation vector (NAV), which is a value that indicates to the station the amount of time remaining before the medium will become available. NAV is the primary mechanism used to prevent stations from accessing the medium during the CFP. The PC announces the end of the CFP by broadcasting the contention-free end (CF-end) frame that signals the end of the CFP and the beginning of the DCF where the stations independently contend for the channel.

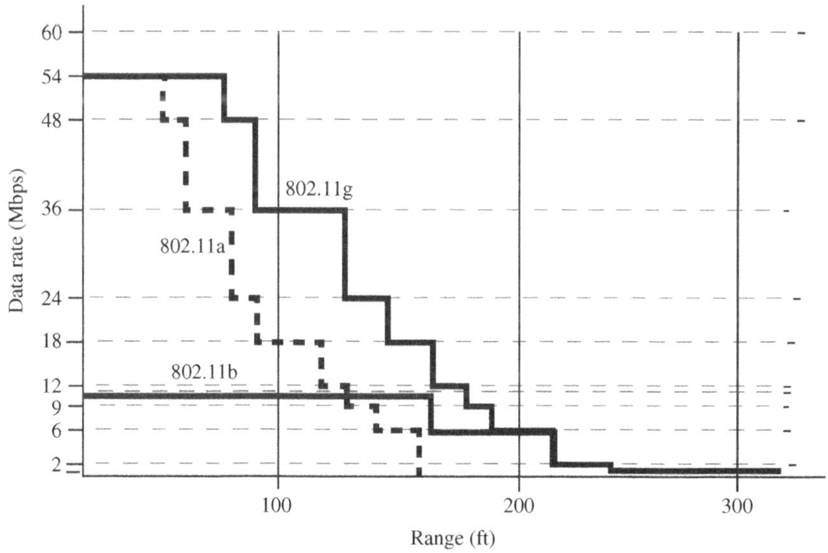

Figure 5.16 Maximum Data Rate Variation with Distance.

5.5.11 Range and Data Rate Comparison in the PCF Environment

The more popular implementation of the IEEE 802.11 WLAN is the infrastructure mode. As distance from the access point increases, 802.11-based products provide reduced data rates to maintain connectivity. Since 802.11b and 802.11g use the same spectrum, they share the same propagation characteristics. In general, radio signals do not propagate as well in the 5 GHz band as they do in the 2.4 GHz band, which means that the 802.11a product range is limited compared to the 802.11b or 802.11g product range. Figure 5.16 illustrates the range performance of the IEEE 802.11a, b, and g. Note that the figure is not drawn to scale; it merely shows approximately how data rate varies with distance.

5.6 Token Ring Network

A token ring network is a network with the ring topology that uses the token passing access control scheme. The IEEE 802.5 token ring network is a standards-based ring network that we will discuss in this section. It supports two basic types of frames:

- Tokens
- Data/command frames.

Tokens are 3 bytes in length and consist of a start delimiter, an access control byte, and an end delimiter. Data frames carry information for upper-layer

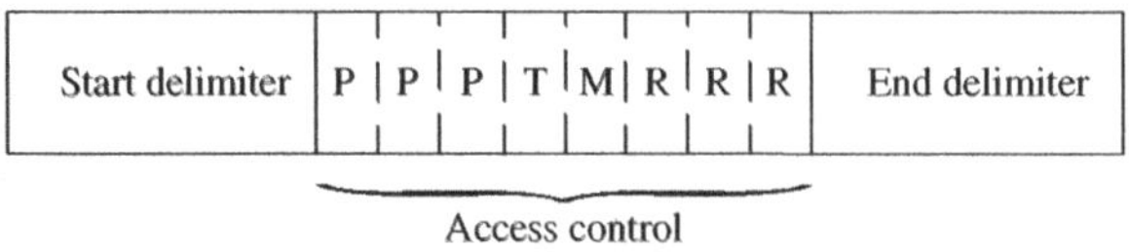

Figure 5.17 Token Frame Format.

protocols and command frames contain control information and have no data for upper-layer protocols.

5.6.1 Token Frame Fields

Tokens are 3 bytes (24 bits) in length and consist of a start delimiter, an access control byte, and an end delimiter. The token frame format is shown in Figure 5.17.

The fields of the token are as follows:

- The *start delimiter* serves to alert each station to the arrival of a token (or data/command frame). This field includes signals that distinguish the byte from the rest of the frame by violating the encoding scheme used elsewhere in the frame.
- The *access control* byte contains the priority and reservation fields, as well as a token bit (used to differentiate a token from a data/command frame) and a monitor bit (used by the active monitor to determine whether a frame is circling the ring endlessly). Specifically,
  - Priority bits (P) are the first three bits used to indicate the priority of the token.
  - Token indicator (T) bit is the fourth bit and it indicates whether this is a token or a data frame.
  - Monitor count bit (M) is used to monitor for packets that continually loop through the network.
  - Priority reservation bits (R) are the last three bits and are used to reserve a token on a priority basis.
- Finally, the *end delimiter* signals the end of the token or data/command frame. It also contains bits to indicate a damaged frame and a frame that is the last in a logical sequence.

5.6.2 Token-Passing Access Method

When the first station comes online, the network generates a token, which is a predetermined bitstream that permits a station to access the channel. The token travels around the ring "polling" each station. When a station receives the token and has a frame to transmit, it toggles the token indicator bit in the access control field to indicate that the frame is a data frame. Thus, the toggled token now becomes the header of a data frame. The station then appends the data information and forwards the frame. When the frame reaches the destination station, the station sets the address-recognized bit in the frame status field. If the station accepts the frame, it also copies the frame into its receive buffer and sets the frame-copied indicator bit without deleting the frame from the ring; this indicates to the sender that the receiver accepted the message.

When the frame returns to the sender, the latter resets the token bit, removes the payload thereby making it a token frame that will first be seen by the next station downstream from the current station. If a station that has no frame to transmit receives the token, it does nothing and allows the token to go to the next station downstream from it. Thus, each station that has data to transmit has an opportunity to transmit a frame in each round of token passage.

5.6.3 Data/Command Frame Fields

A data/command frame is shown in Figure 5.18.

The fields are as follows:

- *Start delimiter* alerts each station of the arrival of a token (or data/command frame). This field includes signals that distinguish the byte from the rest of the frame by violating the encoding scheme used elsewhere in the frame.
- *Access control* contains the priority field (the most significant 3 bits) and the reservation field (the least significant 3 bits), as well as a token bit (used to differentiate a token from a data/command frame) and a monitor bit (used by the active monitor to determine whether a frame is circling the ring endlessly).
- *Frame control* indicates whether the frame contains data or control information. In control frames, this byte specifies the type of control information.
- *Destination address* and *source address* consist of two 6-byte address fields that identify the destination and source station addresses.
- *Data* indicates that the length of field is limited by the ring token holding time, which defines the maximum time a station can hold the token.
- *Frame-check sequence* (FCS) is filed by the source station with a calculated value dependent on the frame contents. The destination station recalculates the value to determine whether the frame was damaged in transit. If so, the frame is discarded.
- *End delimiter* signals the end of the token or data/command frame. The end delimiter also contains bits to indicate a damaged frame and identify the frame that is the last in a logical sequence.

Start delimiter	Access control	Frame control	Destination address	Source address	Data	Frame check sequence	End delimiter	Frame status

Figure 5.18 Frame Format.

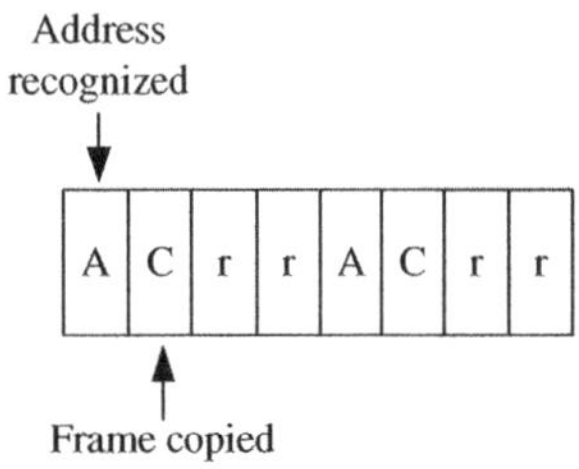

Figure 5.19 Format of Frame Status Field.

- *Frame status* is a 1-byte field that terminates a command/data frame. It includes the one-bit *address-recognized indicator* and *frame-copied indicator* fields. These one-bit fields, if set, provide confirmation that the frame has been delivered to the source address and the data read. Both fields are duplicated within the frame status byte. The format of the frame status field is shown in Figure 5.19.

When a frame arrives at the interface of a station with the destination address, the interface sets the A bit (=1), as it passes through. If the interface copies the frame to the station, it also sets the C bit (=1). A station might fail to copy a frame due to lack of buffer space or other reasons. When the station that sent the frame strips it from the ring, it examines the A and C bits. The three possible combinations are as follows:

1. A = 0 and C = 0; destination not present or powered up.
2. A = 1 and C = 0; destination present, but frame not accepted.
3. A = 1 and C = 1; destination present and frame copied.

This arrangement provides an automatic acknowledgment of the delivery status of each frame. The other bits labeled "r" are reserved and set to 0 in the field.

5.6.4 Token Access Priority

Token ring networks use a priority system that grants certain user-designated, high-priority stations the opportunity to use the network more frequently. The token frame contains two fields that control priority: The *priority field* and the *reservation field.*

The priority field indicates the current priority of the token. This operates as follows: When a station wants to transmit a priority *n* frame, it must wait until it can capture a token whose current priority is less than or equal to *n.* When a data frame goes by, a station can try to reserve the next token by writing the priority of the frame it wants to send into the frame's *reservation bits.* This can only be done, however, if a higher priority has not already been reserved there. When the current frame has been stripped, the next token is generated at the priority that has been reserved.

Figure 5.20 Logical Architecture of the Token Ring Network.

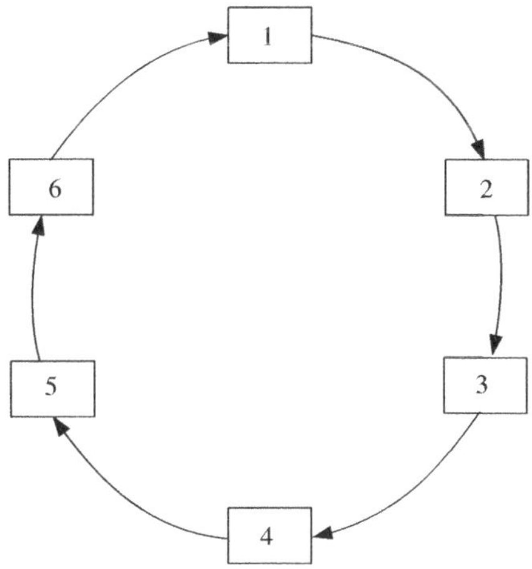

5.6.5 Logical and Physical Implementation

Figure 5.20 shows the logical architecture of the token ring network. A group of six stations are connected in a directed point-to-point manner with station 1 connected to station 2, station 2 connected to station 3, station 3 connected to station 4, station 4 connected to station 5, station 5 connected to station 6, and station 6 connected to station 1 to close the loop. The arrows indicate the direction of frame flow in the network.

The physical realization of the network is usually different from the logical structure. It is based on every station being connected to a panel in a wiring closet, as shown in Figure 5.21. This implementation enables the network to overcome one of its early criticisms: that the failure of any one station brings

Figure 5.21 Physical Architecture of the Token Ring Network.

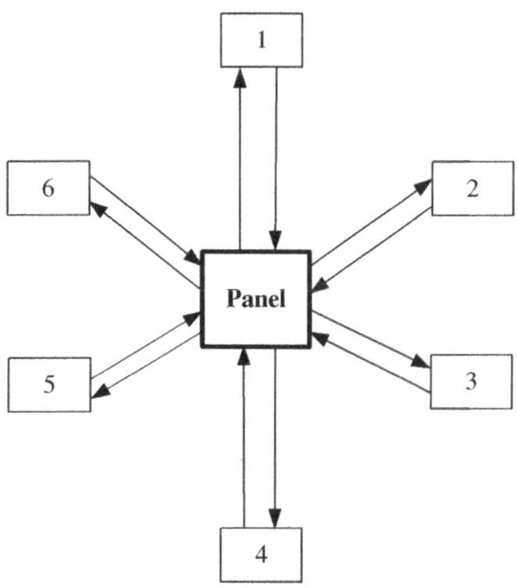

the network down. Thus, by having every station to connect to the panel, it is easy to use a bypass switch at the panel to switch out any station that fails.

5.7 Summary

This chapter has discussed the basic LAN architectures, which include the IEEE 802.3 class of LANs, the Ethernet, the gigabit Ethernet, wireless LANs, and the IEEE 802.5 token ring network. The emphasis is on the architectures of these different networks. No attempt has been made to compare their performance. Getting into the issue of performance comparison requires sophisticated knowledge of stochastic modeling that is outside the scope of this book.

Exercises

1. What is the difference between a Layer 2 LAN switch and a Layer 3 LAN switch?

2. What is the difference between the ad hoc and the infrastructure modes of operation of the IEEE 802.11 wireless LAN?

3. What is the difference between distributed coordination function (DCF) and point coordination function (PCF) in wireless LAN?

4. Name one difference between the IEEE 802.11a and IEEE 802.11b WLANs.

5. Give one difference between the IEEE 802.11a WLAN and the IEEE 802.11g WLAN.

6. What is the major difference between the IEEE 802.11b WLAN and the IEEE 802.11g WLAN?

7. What does gated service mean in a token-passing network?

8. Which node is responsible for removing a frame from a token ring network?

9. How does a node indicate that it has accepted a frame in a token ring network?

10. What does frame bursting mean and where is it used?

11 What is a VLAN?

12 What is the difference between a Layer 2 VLAN and a Layer 3 VLAN?

13 Name one advantage of a port-based VLAN over a MAC address-based VLAN.

14 Name one advantage of a MAC address-based VLAN over a port-based VLAN.

6

Network Layer Part I – IP Addressing

6.1 Introduction

The purpose of this chapter is to discuss IP addressing, which is one of the network layer services. The specific topics to be discussed include the following:

- IP addresses
- IPv4
- IP network subnetting
- IP quality of service (IP QoS)
- Address resolution protocol (ARP)
- Dealing with IPv4 address shortage
- IPv6.

Recall that the network layer is concerned with addressing and routing. The Internet layer is the top part of the network layer. The Internet protocol, which is defined for this layer, is a simple connectionless datagram protocol that provides no error recovery and no delivery guarantee. In this section, we consider the addressing aspect of the layer.

6.2 IP Address

The current version of IP is IP version 4 (IPv4). The IPv4 header has the format shown in Figure 6.1.

The length of IPv4 header is variable, and the fields are measured in terms of 32-bit words. The shortest IPv4 header is 20 bytes, and IP header length (IHL) would be 5 since $20 \times 8/32 = 5$. The fields are defined as follows:

- *Version*: (4 bits) indicates the IP version number, which is 4
- *IHL* (IP Header Length): (4 bits) indicates the number of 32-bit words that form the packet header
- *DSCP* (Differentiated Services Code Point): (6 bits) used to define the quality of service (QoS) for different network applications

Fundamentals of Data Communication Networks, First Edition. Oliver C. Ibe.

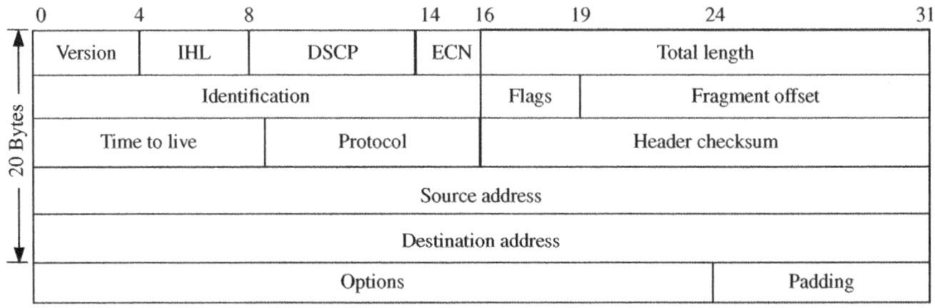

Figure 6.1 IPv4 Header Structure.

- *ECN* (explicit congestion notification): (2 bits) used to practice end-to-end notification of network congestion without dropping packets. ECN is an optional feature that is used only when both endpoints support it and are willing to use it. Also, it is effective only when it is supported by the network.
- *Total length*: (16 bits) specifies packet length in bytes of both the header and the data
- *Identification*: (16 bits) indicates which packet fragments belong together to avoid mismatch in situations where packet fragmentation has been done. The value is incremented every time an IP datagram is sent from source to the destination. The field is particularly useful in the reassembly of fragmented IP datagrams.
- *Flags*: (3 bits) used to control or identify fragments. They are as follows (in order, from most significant to least significant):
  - bit 0: reserved and must be set to zero
  - bit 1: do not fragment (DF)
  - bit 2: more fragments (MF)

  If the DF flag is set, and fragmentation is required to route the packet, then the packet is dropped.
- *Fragment offset*: (13 bits) indicates where this fragment belongs in the original packet, if packet fragmentation has been done. It is measured in 64-bit units (or 8-byte blocks) and specifies the offset of a particular fragment relative to the beginning of the original unfragmented IP datagram. The first fragment has an offset of zero.
- *Time to live* (TTL): (8 bits) indicates amount of time the packet is allowed to stay in network. When the datagram arrives at a router, the router decrements the TTL field by one. When the value of the TTL becomes zero and the packet has not reached its destination, the router discards the packet and sends an ICMP message to the source to let it know that the packet has been dropped. (ICMP stands for *Internet control message protocol*; it is used by routers to report problems with delivery of IP datagrams within an IP network.)
- *Protocol*: (8 bits) indicates ID number of the transport protocol that is being carried. For example, 1 = ICMP; 2 = IGMP; 6 = TCP; 17 = UDP.

(IGMP stands for *Internet group message protocol* and is used by hosts to indicate their multicast group membership to an adjacent router.)

- *Checksum*: (16 bits) used for error detection in the header. It is updated whenever the packet header is modified by a router.
- *Source address* (32 bits): indicates the IP address of the originator of the packet
- *Destination address* (32 bits): indicates the IP address of the final destination of the packet
- *Options* (variable): used to encode options requested by sender and is particularly used for source routing in which the IP addresses to visit are placed in the field. (Field may be empty; the field is not often used.)
- *Padding* (variable): used to ensure that packet header is a multiple of 32 bits.

As stated earlier, the minimum header length is 20 bytes. The header is followed by variable data field that is a multiple of 8 bits with a maximum length of 65,535 bytes.

6.3 Maximum Transmission Unit

Every network specifies the maximum length of packets that pass through it called the *maximum transmission unit* (MTU). Thus, as an IP packet moves through the Internet, it might need to cross a route that cannot handle the size of the packet. If so, the packet will be divided, or fragmented, into smaller packets and reassembled later, if fragmentation is permitted. The field's identification, flags, and fragment offset are used to fragment and reassemble packets.

Example 6.1 Consider a datagram that is 1500 bytes long that needs to travel through a network with MTU of 620 bytes. Assume that there are no options used, which means that the header is 20 bytes long and at most 600 bytes of the data can be accommodated in a packet. Thus, the original packet will be divided into three fragments:

- Fragment 1 (offset 0) has 600 bytes.
- Fragment 2 (offset $600/8 = 75$) has 600 bytes.
- Fragment 3 (offset $1200/8 = 150$) has 300 bytes.

Each fragment contains a header that duplicates most of the original header (except a bit in the flags field that shows that it is a fragment), followed by as much data as can be carried in the network as specified by the MTU. Fragment offset specifies the offset in the original packet of the data being carried in the current packet, measured in units of 8 bytes, starting with zero through the fragment with highest offset. The three fragments are shown in Figure 6.2.

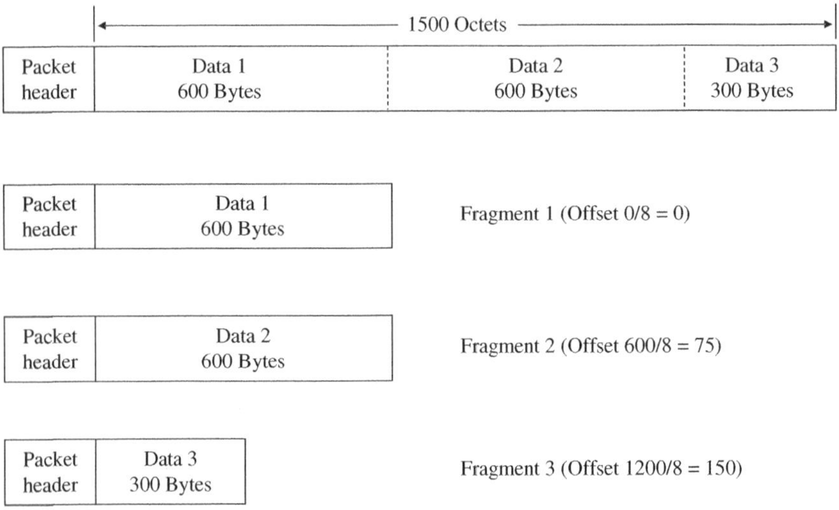

Figure 6.2 MTU Fragments.

6.4 IP Version 4 Addressing

One of the functions of the Internet layer is addressing. IP addressing is based on the concept of *networks* and hosts. A host is any device on the network that can transmit and receive IP packets. An IP address permits both a network and hosts in the network to be uniquely identified. The current version of IP addresses is IPv4. A new version, IPv6, has been defined and is currently being deployed.

IPv4 addresses consist of 32 bits that are represented in the dotted-decimal notation. Each byte of the address is represented by its decimal value and the different values are separated by dots in the form a.b.c.d. In this scheme, each octet is converted into a number that ranges from 0 to 255, where 255 is the value obtained when all the bits in the octet are 1; that is,

$$\begin{aligned}255 &= (1\times 2^7)+(1\times 2^6)+(1\times 2^5)+(1\times 2^4)+(1\times 2^3)+(1\times 2^2)\\&\quad+(1\times 2^1)+(1\times 2^0)\\&= 2^7+2^6+2^5+2^4+2^3+2^2+2^1+2^0 = 128+64+32+16+8\\&\quad+4+2+1\\0 &= (0\times 2^7)+(0\times 2^6)+(0\times 2^5)+(0\times 2^4)+(0\times 2^3)+(0\times 2^2)\\&\quad+(0\times 2^1)+(0\times 2^0)\end{aligned}$$

Thus, valid IP addresses range from 0.0.0.0 to 255.255.255.255. The first part of an IP address designates the network address and second part designates host address within the network, as shown in Figure 6.3.

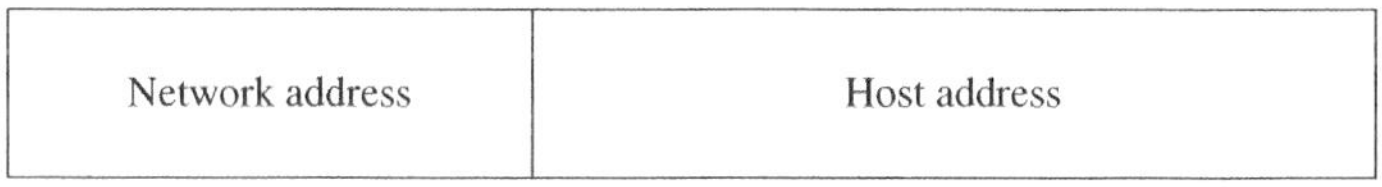

Figure 6.3 IPv4 Address Structure.

There are five classes of IPv4 addresses: Classes A, B, C, D, and E.

6.4.1 Class A IPv4 Addresses

Class A IP addresses are used for large networks (i.e., networks with very large number of hosts). The first octet is used for the network address and the remaining three octets are used for the host address. Also, the first bit of the first octet of a Class A address is 0, which means that the network address is of the form 0xxxxxxx, where x is 0 or 1. Thus, Class A networks are those networks whose first octet is a decimal number from 0 to 127. This means that there are 127 Class A networks addresses, excluding the network with all eight 0s, and each network can support up to 2^{24} hosts, or nearly 17 million hosts. The structure of the Class A address is shown in Figure 6.4.

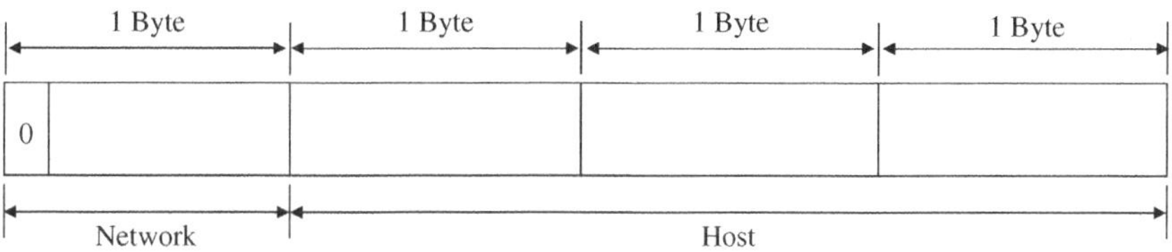

Figure 6.4 Structure of Class A IPv4 Address.

6.4.2 Class B IPv4 Addresses

Class B IPv4 addresses are designed for medium-sized networks. In this case the first two octets are used for the network address and the last two octets are used for the host address. Also, the first two bits of the first octet are 10, which means that the first octet of a class B address is of the form 10xxxxxx, where x is 0 or 1. Thus, Class B networks are those networks whose first octet is a decimal number from 128 to 191. There are 2^{14} or 16,384 Class B network addresses each of which can support 2^{16} or 65,636 hosts. The structure of Class B addresses is shown in Figure 6.5.

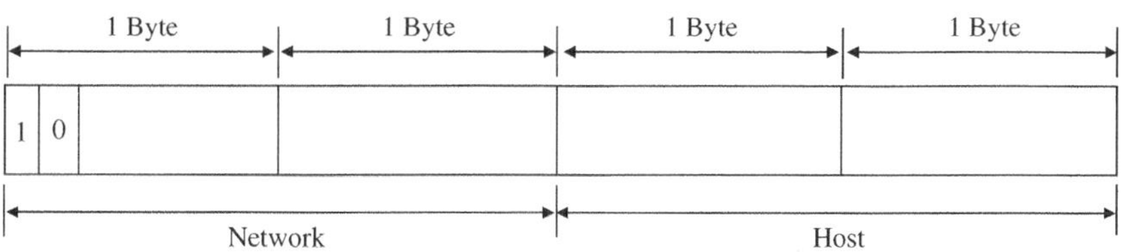

Figure 6.5 Structure of Class B IPv4 Address.

6.4.3 Class C IPv4 Addresses

Class C IPv4 addresses are used for small networks, especially those with less than 250 hosts. The first three octets are used for the network address and the last octet is used for the host address. Also, the first three bits of the first octet are 110, which means that the first octet of a Class C address is of the form 110xxxxx, where x is 0 or 1. Thus, Class C networks are those networks whose first octet is a decimal number from 192 to 223. There are 2^{21} or slightly more than 2 million Class C network addresses each of which can support $2^8 = 256$ hosts. The structure of Class C addresses is shown in Figure 6.6.

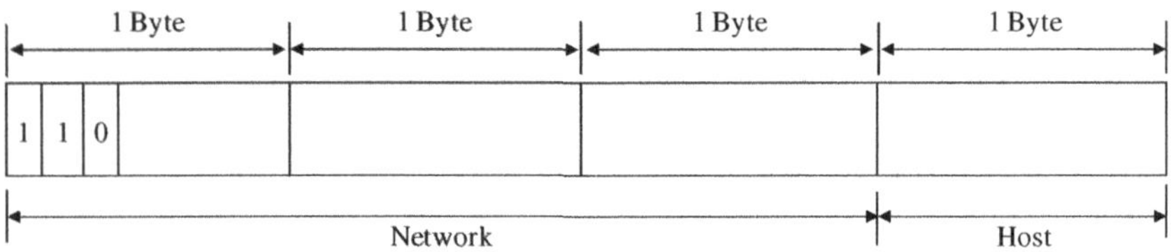

Figure 6.6 Structure of Class C IPv4 Address.

6.4.4 Class D IPv4 Addresses

Class D IPv4 addresses are reserved for multicast groups. There is no network portion of the address, which means that all four octets are used for multicast addresses. Also, the first four bits of first octet are 1110. Thus, multicast addresses are those whose first octet is of the form 1110xxxx, where x is 0 or 1. These are those networks whose first octet is a decimal number from 224 to 239. Multicast groups are not based on any particular location; they subscribe for services that can be accessed from any part of the world, which is why there no network address field in the address. Class A, Class B and Class C addresses are location-based; that is, there is only one network attached to the Internet that has a particular Class A, Class B or Class C address. The structure of Class D addresses is shown in Figure 6.7.

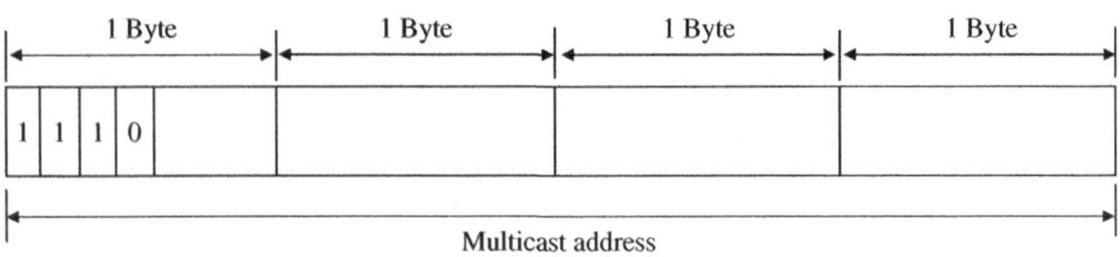

Figure 6.7 Structure of Class D IPv4 Address.

6.4.5 Class E IPv4 Addresses

Class E IP addresses are reserved for future use and may never be used as a result of IPv6. The first five bits of first octet are 11110; thus, first octet starts from 240.

6.5 IP Subnetting

Flat addressing presents an entire network as one gigantic entity to the network administrator. Thus, when a network problem occurs in a network with flat addresses, the administrator can only isolate the problem by the process of trial and error because the structure of the network does not lend itself to easy location of the problem. Subnetting is a networking technique that is used to simplify network administration.

It works by borrowing bits from the host address to create two or more subnetworks (or subnets), leaving the network address space untouched since network addresses must be unique. Thus, subnetting permits us to divide the original host address space into two subspaces: a subnet address space and a new host address space. In this way when a problem occurs in the network, the network administrator will know the affected subnet and can thus narrow the area to look for the problem rather than groping through the entire network. The structure of the IPv4 address for a subnetted network is shown in Figure 6.8.

In addition to easing network administration, subnetting simplifies routing and creates partitions within an enterprise network.

The number of bits borrowed from the host address space to define subnets varies from network to network. The method used to convey this number for a given network is the *subnet mask*. Thus, the subnet mask describes how the host address bits have been partitioned: how many highest order bits of the host address space are used for the subnet and how many bits are used for the actual host address. The subnet mask operates in the following manner:

- It uses the same dotted-decimal notation as the IP address.
- Each bit in the IP address has a corresponding mask bit.
- If a bit in the IP address is part of the network address, including the subnet, its mask bit is set to 1; otherwise, it is set to 0.

Thus, only the bits used for the actual host address have their mask bits set to 0; all other bits have their mask bits set to 1.

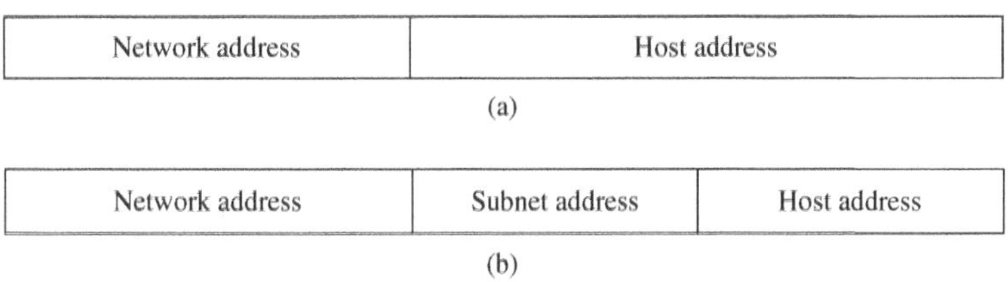

Figure 6.8 Header Structure for Subnetted Networks. (a) Header Format for No Subnetting; (b) Header Format for Subnetted Network.

For example, consider the IP address: 180.200.16.6. Because 180 lies between 128 and 191, the address is a Class B address, and the network address is in the first two octets. Thus, the network prefix (or the network portion of the address) is 180.200. Assume we use the first four highest order bits of the third octet for subnetting, then we obtain the subnet mask as 255.255.240.0 in the following manner:

$$10110100.11001000.00010000.00000110 = 180.200.16.6$$

$$11111111.11111111.11110000.00000000 = 255.255.240.0$$

Note that the value 240 is the result of the structure of the third octet, 11110000, whose decimal value is $2^7 + 2^6 + 2^5 + 2^4 = 128 + 64 + 32 + 16 = 240$. When a packet with a given IPv4 address arrives at such a network, its address is bitwise ANDed with the subnet mask to yield the particular subnet in which the host is located thereby simplifying packet delivery process. For example, when the address 180.200.16.6 is bitwise ANDed with the subnet mask 255.255.240.0, we obtain 180.200.16.0. Thus, the host is in subnet 16.

As an example, consider a company that uses the class C IP address 194.148.1.0 with the subnet mask 255.255.255.0 and needs to split the address range into six subnetworks. To accomplish this, the network administrator needs to use three bits in the last octet to create eight subnets with the subnet mask 255.255.255.224, which is obtained from the mask

11111111.11111111.11111111.11100000

Note that the decimal value of the last octet is obtained as follows: $2^7 + 2^6 + 2^5 = 128 + 64 + 32 = 224$. Figure 6.9 shows the address ranges in the different subnets.

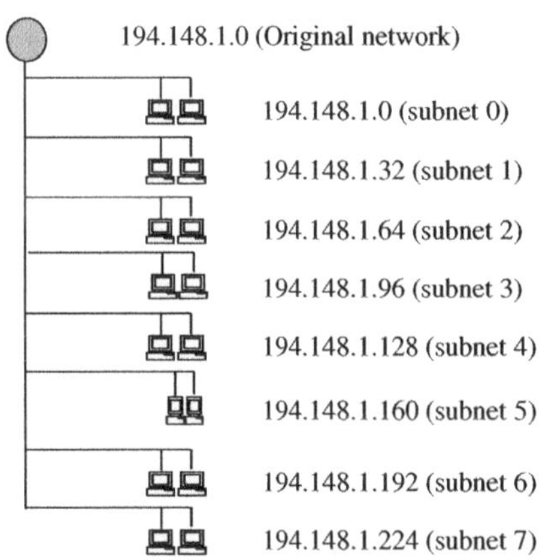

Figure 6.9 Address Ranges in Different Subnets.

Note also that Class D networks do not have any subnet mask because the data in a multicast packet is not destined for any particular host. Thus, there is no need to extract the host address from the IP address.

6.6 Variable Length Subnet Mask Networks

There are two types of subnet masks: fixed length subnet mask and variable length subnet mask (VLSM). In the fixed length subnet mask, all the subnets in the network use the same subnet mask. Native IP and RIP version 1 routing protocol only support the fixed length subnet mask. The main feature of fixed length subnet mask networks is that the maximum number of hosts in each subnet is the same.

VLSM allows the subnets that make up the network to use different subnet masks. The rationale behind a VLSM concept is that some subnets may have a small number of hosts while others have a relatively greater number of hosts. As stated earlier, when the same subnet mask is used for all these subnets, all the subnets will have the same number of IP addresses and there will be a waste of IP addresses in those subnets with smaller numbers of hosts than the available number of IP addresses. A better policy will be to allocate fewer IP addresses to subnets with only a few hosts and more IP addresses to subnets with greater numbers of hosts. This is precisely what the VLSM does in an IP network. Thus, the VLSM permits the network to be divided in such a way that it is possible to limit number of hosts that can be in each subnet by appropriately choosing the subnet mask for each network. Routing protocols RIP version 2 and OSPF support both the VLSM as well as the fixed length subnet mask.

This is accomplished in the following way. Consider a company that uses the IP address 194.148.1.0, which is a Class C address. Thus, the subnet mask of the "unsubnetted" network is 255.255.255.0. Assume that we need to create six subnetworks. To accomplish this, the network administrator needs to use three bits in the last octet to create eight subnets with the subnet mask 255.255.255.224, which is obtained from the mask

11111111.11111111.11111111.11100000

(Note that the decimal value of the last octet of the mask is $2^7 + 2^6 + 2^5 = 128 + 64 + 32 = 224$.) Since we have reserved 5 bits for the host address, each subnet can theoretically support 32 devices. The address ranges of the different subnets are as follows:

- 194.148.1.0 to 194.148.1.31 (subnet 0)
- 194.148.1.32 to 194.148.1.63 (subnet 1)
- 194.148.1.64 to 194.148.1.95 (subnet 2)
- 194.148.1.96 to 194.148.1.127 (subnet 3)
- 194.148.1.128 to 194.148.1.159 (subnet 4)
- 194.148.1.160 to 194.148.1.191 (subnet 5)
- 194.148.1.192 to 194.148.1.223 (subnet 6)
- 194.148.1.224 to 194.148.1.255 (subnet 7).

Note that subnet 0 includes the address 192.168.1.0, which is the address of the entire network, and subnet 7 includes the broadcast address 192.168.1.255. Thus, in some network implementations these two subnets may not be used, which would leave us with the remaining subnets (subnets 1–6).

Next, assume now that the network administrator wants to create two subnets that can handle only 8 IP addresses for some special equipment and would not like to waste the other 24 addresses. He would create these subnets from one of the existing subnets, say subnet 5, which has IP addresses 194.148.1.160 to 194.148.1.191. To do this, he would borrow two bits from the five bits used to address devices in the subnet to define subnets with the mask 255.255.255.248, where the last value is obtained from $2^7 + 2^6 + 2^5 + 2^4 + 2^3 = 128 + 64 + 32 + 16 + 8 = 248$, since the five highest order bits are used for subnetting. The four new subnets have the following IP address ranges:

- 194.148.1.160 to 194.148.1.167 (subnet 50)
- 194.148.1.168 to 194.148.1.175 (subnet 51)
- 194.148.1.176 to 194.148.1.183 (subnet 52)
- 194.148.1.184 to 194.148.1.191 (subnet 53).

The new subnets are shown in Figure 6.10. As before, the usable subnets in some network implementations may be subnets 51 and 52. This example illustrates how the need for variable length subnets arises and the mechanism

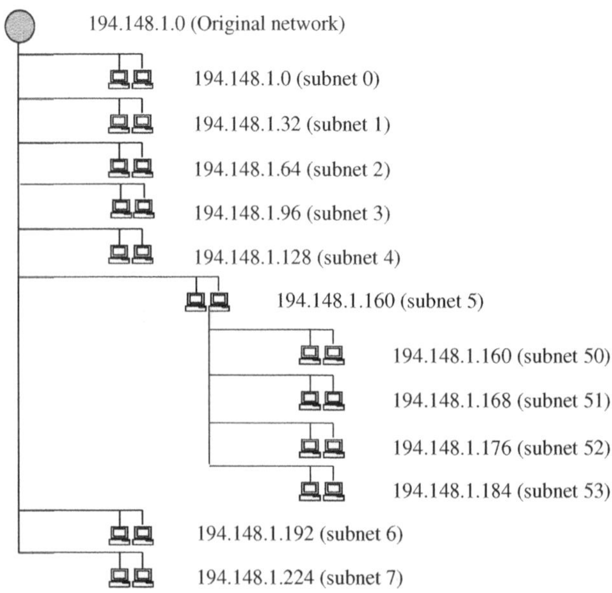

Figure 6.10 Illustration of VLSM Network Creation.

for creating such subnets. Unfortunately, not only is the VLSM network complex to manage, some routing protocols, such as RIPv1, do not support it.

6.7 IP Quality of Service

Quality of service (QoS) is a term used to characterize the performance of a network as seen by the users of the network. It is quantitatively measured by several parameters related to different aspects of the network performance. These include bit error rate, throughput, end-to-end delay, and jitter (or delay variation). The initial efforts on QoS in IP networks were based on the inclusion of the type of service (ToS) field in the original IP header specifications in RFC 791 in 1981. ToS field is an 8-bit field in which the three most significant bits are called the IP precedence bits. The three IP precedence bits can have one of eight settings. The larger the IP precedence value, the more important the packet and the higher the probability of timely forwarding.

However, in the early days of IP networks there was no need for traffic classification as the networks were used exclusively for file transfer and there was no incentive to treat any traffic differently. Thus, the ToS field was essentially grayed out. With the Internet used to carry different types of traffic, ranging from the delay-sensitive real-time voice and video traffic to the delay-tolerant non-real-time data traffic, the need to treat the different types of traffic differently has become obvious. For this reason, the ToS field has been renamed the differentiated services (DiffServ) field.

The structures of both the ToS and DiffServ fields are shown in Figure 6.11. The first three bits of the ToS field are used as the precedence subfield, which defines eight different control actions. The fourth bit is the delay bit; when set it means to minimize delay. The fifth bit is the throughput bit; when set it means to maximize throughput. The sixth bit is the reliability bit; when set it means to maximize reliability. The seventh and eighth bits are reserved.

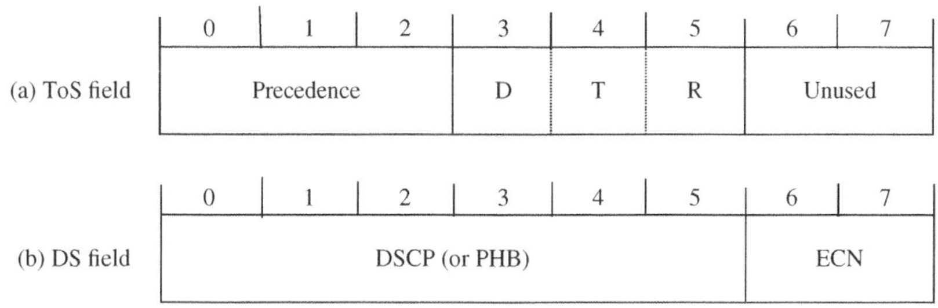

Figure 6.11 ToS and DiffServ Field Structures.

The DiffServ field is subdivided into two parts:

(a) The six most significant bits are called the DiffServ code point (DSCP) and are used to provide QoS.
(b) The last two bits are called the "ECN" field that is used for flow control.

The six DSCP bits can be combined in different ways to define different QoS treatments as a packet moves from one node to the other. Thus, with the DiffServ bits, we can set the DSCP values on the IP packet header to define a per-hop behavior (PHB), where PHB is an observable forwarding behavior of a network node toward a group of IP packets that have the same DSCP value.

Three PHBs have been defined by the IETF:

- *Default* PHB: With the three most significant bits of the DSCP field set to 000, the default PHB is used for best effort service, which is the traditional IP traffic service. Any packet with an unrecognizable PHB value is treated as if it is marked with default PHB.
- *Assured forwarding* (AF) PHB: With the most significant 3 bits of the DSCP field set to 001, 010, 011, or 100 (these are also called AF1, AF2, AF3, and AF4), the AF PHB is used for guaranteed bandwidth service. Thus, it is defined for customers who need reliable service from their service providers even in times of network congestion.
- *Expedited forwarding* (EF) PHB: With the most significant three bits of the DSCP field set to 101 (the whole DSCP field is set to 101110, decimal value of 46), the EF PHB provides low delay service.

The default PHB is the only required behavior. Other PHBs are optional. Any traffic that does not meet the requirements of any of the other PHBs is classified as a default PHB whose value is 000000.

The expedited forwarding PHB whose value is 101110 is used for a traffic class that requires low delay, low loss rate, and low jitter. These are the characteristics of voice, video, and other real-time services. The traffic is often given the highest priority among all the traffic classes.

Assured forwarding PHB allows the operator to provide assurance of delivery as long as the traffic does not exceed some subscribed rate. Traffic that exceeds the subscription rate may be dropped if congestion occurs. The AF behavior group defines four separate AF classes. Within each class, packets are given a drop precedence (high, medium, or low, where higher precedence means *more* dropping). The combination of classes and drop precedence yields 12 separate DSCP encodings from AF11 through AF43, as shown in Table 6.1. The first three bits (bits 0, 1, and 2) define the priority class, the next two bits (bits 3 and 4) specify the drop percentage, and the last bit (bit 5) is always 0.

Thus, AF PHB allows different priority classes each of which can have further differentiation via the percentage of packets that may be dropped. The DSCP values are the decimal values of the PHB values. For example, AF11 has DSCP

Table 6.1 Different Code Points for Assured Forwarding PHB.

Precedence	Class 1	Class 2	Class 3	Class 4
Low drop	AF11 = 10 (001010)	AF21 = 18 (010010)	AF31 = 26 (011010)	AF41 = 34 (100010)
Medium drop	AF12 = 12 (001100)	AF22 = 20 (010100)	AF32 = 28 (011100)	AF42 = 36 (100100)
High drop	AF13 = 14 (001110)	AF23 = 22 (010110)	AF33 = 30 (011110)	AF43 = 38 (100110)

10, which is the decimal value of the PHB value 001010, which is the value in parentheses. Note that each class will be placed in a different queue. When the queue is full, packets with "high drop" probability will be deleted from the queue before other packets. Also, if congestion occurs between classes, the traffic in the higher class is given priority.

6.8 Operation of the Explicit Congestion Notification

The ECN field allows end-to-end notification of network congestion without dropping packets. The two bits are used to encode four different code points:

(a) 00: This indicates that the node is non-ECN capable or non-ECN transport (non-ECT), which means that the packet does not use ECN
(b) 01: ECN capable transport, ECT(0); that is, the packet uses ECN
(c) 10: ECN capable transport, ECT(1); that is, the packet uses ECN
(d) 11: Congestion encountered (CE); that is, the packet has experienced congestion.

When both endpoints support ECN, they mark their packets with ECT(0) or ECT(1). If the packet arrives at a router that is experiencing congestion, and the router supports ECN, it may change the code point to CE instead of dropping the packet. This action is referred to as "marking," and its purpose is to inform the destination of impending congestion in the network. At the destination, this congestion information is handled by TCP, which takes steps to alert source node of the need to reduce its transmission rate.

6.9 Address Resolution Protocol

Although IP addresses are popularly known, message frame delivery is made over a LAN that uses MAC addresses. Thus, before a frame can be delivered to a host, the host's MAC address must be known. The process of mapping the

IP address to the MAC address for actual delivery of frames is called *address resolution*. Address resolution would be trivial if the MAC address is configured to be a part of the host's IP address. In most cases, address resolution is done through *dynamic binding*, which is required because the network interface card (NIC) that is identified with the MAC address may be changed.

The ARP consists of the following steps:

(a) A source that has a packet to transmit broadcasts the IP address of the destination over a LAN.
(b) The destination replies with its MAC address.
(c) The source maintains a cache of IP-to-MAC address bindings that it can use for subsequent communication with the other devices.

We consider the following examples of ARP use.

6.9.1 Source and Sink in Same LAN

Consider two Ethernet LAN segments E1 and E2 linked by router R1. Let the IP addresses of the segments be as shown in Figure 6.12.

Assume that host H1 wants to send a packet to host H3 (IP address 214.124.21.3), which is also on LAN E1. The sequence of steps required for the desired data transfer is as follows:

(a) IP layer of H1 checks whether the (*IP address, MAC address*) *pairs* stored in its cache includes a pair with the desired IP address.
(b) If the search is successful, H1's data link layer (DLL) frames the packet with H3's MAC address as the destination address.
(c) If the search fails, H1 broadcasts an ARP message, which is received by all hosts on LAN E1 including H3, which returns its MAC address.
(d) H1 receives the ARP message, which now contains H3's MAC address, and its DLL frames the packet appropriately.
(e) H1 sends the properly framed packet to H3.

6.9.2 Source and Sink in Different LANs: Proxy ARP

With respect to Figure 6.12, assume that H1 wants to send a packet to host H5 (IP address 192.160.25.8), lying on LAN E2. Assume that H1 does not have a binding on H5; then it sends an ARP request. Since R1 is on the same LAN segment as H1 and is the default gateway for the LAN segment, it responds

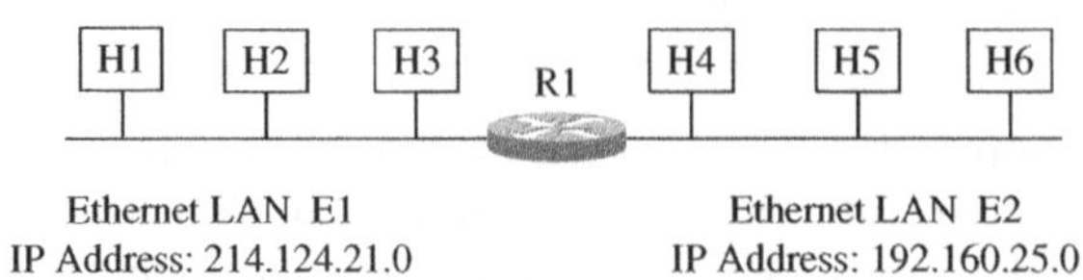

Figure 6.12 ARP Example.

to the request with its own MAC address; this is called *proxy ARP*. The steps involved in the process are as follows:

(a) H1's DLL frames the packet with the MAC address of R1 as the destination address and transmits it through E1.
(b) R1's DLL receives the frame, removes the DLL header and trailer, and passes on the original packet to R1's IP layer.
(c) Since R1 is also on the same LAN segment (which can be an IP subnet) as H5, it broadcasts an ARP request on LAN E2 for H5's MAC address.
(d) H5 responds to the ARP request with its own MAC address to H1 via R1, and R1 forwards the packet with the MAC address of H5 to H1.

6.9.3 Source and Sink in Different Remote LANs

Consider two Ethernet LAN segments E1 and E2 linked through routers (R1, R2) to a network, with IP address ranges as indicated below. Let the IP addresses of the routers be as follows:

- R1: 214.124.21.100 on E1 and 144.16.251.100 on the network
- R2: 192.160.25.200 on E2 and 144.16.251.200 on the network.

The network is illustrated in Figure 6.13.

We consider the case when host H1 wants to send a packet to host H5 (IP address 192.160.25.8), located in LAN E2. Assume that H1 does not have a binding on H5; then it sends an ARP request. Since R1 is on the same LAN segment as H1 and is the default gateway for the LAN segment, it responds to the request with its own MAC address; this again is proxy ARP and the process operates as follows:

- H1's DLL frames the packet with the MAC address of R1 as the destination address and transmits it through E1.
- R1's DLL receives the frame, removes the DLL header and trailer, and passes on the original packet to R1's IP layer.
- R1 then broadcasts the request on the network.
- Since R2 and H5 are in the same subnet, R2 responds to the ARP request with its network address; this is another proxy ARP.

Figure 6.13 Proxy ARP Example.

- R1's DLL frames the packet with R2's network address as the destination address and transmits the frame.
- R2's DLL receives the frame, strips off the frame header and trailer, and passes on the packet to R2's IP layer, which finds from its routing table that the desired IP address refers to LAN E2.
- R2 broadcasts an ARP message with H5's IP address as the target IP address, which finally enables it to obtain from H5 its MAC address.
- H5 returns an ACK to H1 with its MAC address, which is received by R2 and sent to R1, which in turn sends it to H1.

6.10 Dealing with Shortage of IPv4 Addresses

The popularity of the Internet has led to an explosion in the demand for and depletion of IPv4 addresses. The Internet engineering task force (IETF), which is an organization that is responsible for the development and smooth operation of the Internet, has proposed three short-term solutions to deal with the IPv4 address shortage. These are as follows:

(a) Private Internets
(b) Network address translation (NAT)
(c) Classless inter-domain routing (CIDR).

The long-term solution is the introduction of IPv6, which uses 128 bits for addressing and is discussed later in this chapter.

6.10.1 Private Internets

The rationale for developing the private Internets scheme is that many hosts in a network do not always require access to hosts in other networks. Moreover, when they do, such an access can be handled by a mediating gateway. The IETF has set aside three blocks of IP addresses to be used by any organization in an unrestricted manner as long as they are not advertised outside that organization's network:

10.0.0.0	-	10.255.255.255
172.16.0.0	-	172.31.255.255
192.168.0.0	-	192.168.255.255

Observe that these address blocks are from Class A, Class B, and Class C networks, respectively. These IP addresses are defined to be *nonroutable* or *private* IP addresses, and any packet with any of these addresses will be discarded by a router outside the network where they originated. All IP addresses outside these blocks are defined to be *routable* or *public* IP addresses.

6.10.2 Network Address Translation

NAT is a method of connecting multiple computers to a network using one routable IP address. It is popularly used in conjunction with the private Internets and is particularly useful in home networks because it allows all devices in the home network to access the Internet using a single IP address. It operates in a border router that has one interface to a network with private IP addressing and another interface to a network with public IP addressing. Packets from the private network are assigned the IP address of the router in the public network when they need to enter the public network.

NAT can also operate in a dynamic mode where a small set of public IP addresses is set aside for a much larger group of users such that when a member of the group attempts to get outside the private IP network, they will be assigned one of the public IP addresses. In this case, requests that are received after the IP addresses are used up are denied.

Figure 6.14 illustrates the combined use of the private Internet and NAT. The private network with the IP address 192.168.1.0 is connected to a public network via a NAT-capable router. Thus, the router straddles the private network to which it is connected with the port with IP address 192.168.1.1, and the public network to which it is connected via the port with the IP address 180.240.16.6. All packets that are generated in the private network and want to get into the public network will have their source addresses overwritten to the address 180.240.16.16, the public IP address of the NAT router. In this way, the public network does not see the private addresses of the devices inside the private network. The NAT router is also responsible for directing a response from the public network to the appropriate owner in the private network by replacing the destination address of the packet with the private IP address of the owner and forwarding the packet to the latter via the private network.

6.10.3 Classless Inter-Domain Routing

Given an IPv4 address, one can easily know to which class of IP addresses it belongs. Thus, we refer to the IPv4 addresses we discussed so far as *classful*

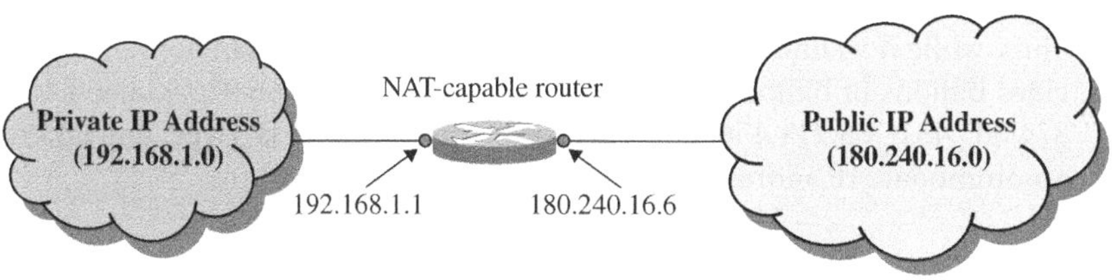

Figure 6.14 Illustration of NAT.

addresses. CIDR replaces classful IP addresses with the more general notion of network prefix with a specified length. A CIDR address includes the standard 32-bit IP address and the number of bits in the network prefix. The structure of a CIDR address is a.b.c.d/*n*, where *n* is the number of bits in the network address space and ranges between 13 and 27. For example, the classless IP address 180.32.10.48/24 means that the first 24 bits are used to identify the network (i.e., the network address), while the remaining 8 bits are used to identify the hosts. "24" is defined as the prefix length. Observe that the address 180.32.20.48 is traditionally a Class B address. With CIDR it now behaves like a Class C address.

6.11 IPv6

The continuous growth of the global Internet requires that its overall architecture evolve to accommodate the new technologies that support the growing numbers of users, applications, appliances, and services. The current IP address space is unable to satisfy the potential large increase in the number of users or the geographical needs of the Internet expansion. Also, IPv4 cannot meet the requirements of emerging applications such as Internet-enabled personal digital assistants (PDAs), Internet-connected automobiles, integrated telephony services, and distributed gaming.

IPv6 increases the number of network address bits from the current 32 bits in IPv4 to 128 bits, which provides more than enough globally unique IP addresses for every network device that is currently on the planet. The lifetime of IPv4 has been extended using such techniques as private Internet, network address translation, and CIDR. While these techniques appear to increase the address space, they fail to meet the requirements of the new applications. Standard bodies for the wireless data services are preparing for the future, and IPv6 provides the end-to-end addressing required by these new environments for mobile phones and residential Voice over IP (VoIP) gateways. IPv6 provides the services, such as integrated autoconfiguration, QoS, and security that are required by these new services.

The basic features of IPv6 include the following:

(a) *Expanded address space*: IPv6 has a larger address field that is 128 bits long. Thus, while IPv4 limits the number of addresses to about 4 billion, IPv6 provides billions of billion addresses. There are 340,282,366,920,938,463,463,374,607,431,768,211,456 IP addresses. This number is large enough to accommodate IP address demands in the foreseeable future.

(b) *Streamlined header and optional extensions*: Some of the IPv4 fields have been dropped in IPv6, thereby reducing the processing overhead of packet handling and keeping the bandwidth cost as low as possible.
(c) *Efficient routing*: CIDR was introduced in IPv4 to do away with address classes and permit flexible use of variable-length network prefixes. This permits route aggregation in the Internet whereby a backbone router can store a single routing table entry that provides reachability to many lower-level networks. CIDR is an integral part of IPv6 because no address classes are defined in IPv6.
(d) *Stateless autoconfiguration*: In IPv4, the dynamic host configuration protocol (DHCP) is used to manage IP address assignment. DHCP is a "stateful" protocol that maintains static tables to determine the addresses that have been assigned to hosts. DHCP for IPv6 not only provides this stateful feature but also adds a new "stateless" *autoconfiguration* feature that permits hosts to configure their own addresses with the help of a local IPv6 router. This is done by a host combining its 48-bit MAC address, known as the link-local address, with a network prefix obtained from the local router.
(e) *Flexible addressing*: Supports three modes of addressing: *unicast*, *multicast*, and *anycast*.
(f) *Performance improvement*: There is no packet fragmentation in IPv6. Thus, the source router sets the packets to the minimum MTU of the path (called path-MTU) before they are transmitted.
(g) *Class of service support*: Packets are labeled according to their flows and routers handle packets that belong to the same flow in the same manner. This will allow all IPv6 packets with the same label to be routed through specified network nodes and prevent them from straying into unnecessary parts of the network thereby guaranteeing fast delivery of packets. Also, priority bits are used to provide traffic handling order.
(h) *Built-in security feature*: IPv6 includes IPSec and provides authentication and privacy.
(i) *Mobility support*: IPv4 supports mobility through a special Mobile IP protocol. IPv6 has built-in roaming support.
(j) *Broadcast*: There is no broadcast in IPv6; this functionality is taken over by multicast. A consequence of this is that the all 0's and all 1's addresses are legal.

IPv6 defines the concept of *link* as follows: A link is a communication facility or medium over which nodes can communicate at the link layer, such as an Ethernet (simple or bridged). The link layer is the layer immediately below IP. Thus, all nodes connected to a hub or switch or indeed belonging to one broadcast

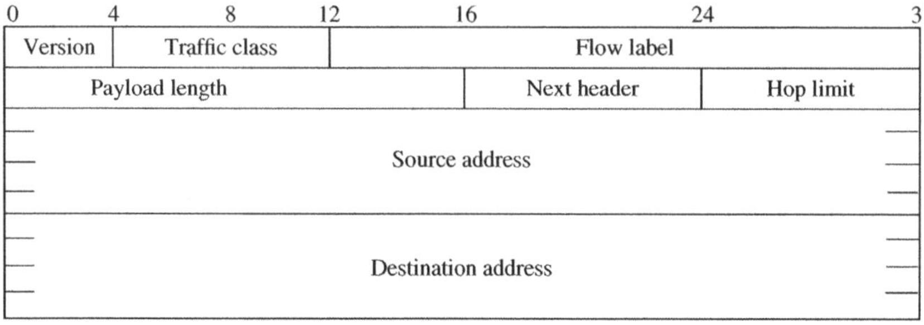

Figure 6.15 IPv6 Header.

domain are on the same link. The concept of a link means that a device can communicate with other devices on the same link without using an IPv6 address. An IPv6 address is necessary only if a device wants to communicate with other devices that are not on the same link as the device. Thus, unlike IPv4 where an IP address is required for every type of communication, an IPv6 address is not required for every type of communication.

6.11.1 IPv6 Header

IPv6 has six fixed fields and two address fields; it has no option field and so no *HLEN* field. It has a total of 40 fixed bytes of header and has no checksum, which reduces header processing cost since there is no need to check and update checksum at each intermediate node. It allows extension headers to be appended after the main header. Figure 6.15 shows the IPv6 header.

The header fields are defined as follows:

- *Version* (4 bits): indicates version number (=6)
- *Traffic class* (4 bits): indicates priority value
- *Flow label* (20 bits): used to label packets that the source wants routers within the network to handle in a special manner. All packets that belong to the same "flow" are assigned the same flow label.
- *Payload length* (16 bits): indicates the length of the remainder of the IP packet following the header
- The *next header* (8 bits): identifies type of header immediately following the main header. Six extension headers are defined as follows:
  - *Hop-by-hop option header*: passes management or debugging information to routers handling the packet
  - *Routing header*: performs source routing
  - *Fragment header*: contains information for reassembling packets that were fragmented before entering the network
  - *Authentication header*: provides packet integrity and authentication

– *Encapsulating security payload header*: provides privacy
– *Destination option header*: contains information that only the destination station may examine.

A packet can carry zero or more extension headers, and each extension is identified in the *next header* field of the preceding header.

- *Hop limit* (8 bits): indicates the remaining number of allowable hops for this packet; it is set by the source and decremented by 1 by each node that forwards the packet.
- *Source address* (128 bits): indicates the IP address of source that originated the packet
- *Destination address* (128 bits): indicates the IP address of the intended recipient of the packet.

6.11.2 Concept of Flexible Addressing in IPv6

As previously discussed, IPv6 supports three modes of addressing: *unicast, multicast*, and *anycast*. Unicast is used to describe communication between one sender and one receiver. Thus, in a unicast transmission, a packet is sent from a single source to a single destination.

Multicast is a type of communication where one packet is simultaneously sent to a group of devices on the network that are members of a multicast group. Any device that is interested in a particular multicast traffic must join that multicast group in order to receive the traffic. IPv6 multicast groups are identified by IPv6 multicast addresses. In multicast, the sender transmits only one copy of packet and it is delivered to any device that is interested in that traffic. The multicast routers in the network are responsible for duplicating the packet based on how many ports on a router are on the paths to the users who are members of the multicast group to which that packet belongs.

Anycast is a type of IPv6 network communication in which a single IPv6 address is assigned to multiple nodes. These nodes are usually servers that perform the same function. When a request for the service performed by anycast address servers is received, it is sent to the first available server.

6.12 Summary

In this chapter, we have discussed the network addressing function of the network layer. In particular, we discussed IPv4 and the different classes of the address. We discussed the concept of subnetting that enables us to partition an IP network into smaller groups called subnets, which eases the management of data networks. We discussed the VLSM that permits the maximum number of IP addresses in some subnets to be different. We also discussed the concept of IP QoS. We discussed the short-term methods used to deal with the shortage of IPv4 addresses. These include the private Internet, network address translation, and CIDR. Finally, we discussed IPv6 as a long-term solution of the IPv4 address shortage.

Exercises

1 Name the two functions of the network layer.

2 What is the special feature of Class A IPv4 addresses?

3 What is the special feature of Class C IPv4 addresses?

4 What is the special feature of IPv4 multicast addresses?

5 Name two limitations of IPv4.

6 Name two features of IPv6 that provide solutions to the two limitations of IPv4 you listed in Question 5 above. (*Make sure that each answer you give addresses a shortcoming you listed above.*)

7 Name two short-term methods used to deal with IPv4 address shortage.

8 What is name of an IP data block?

9 What does the IPv4 address 120.10.15.20/15 mean?

10 What is the subnet mask of the IP address 120.10.15.20/15?

11 What is the subnet mask of the IPv4 address 120.10.15.20 in which 4 bits are used to create subnets?

12 What is a variable-length subnet mask (VLSM) network?

13 What is the function of the address resolution protocol?

14 What is the differentiated services code point field in the IP header used for?

15 What is the explicit congestion notification field in the IP header used for?

7

Network Layer Part II – Routing

7.1 Introduction

This chapter is a continuation of the discussion on the network layer service of IP addressing. The topics to be covered in this chapter include the following:

a. Routing principle
b. Routing algorithms
c. Unicast routing protocols
d. Multicast routing.

7.2 Routing Principle

Routing is the process of forwarding messages from source to destination through a packet-switched network. Thus, before a sending host can transfer packets to the destination host, the network layer must determine the *path* or *route* that the packets are to follow. This route determination is made by the network layer *routing protocol*. At the heart of any routing protocol is the *routing algorithm* that computes the path for a packet.

The purpose of a routing algorithm is simple: given a set of interconnected routers, a routing algorithm finds a "good" path from source to destination. Typically, a "good" path is one that has the "least cost," where cost is defined according to the type of protocol used. That is, the routing protocol specifies the criteria for determining the routes, and the routing algorithm carries out the actual computation based on those criteria.

7.3 Routing Algorithms

A routing algorithm is that part of the network layer software that decides to which output link an incoming packet should be transmitted. Routing

Fundamentals of Data Communication Networks, First Edition. Oliver C. Ibe.

algorithms can be classified in several ways, some of which include the following:

- Static routing versus dynamic routing
- Link-state routing versus distance–vector routing
- Flat routing versus hierarchical routing
- Host-intelligent routing versus router-intelligent routing
- Centralized versus distributed routing algorithms.

7.4 Static Versus Dynamic Routing

In static routing (also called nonadaptive routing), routing decisions are not based on any measurements or estimates of the current traffic levels and changes in network topology. Instead, the choice of routes is computed in advance, offline and downloaded typically by the network administrator. Thus, the routing table is populated in advance by the network administrator. These computations do not change unless the network administrator alters them.

Algorithms that use static routes are simple to design and work well in environments where network traffic is relatively predictable and where network design is relatively simple. Because static routing systems cannot react to network changes, they are generally considered unsuitable for today's large, changing networks.

Dynamic routing algorithms (also called *adaptive routing algorithms*) periodically change their routing decisions to reflect changes in topology and network traffic. If the network update information indicates that a network change has occurred, the routing software recalculates routes and sends out new routing update messages. These messages permeate the network, stimulating routers to rerun their algorithms and change their routing tables accordingly.

7.5 Link-State Versus Distance–Vector Routing

Dynamic routing algorithms can be further divided into link-state algorithms and vector–distance algorithms. Link-state algorithms operate with information about routes over the entire network in the following manner:

- Each link-state router gathers this information from each of its neighbors and in turn passes it on to other neighbors who pass it on to their own neighbors, and so on. Eventually, all the routers have information about all the links on the network.
- Then, each router runs a shortest path algorithm, such as the *Dijkstra shortest path algorithm*, to calculate the best path to each network router and create routing tables.

Distance–vector algorithms require each router to send all or some portion of its routing table to only its neighbors. An example of a distance–vector algorithm is the *Bellman–Ford* algorithm. In essence, link-state algorithms send small updates everywhere, while distance–vector algorithms send larger updates only to neighboring routers.

7.6 Flat Versus Hierarchical Routing

Some routing algorithms operate in a flat space, while others use routing hierarchies. In a flat routing system, the routers are peers of each other. In a hierarchical routing system, some routers form what amounts to a routing backbone. Packets from nonbackbone routers travel to the backbone routers where they are sent through the backbone until they reach the general area of the destination. At this point, they travel from the last backbone router through one or more nonbackbone routers to the final destination.

One way to understand hierarchical routing is to consider a large network that is divided into logical groups of nodes called domains, autonomous systems (ASs), or areas. Routers in one domain can communicate with each other without their packets leaving the domain; these routers are called *intradomain routers*. Some routers in a domain can communicate with routers in other domains and these are called *interdomain routers*. The interdomain routers are the ones that form the backbone network that interconnects the different domains and packets going from one domain to another pass through them.

The primary advantage of hierarchical routing is that it imitates the organization of most companies and therefore supports their traffic patterns well. Most network communication occurs within small company groups (domains). Because intradomain routers need to know only about other routers within their domain, their routing algorithms can be simplified and, depending on the routing algorithm being used, routing update traffic can be reduced accordingly.

Figure 7.1 illustrates the hierarchical routing. There are three domains and five interdomain routers (labeled backbone routers, BBR) that form the backbone network. All the other routers are intradomain routers.

7.7 Host-Based Versus Router-Intelligent Routing

Some routing algorithms assume that the source node will determine the entire route. In these algorithms, called *source routing* algorithms, routers merely act as store-and-forward devices that mindlessly send the packet to the next router listed in the packets' headers. Other algorithms assume that hosts know nothing about routes. In these algorithms, the routers determine the path through the network based on their own calculations. Thus, in source routing, the hosts

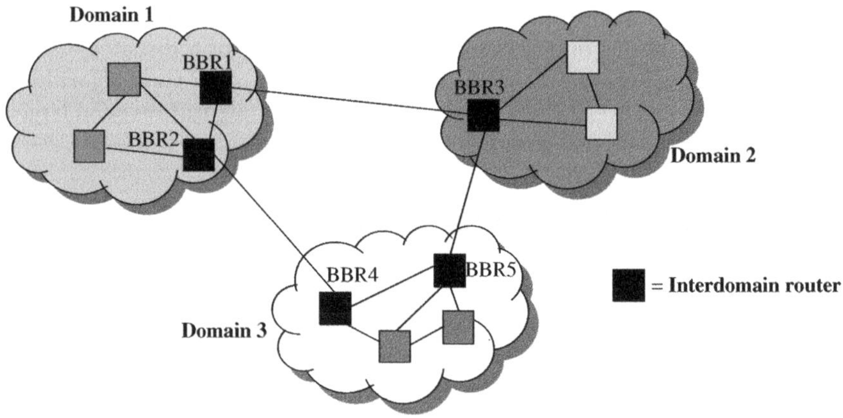

Figure 7.1 A Simple Hierarchical Routing.

have the routing intelligence, while in nonsource routing, the routers have the routing intelligence. The problem with source routing is that if the defined path becomes unavailable due to one of the routers failing, there is usually no provision for an alternate path and so the packet is discarded.

7.8 Centralized Versus Distributed Routing

In centralized routing algorithms, a special node has full view of the network and this node makes all routing computations. The advantage in this case is that only one node is required to maintain routing information. The disadvantage is that it is a single point of failure problem because if the computing node fails, network routing will be disrupted. In particular, no new paths can be set up and updates of existing routing paths cannot be made.

In distributed routing algorithms, each router computes the best path to all destinations using shortest paths that are computed based on exchange of link information between neighbors.

7.9 Routing Metrics

Routing tables contain information that routers use to select the best route. But how, specifically, are routing tables built? What is the specific nature of the information they contain? How do routing algorithms determine that one route is more preferable than others?

Routing algorithms use many different metrics to determine the best route. Sophisticated routing algorithms can base their route selection decisions on multiple metrics, combining them in a single (hybrid) metric. The following metrics are often used in route computation:

- Path length
- Reliability
- Delay
- Bandwidth
- Load
- Communication cost.

7.9.1 Path Length

Path length is the most common routing metric. Some routing protocols allow network administrators to assign arbitrary costs to each network link. In this case, path length is the sum of the costs associated with each link traversed. Other routing protocols define *hop count*, which is a metric that specifies the number of routers that a packet must traverse on its route from the source to the destination.

7.9.2 Reliability

Reliability, in the context of routing algorithms, refers to the dependability (usually described in terms of the bit-error rate) of each network link. Some network links might go down more often than others. After a network fails, certain network links might be repaired more easily or more quickly than other links. Any reliability factors can be taken into account in the assignment of the reliability ratings, which are arbitrary numeric values usually assigned to network links by network administrators.

7.9.3 Delay

Routing delay refers to the length of time required to move a packet from the source to the destination through the network. Delay depends on many factors, including the bandwidth of intermediate network links, the port queues at each router along the way, network congestion on all intermediate network links, and the physical distance to be traveled. Because delay is a composite of several important variables, it is a common and useful metric.

7.9.4 Bandwidth

Bandwidth refers to the available traffic capacity of a link. All other things being equal, a 10-Mbps link would be more preferable than a 64-kbps link. Although bandwidth is a rating of the maximum attainable throughput on a link, routes through links with greater bandwidth do not necessarily provide better routes than routes through slower links. If, for example, a faster link is busier, the actual time required to send a packet to the destination could be greater.

7.9.5 Load

Load refers to the degree to which a network resource, such as a router, is busy. Load can be calculated in a variety of ways, including CPU utilization and packets processed per second. Monitoring these parameters on a continual basis can be resource-intensive itself.

7.9.6 Communication Cost

Communication cost is another important metric, especially because some companies may not care about performance as much as they care about operating expenditures. Even though link delay may be greater, a company may prefer to send packets over its own data communication network rather than through a service provider's network that is faster but costs a lot of money for usage time.

7.10 Flooding Algorithm

Flooding is a routing algorithm in which each incoming packet is sent on every outgoing link except the one through which it was received. Flooding generates a large number of duplicate packets, which can in fact be an infinite number if no measures are taken to dampen the process. To dampen the flooding process, some methods such as hop counter and selective flooding are used. Because it uses all the possible paths, flooding always produces shortest path routing; thus, no algorithm can produce a shorter delay if overhead generated by flooding is ignored. Figure 7.2 is an illustration of flooding in a network.

7.11 Distance–Vector Routing Algorithms

A distance–vector routing algorithm is sometimes called the *Bellman–Ford* algorithm and the *Ford–Fulkerson* algorithm. It was the original ARPANET

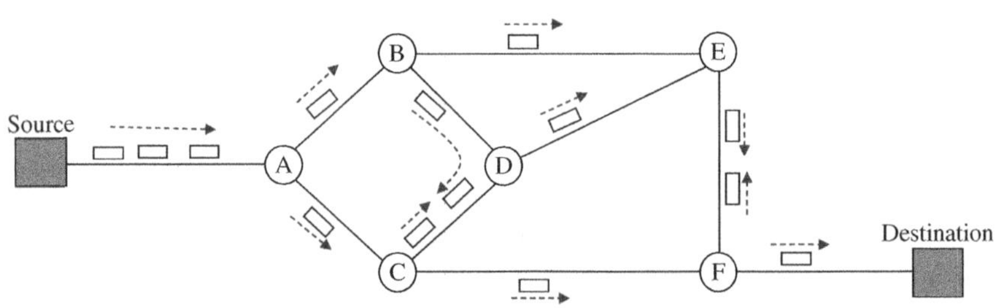

Figure 7.2 Illustration of Flooding.

routing algorithm until 1979 when it was replaced by a link-state routing algorithm. (ARPANET is the predecessor of the current Internet.) It is also used to implement the Routing Information Protocol (RIP).

A distance–vector routing algorithm operates by having each router maintain a table (or vector) that contains an entry for each router in the network. This entry consists of two parts: an estimate of the shortest distance to the router and the preferred outgoing link to use to get to the router. Different metrics can be used to define distance, and they include the following:

- Hop count, which means that the entry is the number routers that are traversed to get to the destination router.
- Queue length, where the router examines the number of packets that are waiting to be transmitted on each output link.
- Delay, which the router measures directly by a special ECHO packet that the receiver just timestamps and sends back as fast as it can.

7.12 Link-State Routing Algorithms

One of the reasons why the distance–vector algorithm was discontinued in the ARPANET in 1979 was that at that time the distance metric was queue length, which did not take bandwidth into consideration when choosing routes. The basic principle of the link-state routing algorithm can be summarized as follows:

- Each router must discover its neighbors and learn their network addresses.
- It must measure the delay or cost on the link to each of its neighbors.
- It must construct a packet detailing all it has just learned and send this packet to all other routers in the network.
- It must compute the shortest path to every other router using similar information it receives from other routers.

Essentially, the complete topology and all delays are experimentally measured and distributed to every router. The *Dijkstra's algorithm* is usually used to run the shortest path to every other router. To identify the routers that are physically connected to a given router and get their IP addresses, the router operates as follows:

- When the router starts working, it first broadcasts a "HELLO" packet over the network.
- Each router that receives this packet replies with a message that contains its IP address.

To measure the delay to neighboring routers, the router sends *echo packets* over all outgoing links. Every router that receives these packets replies with an echo reply packet. By dividing round trip time by 2, the router can estimate the delay time. Note that this time includes both transmission and processing times – the time it takes the packets to reach the destination and the time it takes the receiver to process it and reply to it.

The router shares its knowledge by the reliable flooding of the network with its information to enable every router to know the structure and status of the network. Reliable flooding is the process of making sure that all routers in the network that are participating in the routing protocol get a copy of the link-state information from all other routers. A router first sends the information to its immediate neighbors on each output link. Each router that receives the packet forwards it to all its neighbors except the neighbor from where it was received, and so until the information reaches all other nodes. Each router stores the sequence number of the packet so that it does not forward it again to its neighbors if a copy of the packet makes its way back to the router later.

To share routing information with other routers using reliable flooding, each router creates an update packet called a *link-state packet* (LSP) that contains the following information:

- The ID of the router creating the LSP
- A list of directly connected neighbors of the router along with the cost of each link to each neighbor
- A sequence number that enables a receiving router to know when it is receiving a duplicate copy of an LSP
- The time to live (TTL) of the LSP to ensure that older information is removed from the network; TTL is decreased by one at each node that receives the LSP and discarded when it reaches zero and has not reached its destination.

Consider an LSP that arrives at node X in Figure 7.3. The node forwards the packet to its two neighbors A and C. (After a node has received the LSP, we color it gray.) A and C forward the packet to node B who receives two identical copies of the LSP and thus accepts one (because they have the same sequence number and same source ID). B then forwards LSP to D and all the nodes receive the packet. This is an example of reliable flooding.

7.13 Routing Protocols

As stated earlier, routing protocols define the rules that are used by the routing algorithms. Routing protocols are divided into two broad classes:

- *Intradomain routing protocols,* which are used within an AS (or routing domain); they are also called *interior gateway routing protocols.*
- *Interdomain routing protocols,* which are used between ASs.

Examples of intradomain routing protocols are the *RIP* and the *open shortest path first* (OSPF) protocol, which are discussed later in this chapter. An example

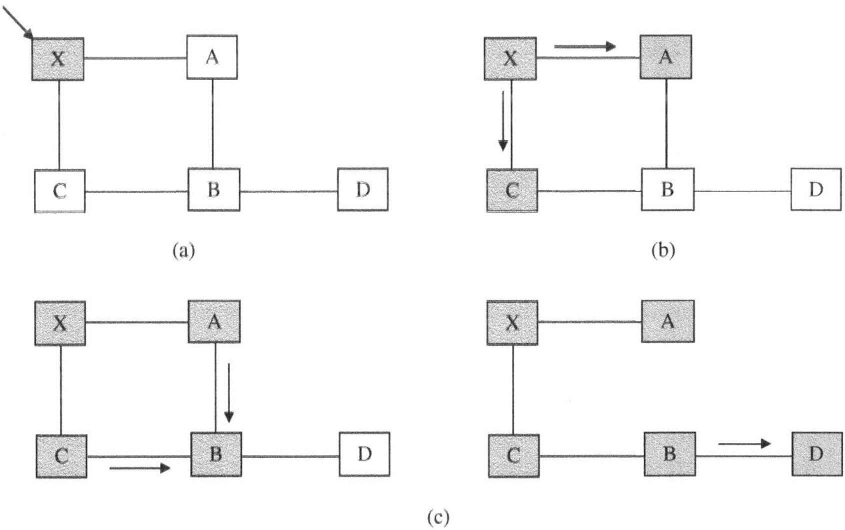

Figure 7.3 Example of Reliable Flooding. (a) LSP Arrives at X; (b) X Floods LSP to A and C; and (c) A and C Flood LSP to B But Not to X.

of interdomain routing protocols is the border gateway protocol, which we will not discuss.

One way to understand the concept of domain is that it is a network in which all the routers are under the same administrative control, such as a single university campus or the network of an Internet service provider. Consider Figure 7.4. In this case, the interdomain routers implement both an intradomain routing protocol so they can communicate with intradomain

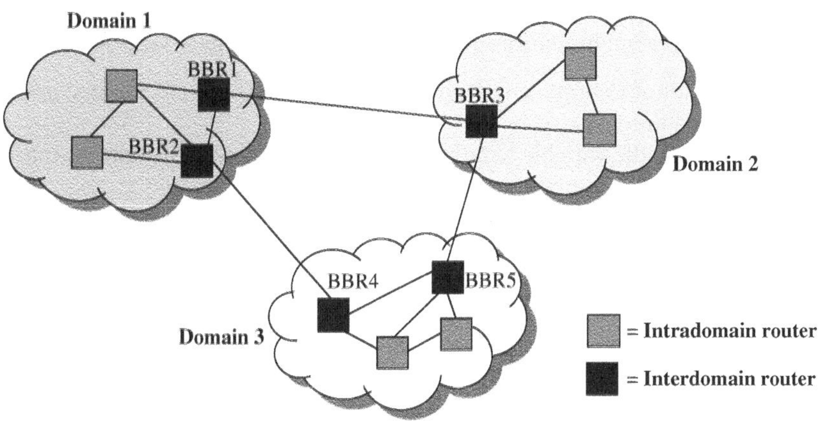

Figure 7.4 Illustration of Domain.

routers in their domains, and also an interdomain routing protocol so they can talk across their domains.

7.14 Routing Information Protocol

RIP is an implementation of a distance–vector routing protocol. It classifies routers as active and passive (silent). Active routers advertise their routes (reachability information) to others; passive routers listen and update their routes based on advertisements but do not advertise. Typically, routers run RIP in active mode, while hosts use passive mode. A router running RIP in active mode broadcasts updates at set intervals. Each update contains paired values where each pair consists of an IP network address and an integer distance to that network address. RIP uses a hop count metric to measure the distance to a destination; thus, a router advertises directly connected networks at a metric of one.

Networks that are reachable through another router are two or more hops away; thus, the number of hops (or hop count) along a path from a given source to a given destination refers to the number of routers that a datagram would encounter along that path. Using hop counts to calculate shortest paths does not always produce optimal results. For example, a path with hop count 3 that crosses three T3 links (with a data rate of 44.736 Mbps) may be substantially faster than a path with a hop count 2 that crosses two T1 links (with a data rate of 1.544 Mbps).

7.15 Routing Information Protocol Version 2

The rapid growth and expansion of today's networks have pushed RIP to its limits. RIP has certain limitations that could cause problems in large networks:

- RIP has a limit of 15 hops; a node that is more than 15 hops (15 routers) away is considered unreachable.
- RIP cannot handle variable length subnet masks (VLSM). All subnets must have the same subnet mask length.
- Periodic broadcasting of the full routing table consumes a large amount of bandwidth, which is a major problem with large networks especially on slow links and WAN clouds.
- RIP converges slowly. This means that we may not finish running one round of updates before it is time for the next one. Thus, it takes a long time to get a complete update using the algorithm.
- RIP has no concept of network delays and link costs; routing decisions are based on hop counts. The path with the lowest hop count to the destination is

always preferred even if the longer path has a better aggregate link bandwidth and smaller delays.
- RIP networks are flat networks; there is no concept of areas or boundaries.

Some enhancements were introduced in a new version of RIP called RIP2 to address the issues of VLSM, authentication, and multicast routing updates. However, RIP2 is not a big improvement over RIP (now called RIP1) because it still has the limitations of hop counts and slow convergence; these are issues that are not expected in today's large networks.

7.16 Open Shortest Path First Protocol

OSPF is a link-state protocol that distributes routing information between routers in a single AS; thus, it is an interior gateway routing protocol. OSPF chooses a least-cost path as the best path and is suitable for complex networks with a large number of routers. It provides equal cost multipath routing where packets to a single destination can be sent via more than one interface simultaneously.

As a link-state protocol, each router maintains a database describing the entire AS topology, which it builds out of the collected link-state advertisements of all routers. Each participating router distributes its local state (i.e., the router's usable interfaces and reachable neighbors) throughout the AS by flooding. Each network that has at least two attached routers has a designated router and a backup-designated router. The designated router floods a link-state advertisement for the network and has other special responsibilities.

7.16.1 OSPF Routing Hierarchy

An AS is a collection of networks under a common administration, sharing a common routing strategy. OSPF enhances the management of large ASs by further dividing the system into logical areas. An *area* is a group of contiguous networks and attached hosts. Routers with multiple interfaces can participate in multiple areas. Such routers are called *area border routers* (ABRs) and they maintain separate topological databases for each area. A topological database is an overall picture of networks in relationship to routers. An area's topology is invisible to entities outside the area, which enables OSPF to pass less routing traffic than if the AS is not partitioned. Figure 7.5 is an illustration of the OSPF routing hierarchy.

7.16.2 OSPF Routers

With the hierarchical organization described, the OSPF architecture includes four types of routers, where a router's classification is determined by its

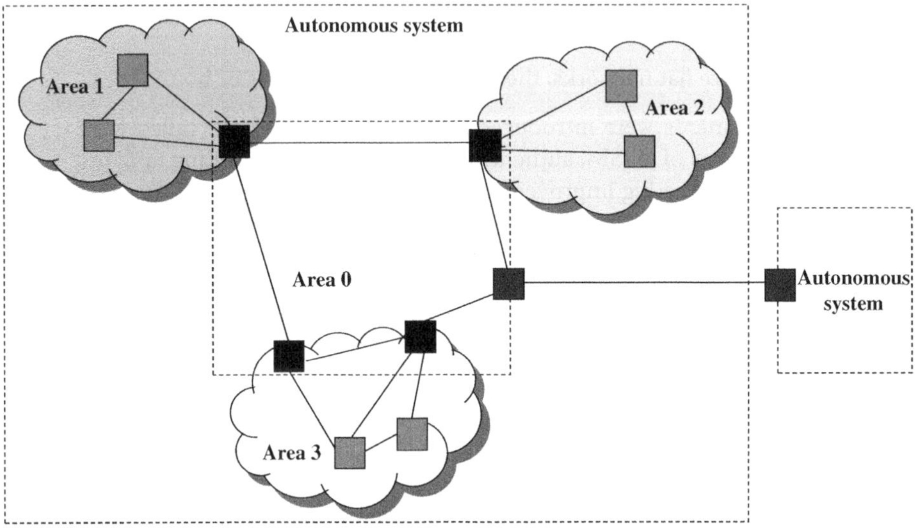

Figure 7.5 Illustration of OSPF Routing Hierarchy.

function and location within an OSPF area. These are as follows:

- *Internal routers* (IRs) interconnect networks that are internal to a single area. IRs run only one copy of the OSPF routing algorithm and maintain a topological database for only their area.
- ABRs are connected to more than one area. These routers run multiple copies of the basic algorithm, with one copy of the topological database for each area to which they are connected, and one copy for the AS backbone. Routers attached to the backbone network forward the routing information to other areas.
- *BBRs* are routers that have an interface to the backbone. This includes all the ABRs. However, backbone routers do not have to be ABRs. Routers with all interfaces connected to the backbone are considered to be IRs. By convention the area number of the backbone is 0.
- *Autonomous system boundary routers* (ASBRs), which are routers that exchange routing information with non-OSPF networks. For example, a router that connects to a network running RIP is an AS boundary router.

These routers are illustrated in Figure 7.6.

7.16.3 OSPF Routing

AS partitioning creates three different types of OSPF routing:

- Intra-area routing, which occurs when the source and destination are in the same area.

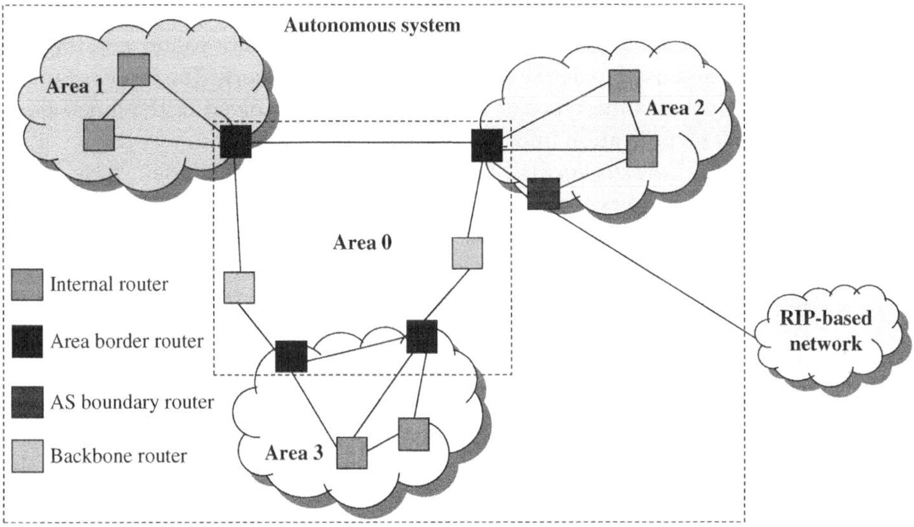

Figure 7.6 Illustration of Different OSPF Routers.

- Inter-area routing, which occurs when the source and destination are in the different areas in the same AS.
- Inter-AS routing, where the source and destination are in different ASs.

Routing packets outside of the source area is handled in a manner similar to the techniques applied to routing packets across the Internet. In this case, three routes are managed in transferring the packet. First, the shortest path to the AS backbone is identified, and the packet is forwarded to the source host's nearest border router. Once the packet has reached the border router, the backbone routing determines the shortest path to the destination area. Once the backbone routes the packet to one of the destination node's ABRs, the third shortest path route is calculated, and the packet is delivered to its intended destination.

7.16.4 Maintaining the Topological Database

One of the critical services provided by OSPF is a set of messages that are used to define and control the topological database, which can be broken into three primary categories:

- *Hello messages* that are periodically sent to validate neighbor reachability. Each router maintains a list of timers that are used to determine the presence of neighbor routers. As OSPF hello messages are received, the router resets its timer. Failure to receive the hello messages from neighbor routers is used to determine the loss of a link to a router.

- *Database description messages* that are transferred to initialize the topology database. These messages are sent on the request of a router entering the network and used to define the topology of the network. The topology information includes the types of links, the routers involved in the links, along with link identification information.
- *Link-state advertisements*: Once the topology databases have been exchanged, routers may determine that some of the information in the database is no longer accurate. To resolve these changes, the router will send a link status request message to its neighbor. The reply message is the link status update message that contains a list of link status advertisements.

7.17 Advantages of OSPF Over RIP

As stated earlier, some enhancements were introduced in RIP2 to address the issues of VLSM, authentication, and multicast routing updates. However, RIP2 is not a big improvement over RIP1 because it still has the limitations of hop counts and slow convergence which are essential in today's large networks.

OSPF, on the other hand, addresses most of the limitations of RIP:

- With OSPF, there is no limitation on the hop count.
- OSPF supports VLSM, which is very useful in IP address allocation.
- OSPF uses IP multicast to send link-state updates, which ensures less processing on routers that are not listening to OSPF packets.
- Also, updates are only sent in case routing changes occur instead of periodically, which ensures a better use of bandwidth.
- OSPF has better convergence than RIP because routing changes are propagated instantaneously and not periodically.
- OSPF allows for a logical definition of networks where routers can be divided into areas; this limits the explosion of link-state updates over the whole network and also provides a mechanism for aggregating routes and cutting down on the unnecessary propagation of subnet information.
- OSPF allows for routing authentication using different methods of password authentication.

7.18 The Dijkstra's Algorithm

A communication network is a graphical structure and is usually analyzed with graph theoretic tools. A graph G is a set of points (called *vertices* or *nodes*) that are interconnected by a set of lines (called *edges* or *arcs*). We denote the set of nodes by V and the set of edges by E and thus write $G = (V, E)$. An edge is specified by the two nodes that it interconnects; the two nodes are the

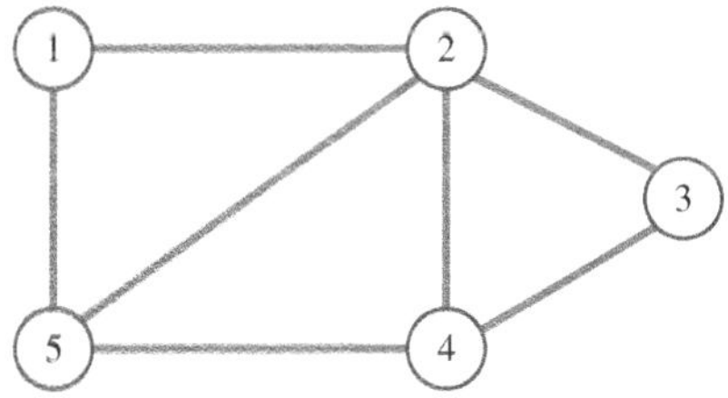

Figure 7.7 Example of a Graph.

endpoints of the edge. Thus, an edge that connects nodes a and b is denoted by (a, b). For example, consider the graph shown in Figure 7.7. We have that

$$V = \{1, 2, 3, 4, 5\}$$
$$E = \{(1, 2), (1, 5), (2, 3), (2, 4), (2, 5), (3, 4), (4, 5)\}$$

Sometimes there are weights associated with the edges, which leads to what is called a *weighted graph*. A weight on an edge can be the cost associated with using the edge. For example, a weighted version of the graph of Figure 7.7 is shown in Figure 7.8. In this case, the cost of edge (1, 2) is 5, the cost of edge (1, 5) is 7, and so on. The weight on edge (a, b) is usually denoted by $w(a, b)$. Thus, for example, $w(1, 2) = 5$ and $w(1, 5) = 7$.

The Dijkstra's algorithm is used in graphs and networks to find the shortest paths from a source node $\boldsymbol{s}$ to all other nodes in a graph or network that contains no negative cycles. It is a centralized routing algorithm that maintains information in a central location. The algorithm works by visiting vertices (or nodes) in the graph starting with the source node (the starting point). It then repeatedly examines the closest not-yet-examined vertex, adding its neighbors to the set of vertices to be examined. It expands outward from the starting point until it reaches the goal. The original version was used to find the shortest path between any two nodes in a graph. However, the algorithm can also be used to find the minimum spanning tree on a graph that is rooted at a given node.

Dijkstra's algorithm is guaranteed to find a shortest path from the starting point to the target destination node *provided none of the edges (or links) has a negative cost*. We say "a shortest path" because there are often multiple equivalently short paths. The algorithm is an example of a *greedy* algorithm because

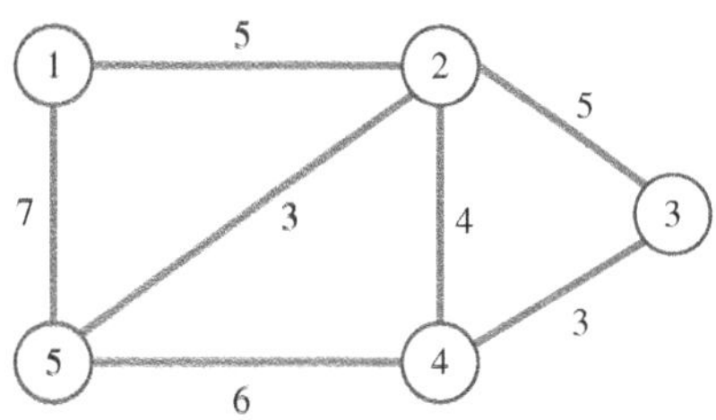

Figure 7.8 A Weighted Version of the Graph in Figure 7.7.

at each step in the algorithm, we pick a "best" node, which is one of the nodes with the smallest cost.

The algorithm works by keeping for each node v the cost $d[v]$ of the shortest path found so far. Initially, this value is 0 for the source node s and infinity for all other nodes, representing the fact that we do not know any path leading to those nodes. When the algorithm finishes, $d[v]$ will be the cost of the shortest path from s to v or infinity if no such path exists.

The basic operation of Dijkstra's algorithm is *edge relaxation*: If there is an edge from u to v, then the shortest known path from s to u can be extended to a path from s to v by adding edge (u, v) at the end. If $w(u, v)$ denotes the weight (or length) of the edge (u, v), then this path will have length $d[u] + w(u, v)$; if this is less than $d[v]$, we can replace the current value of $d[v]$ with the new value. Edge relaxation is applied until all values $d[v]$ represent the cost of the shortest path from s to v.

The algorithm maintains two sets of nodes S and Q. Set S contains all nodes for which we know that the value $d[v]$ is already the cost of the shortest path and set Q contains all other nodes. Initially, set S is empty, while set Q contains all the nodes. Then in each step one node is moved from Q to S, where the node moved is the node with the lowest value of $d[u]$. When a node u is moved to S, the algorithm relaxes every outgoing edge (u, v).

The following is the pseudocode for the algorithm. In the code, the operation $u = \text{Min}(Q)$ searches for the node u in the node set Q that has the least $d[u]$ value. That node is removed from the set Q and added to the set S.

```
for each node v in V[G]                          // Initialization
        do d[v] = infinity
d[s] = 0
S = empty set
Q = set of all vertices
while Q is not an empty set
        do u = Remove{Min(Q)}
        S = S ∪ {u}
        Q = Q \ {u}
        for each edge (u, v) outgoing from u
                do if d[v] > d[u] + w(u, v)               // Relax (u, v)
                        then d[v] = d[u] + w(u, v)
```

Example 7.1 Consider the network shown in Figure 7.9, where it is required to find the shortest path between A and E. The number on each link represents the weight (or cost) of using the link. We will find the shortest path from node A to every other node and extract the shortest path from A to E from that solution. We start by defining 100 to represent infinity and the sets S and Q. Initially, S is empty (or has no members) and Q contains all the nodes with the cost of the source A being zero and the costs of others being "infinity."

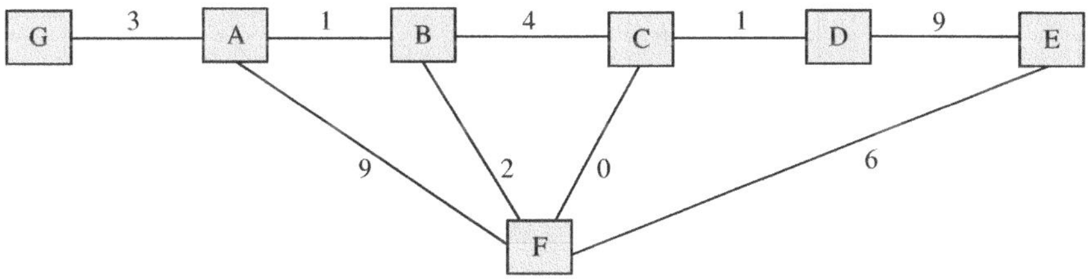

Figure 7.9 Graph for the Example.

We define X(a, Y) to mean node X with minimum cost a measured from node A through its neighboring node Y, where $a = 100$ means "cost is infinity" and Y = "–" means "neighbor unknown."

$S = \{\ \}$

$Q = \{$A(0, A), B(100, –), C(100, –), D(100, –), E(100, –), F(100, –), G(100, –)$\}$

The smallest distance in that graph is the distance to A, so we move A to S and update the distances to all of its neighbors (B, F, and G, which now have actual paths and distances). We now have the following information:

$S = \{$A(0, A)$\}$

$Q = \{$B(1, A), G(3, A), F(9, A), C(100, –), D(100, –), E(100, –)$\}$

Because B has the lowest cost, we move B to S and update its neighbors. We now know a path to C and a better path to F (A → B → F has cost $1+2=3$).

$S = \{$A(0, A) B(1, A)$\}$

$Q = \{$F(3, B), G(3, A), C(5, B), D(100, –), E(100, –)$\}$

We could then move F or G to S since the two nodes have the same cost. Let us say that we move F. We now know a shorter path to C (A → B → F → C has cost 3, A → B → C has cost 5) and a path to E.

$S = \{$A(0, A), B(1, A), F(3, B)$\}$

$Q = \{$G(3, A), C(3, F), E(9, F), D(100, –)$\}$

We could then move G or C to S since the two nodes have the same cost. Assume that we move G. Since G has no neighbors besides A, there are no other changes in costs.

$S = \{$A(0, A), B : 1, A), F(3, B), G(3, A)$\}$

$Q = \{$C(3, F), E(9, F), D(100, –)$\}$

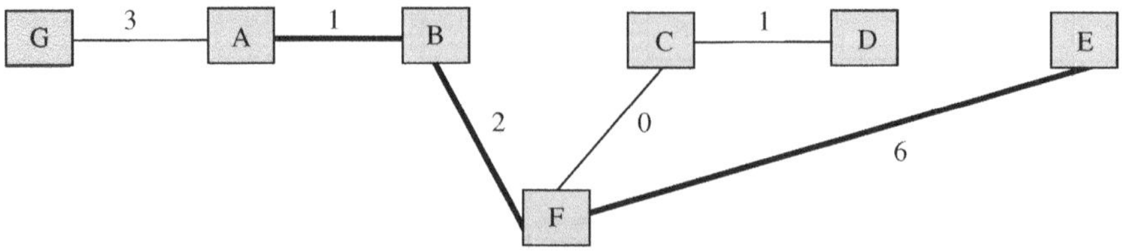

Figure 7.10 The Minimum Spanning Tree Rooted at Node A.

Next we move C. We can then update the distance to D.

$$S = \{A(0, A), B(1, A), F(3, B), G(3, A), C(3, F)\}$$
$$Q = \{D(4, C), E(9, F)\}$$

We next move D. Note that even though there is an edge from D to E, it does not give us any better path; so we do not change the entry for E.

$$S = \{A(0, A), B(1, A), F(3, B), G(3, A), C(3, F), D(4, C)\}$$
$$Q = \{E(9, F)\}$$

We move E and obtain the following final entries with the set Q now empty. After this, the algorithm is terminated.

$$S = \{A(0,A), B(1,A), F(3,B), G(3,A), C(3,F), D(4,C), E(9,F)\}$$
$$Q = \{\ \}$$

Thus, we have obtained a minimum spanning tree rooted at node A, and the shortest path from A to E has cost 9 and is obtained in the reverse order (E → F → B → A) by going to E to see the node that is the last hop to E, and that is F; going to F to see the node that is the last hop to F, which is B; and going to B to see the last hop, which is A. Figure 7.10 shows the minimum spanning tree rooted at A. The shortest path between nodes A and E is shown by the bold edges.

7.19 Multicast Routing

Recall that IP multicast addresses are the Class D addresses where the first four bits of first octet are 1110. IPv4 multicast addresses are in the range 224.0.0.0–239.255.255.255. Reserved link-local addresses are in the range: 224.0.0.0–224.0.0.255. Examples of these link-local addresses include the following:

224.0.0.1 All systems on this subnet
224.0.0.2 All routers on this subnet
224.0.0.4 Distance–vector multicast routing protocol (DVMRP) routers
224.0.0.5 OSPF routers

As we discussed earlier, multicast addresses are not location dependent. These addresses are used to subscribe for services and thus members of a multicast group can be located anywhere in the world. To receive a multicast packet, a user must join the group that receives that packet.

7.20 Types of Multicast Systems

When joining a multicast group, a host can opt to receive data sent to the group from any source or to receive data sent to the group from one specific source. To receive data from any source, the host needs to specify only the IP address of the multicast group; this is known as *any source multicast* (ASM). Similarly, to receive data from one particular source only, a host must specify both the IP address of the multicast group and the IP address of the source; this is known as *source-specific multicast* (SSM). Thus, SSM uses the source and group address pair (*S*, G), where S is the source IP address and G is the group multicast address, while ASM uses the group address with the wildcard as follows: (*, G).

One of the advantages that SSM has over ASM is improved security and access control: It is less susceptible to denial of service attacks that can arise when an unauthorized sender sends traffic to the multicast group. One of its disadvantages is that a recipient needs to know the source address of the SSM traffic.

7.21 Host-Router Signaling

The Internet Group Management Protocol (IGMP) is used by hosts to tell routers about their multicast group membership. Multicast routers solicit group membership from directly connected hosts. Specifically, a multicast router periodically sends queries to the IP address 224.0.0.1 to invite users on the particular network to indicate their intention to join the multicast group. Any member that wants to join the multicast group will send a request called *IGMP report*. Once a router receives one member's report, it suppresses subsequent reports from other hosts in the same subnet because when it receives any multicast message for that group, it broadcasts it for all users on the subnet to see. Thus, it takes only one interested member for the users on a subnet to receive multicast messages.

Figure 7.11 illustrates the process of joining a multicast group. The router sends a query to the subnet with hosts H1, H2, H3, and H4 connected. H1 is the first host to respond to the query; it sends a report and the router now knows that at least one host in the subnet wants to join the multicast group. Later host H4 also sends a report that the router suppresses because it has already known that at least one host in the subnet has already joined the multicast group.

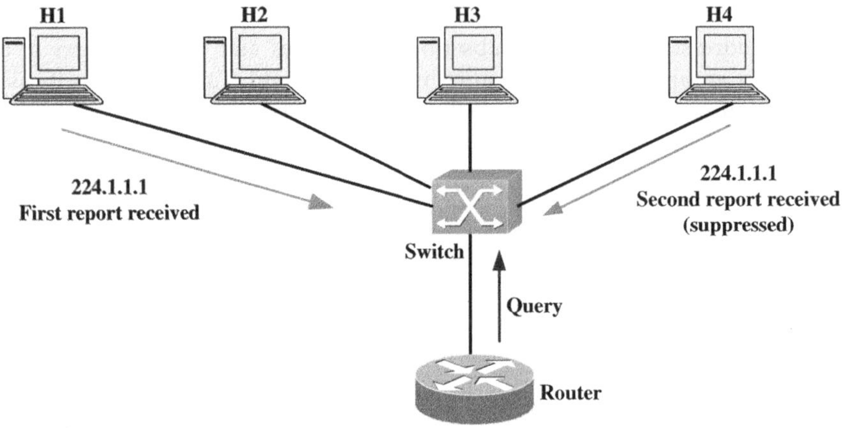

Figure 7.11 Host Registration for Multicast Service.

7.22 Multicast Routing Protocols

Multicast routers exchange information with other multicast routers to form a tree of multicast group recipients. Data sent to the multicast group is forwarded to all branches of the tree. The commonly used multicast routing protocols are as follows:

- Protocol-independent multicast (PIM), which can be used in the dense mode (DM) or sparse mode (SM)
- Distance Vector Multicast Routing Protocol (DVMRP)
- Multicast Open Shortest Path First (MOSPF)
- Core-based Tree (CBT) Multicast Routing.

MOSPF is used in networks that use the OSPF routing protocol as the background routing protocol, while DVMRP is used in a network that uses a distance–vector routing protocol, such as RIP. PIM is used in any network. Thus, with PIM, the background routing protocol can be the RIP, OSPF, or any other protocol. From this, we can refer to both MOSPF and DVMRP as *protocol-dependent multicast protocols* that must be implemented in networks whose unicast routing protocol is either OSPF (for MOSPF) or RIP (for DVMRP).

Three important features of multicast routing protocols are as follows:

(a) Whether they use *opt-in* or *opt-out* routing protocols. Opt-in protocols are also called *sparse protocols* and opt-out protocols are also called *broadcast-and-prune* or *dense protocols.*
(b) Whether they use source-based trees or shared trees.
(c) The method they use to find the upstream router.

7.22.1 Opt-In Protocols

In opt-in protocols, routers indicate which multicast groups they want to receive from in advance of that data flowing. They are designed on the assumption that the receivers for the multicast group are sparsely distributed throughout the network. Thus, most subnets in the network will not want a given multicast data stream. We can characterize an opt-in or sparse-mode protocol as a protocol that uses the "pull" model; that is, receivers explicitly join the multicast tree and traffic is sent only to where it is requested. An example of an opt-in protocol is the PIM-SM.

Figure 7.12 is an illustration of the opt-in multicast scheme. The figure shows receivers R1 and R2 requesting to join the multicast group for a service offered by Source 1. The root of this multicast is Router A. When Router C receives the request from R1, it sends a request to Router A to join the multicast group. Similarly, when Receiver E receives the request from R2, it sends a request to join the multicast group to Router A via Router C. Since Router C has already joined the group, it does not forward the request to Router A since C is already on the multicast tree. Instead it notes that it has received a request on the port

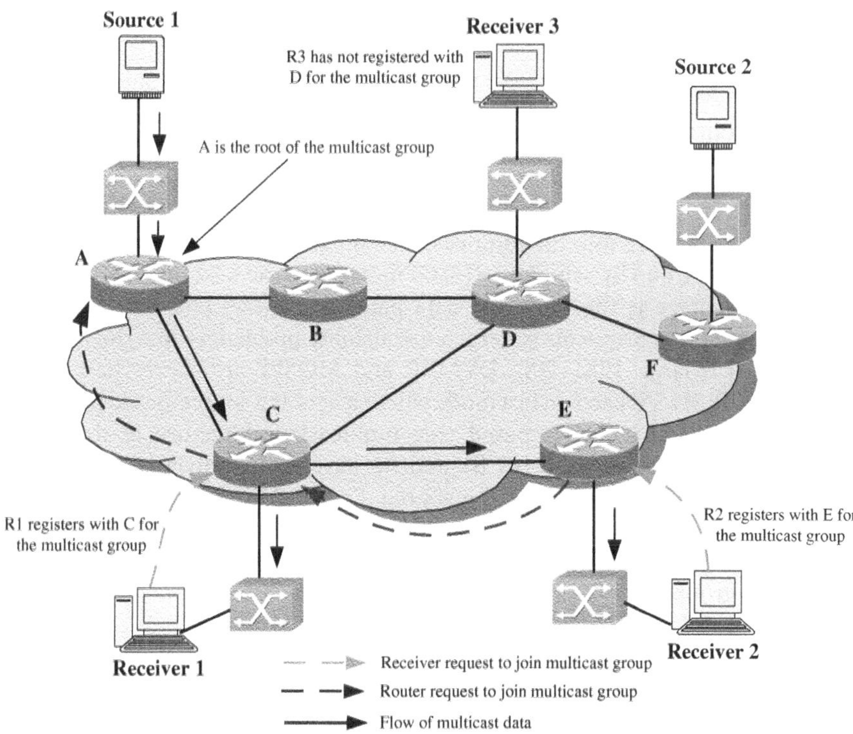

Figure 7.12 An Example of the Opt-In Multicast Process.

to E. Receiver R3 has not joined the multicast group, and so Routers B and D do not join the multicast group. Multicast data from A will not be sent to B; it will only be sent to C, which will send a copy to R1 and a copy to Router E to be sent to R2.

7.22.2 Opt-Out Protocols

In opt-out protocols, every router is initially assumed to want to receive multicast data, and so each tree initially spans every node in the network. Thus, initially data is sent to all routers and any router that does not wish to receive the data will send a *Prune* message upstream to remove itself from the multicast tree. A Prune message is sent when a router receives a multicast data for the group or from a source that it is not interested in. PIM–DM and DVMRP are opt-out protocols. In summary, the opt-out or DM protocol has the following features:

- It uses "push" model; that is, traffic is flooded to all the routers in the network.
- It prunes the multicast tree where the multicast is unwanted.

Figure 7.13 is an illustration of the opt-out protocol, where Routers F, D, and B request to be pruned from the multicast tree after receiving the first multicast packet.

7.22.3 Source-Based Tree Protocols

Source-based tree protocols build a separate tree for each source sending data to a multicast group, with a tree rooted at the router adjacent to the source. Routers that wish to join the multicast tree must specify both the source and the multicast group by sending an (S, G) message to the next upstream router.

One advantage is the multicast data paths are always efficient. However, source-based tree protocols suffer from scalability problems when the number of sources is large. PIM–DM, DVMRP, and MOSPF are source-based tree protocols. PIM–SM can run in a mode where it acts as a source-based protocol.

An example of a source-based tree multicast architecture is shown in Figure 7.14, where the sources S1 and S2 have their individual trees each rooted at the closest router to the source. Each receiver joins the multicast group that it wants to receive from and the multicast routers join the appropriate multicast group in order to be on the multicast tree for the particular multicast group.

7.22.4 Shared Tree Protocols

Shared tree protocols build a single tree for all sources sending to a multicast group. The tree is rooted at some selected node, which is a router called the *rendezvous point* (RP). The sources then use a protocol-specific mechanism to

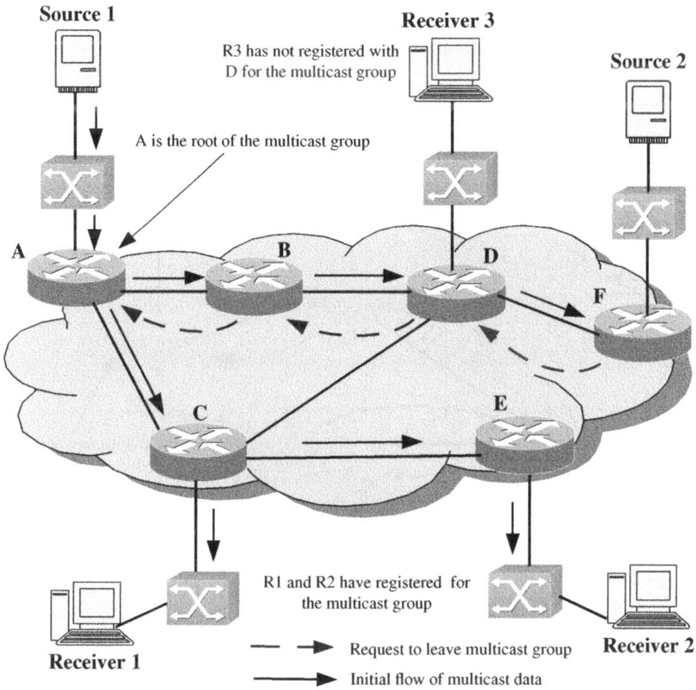

Figure 7.13 An Example of the Opt-Out Multicast Process.

transport the multicast datagrams from the source to the root of the tree. Typically, the data from the source is encapsulated in a unicast packet and sent to the root of the tree from where the multicast is delivered to the users.

Shared tree protocols are used in only opt-in multicast protocols. One example is CBT; PIM–SM can run in a mode where it acts as a shared tree protocol. When a router wants to join a multicast group, it does not specify the source but sends a (*, G) message to the next upstream router. One advantage of this protocol over the source-based protocol is its scalability. There is only one multicast tree as opposed to several multicast trees in source-based tree multicast.

Figure 7.15 illustrates the shared tree multicast architecture. The sources unicast their data to the RP, which is Router B, and this is sent to the users over a common (or shared) multicast tree.

7.23 Multicast Forwarding

Multicast routing operates on a different principle from unicast routing. Specifically, while unicast routing is concerned about where the packet is going,

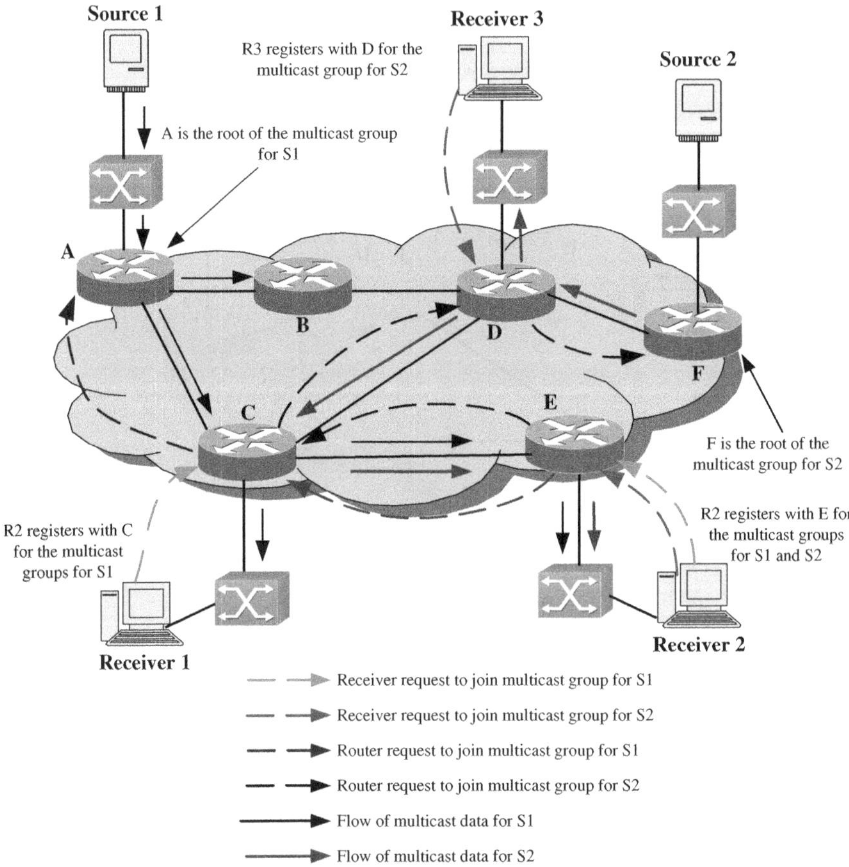

Figure 7.14 An Example of the Source-Based Tree Multicast Architecture.

multicast routing is concerned about where the packet came from because the router forwarding the packet (or more specifically, the port through which a multicast packet was received) must be on the multicast tree.

Multicast routing uses a feature called "reverse path forwarding" (RPF) that permits a router to forward a multicast packet only if it was received on the upstream interface that leads to the source of the multicast packet. This means that the router must verify that the packet is following the multicast distribution tree. Thus, when a multicast router receives a multicast packet it performs the *RPF check*, as follows: The routing table used for multicasting is checked against the "source" IP address in the packet.

(a) If the packet arrived on the interface specified in the routing table for the source address (i.e., the upstream interface through which it was received

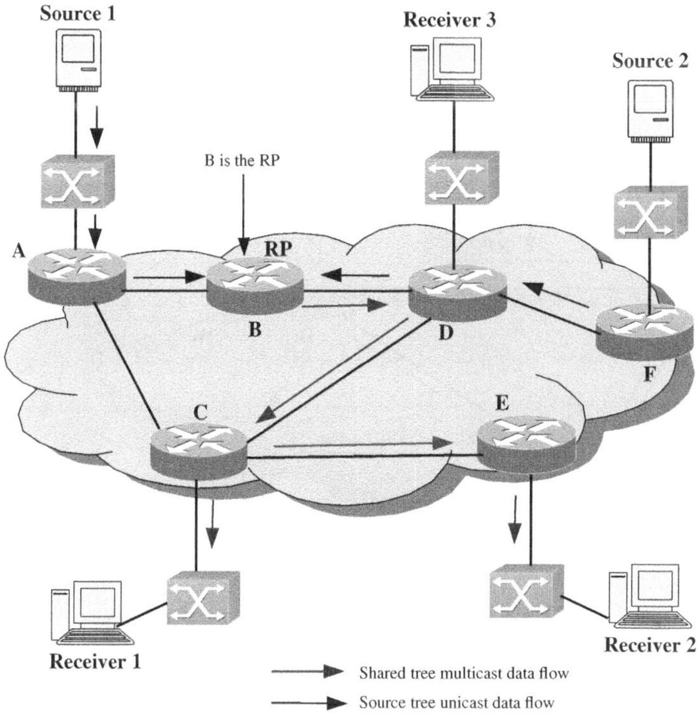

Figure 7.15 An Example of the Shared Tree Multicast Architecture.

is on the multicast distribution tree), then the RPF check succeeds and the packet is forwarded.

(b) Otherwise, the RPF check fails and the packet is discarded.

Figure 7.16 illustrates the RPF checking process. Two multicast packets are received at Router C, one on an interface that is on the multicast tree and the other on a different interface. The RPF check on the correct interface will pass while that on the wrong interface will fail and the packet discarded.

7.24 Summary

This chapter has been devoted to discussions on routing algorithms and routing protocols. These include the routing principle, RIP, OSPF protocol, and multicast routing. There is a brief discussion on the Dijkstra's algorithm. The last part of the chapter deals with multicast routing with discussion on the different types of multicast protocols: opt-in versus opt-out. A distinction between how unicast routers handle packets and how multicast routers handle packets is

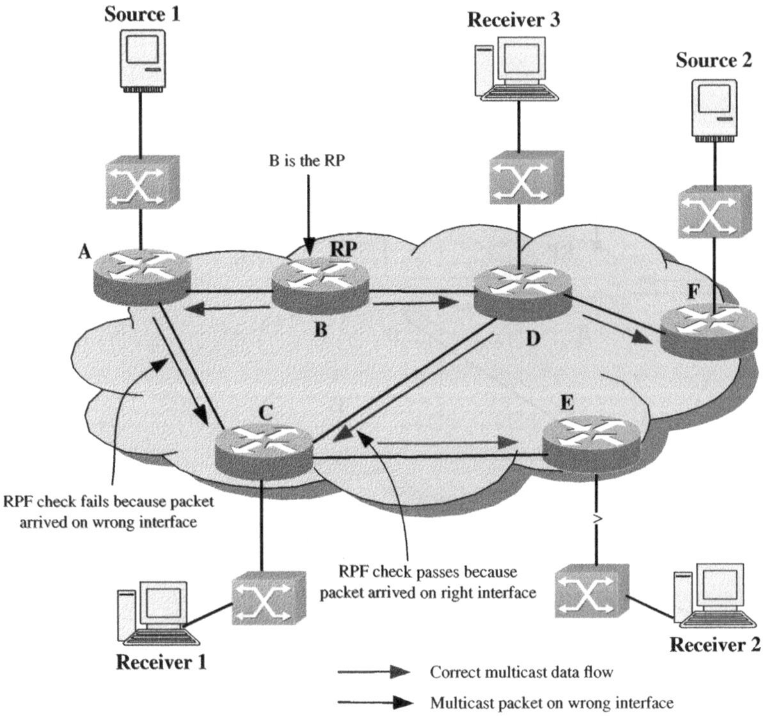

Figure 7.16 RPF Checking at Router C.

also made: unicast routers are concerned with where the packet is going, while multicast routers are concerned with where the packet came from. Finally, the two types of multicast trees are discussed, which are source-based trees and shared trees.

Exercises

1 What is the difference between a distance–vector routing protocol and a link-state routing protocol?

2 Give an example of a distance–vector routing protocol.

3 Give an example of a link-state routing protocol.

4 What is the difference between a static routing protocol and an adaptive routing protocol?

5 Use the Dijkstra's algorithm to find the shortest path from node A to every other node in the following network:

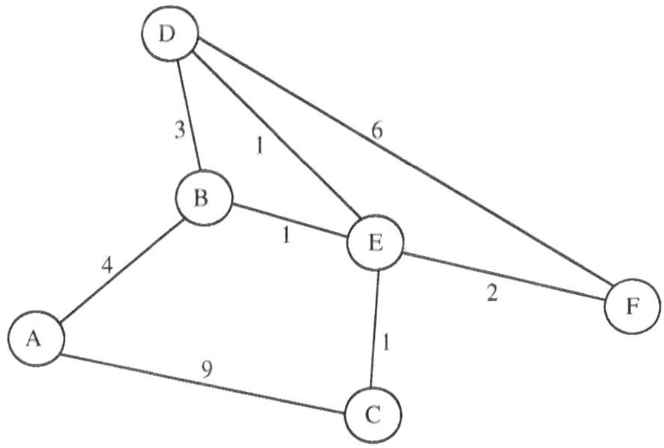

6 Use the Dijkstra's algorithm to find the shortest path from node A to every other node in the following network:

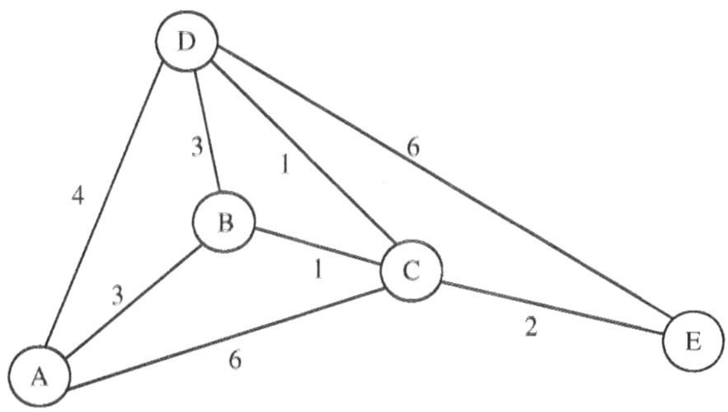

7 When joining a multicast group, how does a node specify the fact that it wishes to receive data sent by any source?

8 What is the major difference between how routers handle unicast routed packets and multicast routed packets in a network?

9 What is the difference between an opt-in multicast routing protocol and an opt-out multicast routing protocol?

10 What is a rendezvous point in multicast routing?

8

Transport Layer – TCP and UDP

8.1 Introduction

In this chapter, we discuss the transport layer protocols. As discussed in Chapter 1, the transport layer provides reliable communication between the source and the sink. Two protocols were originally defined in this layer, namely, the *Transmission Control Protocol* (TCP), which is a connection-oriented protocol that provides reliable end-to-end transfer, and the *User Datagram Protocol* (UDP), which is an unreliable connectionless protocol that was designed for applications that do not need the reliability and overhead of TCP. Two other protocols have recently been defined to address some of the shortcomings of TCP and UDP. These are the *Stream Control Transmission Protocol* (SCTP), which, like TCP, is a connection-oriented protocol that provides reliable end-to-end transfer and the *Datagram Congestion Control Protocol* (DCCP), which was designed to inherit some of the features of TCP and some of the features of UDP. In this chapter, we discuss TCP and UDP and defer the discussion on SCTP and DCCP to Chapter 9.

Transport layer protocols are the first end-to-end protocols in the OSI reference model. As can be seen in Figure 8.1, protocols in the first three lowest layers operate on a hop-by-hop basis.

Transport-layer protocols are usually classified according to three properties:

(a) Whether they are connection-oriented or connectionless
(b) Whether they are stateful or stateless
(c) Whether they are reliable or unreliable.

Connection-oriented protocols require a connection to be established between the communicating parties prior to the commencement of the information transfer. These protocols are generally reliable because they require the acknowledgment of each transmission. However, they have more overhead because of the need to establish and tear down connections. With

Fundamentals of Data Communication Networks, First Edition. Oliver C. Ibe.

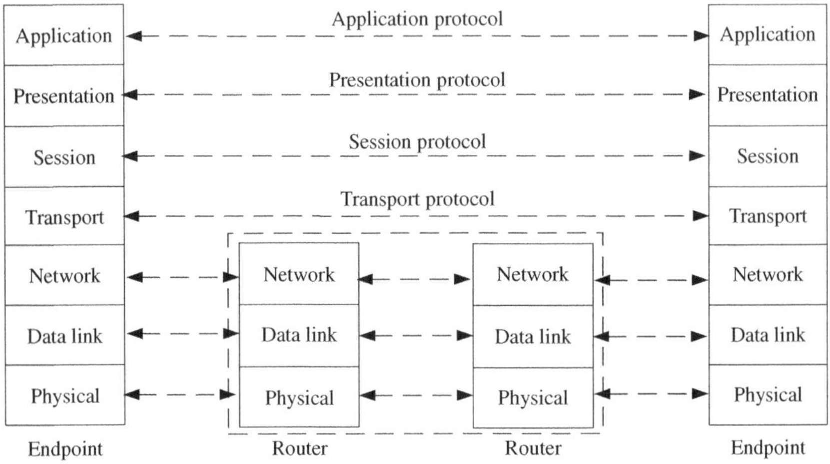

Figure 8.1 Spans of the Protocols in the Different Layers of the OSI Model.

a connectionless protocol, information transmission can proceed without a connection being established. Each transmission is a full message that does not require any acknowledgment.

A stateless protocol is a protocol that does not require the communicating parties to track the state of their session for the duration of the session. A stateful protocol requires the communicating parties to maintain and track the state of their session. Thus, with a stateful protocol data cannot be lost or delivered out of sequence, while data can be lost or delivered out of sequence in a stateless protocol.

A reliable protocol is one that requires that each transmission be acknowledged by the receiving partner. The sender is aware of a lost data packet and retransmits the data packet possibly several times until the receiving partner indicates that it has been correctly received. An unreliable protocol does not require each transmission to be acknowledged by the communicating partner.

As will be seen in the remainder of this chapter, TCP is a connection-oriented, stateful, and reliable protocol, while UDP is a connectionless, stateless, and unreliable protocol. The specific topics to be covered in this chapter include the following:

a. TCP basics
b. TCP window flow control
c. TCP congestion control
d. UDP model.

8.2 TCP Basics

As stated earlier, TCP is a connection-oriented reliable transport protocol that provides end-to-end service. This means that an application that uses the protocol knows that the data it sends is received at the destination. It is a stateful protocol whose responsibilities in data communication process include the following:

- Communicating with the application layer
- Providing error checking and flow control
- Ensuring reliability.

It uses a checksum in the header for error detection. When data is received, TCP sends an ACK back to the sender. If the sender does not receive an ACK within a predefined time interval after sending the data, it retransmits the packet.

TCP provides packet resequencing for packets that arrive out of order. It implements flow control to ensure that the receiver is not overwhelmed by the sender. It uses the service of IP to send data in blocks called *segments* whose length is defined by the protocol. The amount of data placed in a segment is limited by the *maximum segment size* (MSS), which is in turn determined by the maximum transmission unit (MTU) of the network.

8.2.1 TCP Ports

The concept of ports and sockets provides a way to uniquely identify connections and the processes that use them. Each process that wants to communicate with another process identifies itself to the TCP/IP suite through one or more ports. A port is a 16-bit number (permitting 65,536 port addresses) used to identify the higher-layer protocol or application program (or process) to which an incoming message will be delivered.

In the past, there were two types of ports: *well-known* ports and *ephemeral* ports. Well-known ports belonged to standard servers and had numbers that range between 1 and 1023 and were controlled by the Internet Assigned Numbers Authority (IANA). They are typically odd numbers because early systems required an odd/even pair of ports for duplex operation. Ephemeral ports were assigned on the fly to clients that need to communicate with servers and the allocation lasts as long as the client needs it. Ephemeral port numbers had values that range between 1024 and 65,535; they were not controlled by the IANA and could be used by any application.

According to the IETF RFC 6335, these ports have recently been reclassified, and there are currently three types: *System* (or well-known) ports, *user* ports, and *dynamic* (or *private* or ephemeral) ports. The basic difference between the current classification and the previous one is that the ports that were previously

Table 8.1 Examples of System Ports

Port	Protocol	Use
21	FTP	File transfer
23	Telnet	Remote login
25	SMTP	E-mail
69	TFTP	Unreliable file transfer
79	Finger	Look up information about a user
80	HTTP	World Wide Web
110	POP-3	Remote e-mail access

classified as ephemeral have now been divided into two classes: user ports and dynamic ports.

Ports 1 through 1023 are system ports that are available for assignment through IANA. They cover the previous well-known ports and are assigned to the well-known applications, such as those listed in Table 8.1.

User ports are those in the range from 1024 to 49,151. They are also available for assignment through IANA and may be used as service identifiers upon successful assignment. The IANA requires that the requester documents the intended use of the port number. The applicant for a user port is required to explain why using a port number in the dynamic ports range is unsuitable for the given application.

Dynamic ports are those in the range 49,152–65,535. They have been specifically set aside for local and dynamic use and cannot be assigned through IANA. Application software may simply use any dynamic port that is available on the local host, without any sort of assignment.

8.2.2 TCP Sockets

Any higher-layer application that communicates across a TCP connection with another higher-layer application maintains an association that takes the following form:

[protocol, source IP address, source port, destination IP address, destination port]

This association uniquely identifies a connection in the network and is made up of two half-associations that take the form:

- [*protocol, source IP address, source port*] or
- [*protocol, destination IP address, destination port*].

For example, [tcp, 190.20.4.10, 23] specifies a half-association that involves a TCP port 23, which is a Telnet session, on a machine with the IP address

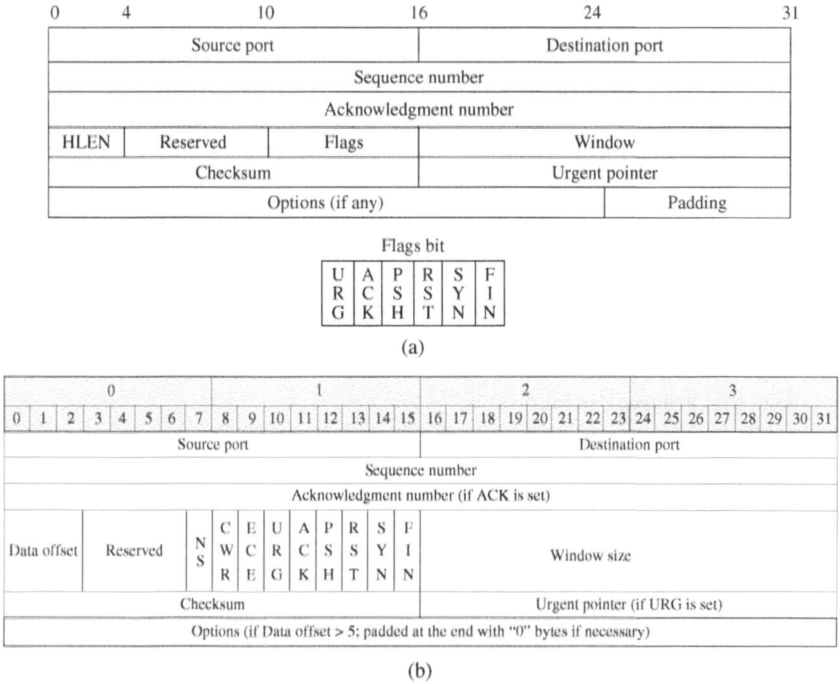

Figure 8.2 The TCP Header. (a) Old Format and (b) New Format.

190.20.4.10. A half-association is called a ***socket***. Thus, communication between two processes over a TCP connection is carried out over a TCP socket.

8.2.3 TCP Segment Format

As stated earlier, the unit of transfer across a connection is the TCP segment. Each TCP segment is divided into two parts: a header followed by data or payload. The header, known as the TCP header, is 20 octets long and carries identification and control information. The data is the actual data to be transmitted. The TCP header has been reformatted; the old format is shown in Figure 8.2(a) and the new format is shown in Figure 8.2(b).

The different fields of the new format are as follows:

(a) *Source port* (16 bits) is the port number of the sending port; up to 65,536 port numbers can be defined.
(b) *Destination port* (16 bits) is the port number of the receiving port.
(c) *Sequence number* (32 bits) has a dual role: If the SYN flag is set (i.e., SYN = 1), then this is the initial sequence number. If the SYN flag is not set

(i.e., SYN = 0), then this is the accumulated sequence number of the first data byte of this segment for the current session.

(d) *Acknowledgment number* (32 bits) is used as follows. If the ACK flag is set, then the value of this field is the next sequence number that the sender is expecting. This acknowledges receipt of all prior bytes (if any).

(e) *Data offset* (4 bits), which was originally called header length, specifies the size of the TCP header in 32-bit words. The minimum size of the header is 5 words and the maximum is 15 words, which gives a minimum length of 20 bytes and a maximum of 60 bytes, allowing for up to 40 bytes of options in the header.

(f) *Reserved* (3 bits), which is for future use and the field is set to zero.

(g) *Flags* (9 bits), which are sometimes called *control bits*. It contains nine 1-bit flags that are used as follows:

- NS (1 bit) – explicit congestion notification (ECN) nonce sum flag, which is used to guard against the malicious or accidental concealment of marked packets from the TCP sender.
- CWR (1 bit) – congestion window reduced (CWR) flag is set by the sending host to indicate that it received a TCP segment with the ECE flag set and had responded to the condition using the congestion control mechanism.
- ECE (1 bit) – ECN-Echo has a dual role, depending on the value of the SYN flag. It operates as follows:
  - If the SYN flag is set (i.e., SYN = 1), then the TCP peer is ECN capable.
  - If the SYN flag is clear (i.e., SYN = 0), then a packet with congestion experienced (CE) flag set (ECN = 11) in IP header was received during normal transmission, which is an indication of an impending network congestion to the TCP sender.
- URG (1 bit) – indicates that the urgent pointer field is significant.
- ACK (1 bit) – indicates that the acknowledgment field is significant. All packets after the initial SYN packet sent by the client should have this flag set.
- PSH (1 bit) – push function. Asks to push the buffered data to the receiving application.
- RST (1 bit) – reset the connection.
- SYN (1 bit) – synchronize sequence numbers. Only the first packet sent from each end should have this flag set.
- FIN (1 bit) – last packet from the sender.

(h) *Window size* (16 bits), which indicates the size of the *receive window*; this value specifies the number of window size units (by default, bytes) that the sender of this segment is currently willing to receive.

(i) *Checksum* (16 bits), which is a 16-bit checksum field that is used for error checking of the header and data.

(j) *Urgent pointer* (16 bits), which is such that if the URG flag is set, then this 16-bit field is an offset from the sequence number indicating the last urgent data byte.
(k) *Options* (variable 0–320 bits) whose length is determined by the data offset field.
(l) *Padding*, which is used to ensure that the TCP header ends and data begins on a 32-bit boundary. The padding is composed of zeros.

8.3 How TCP Works

TCP uses several methods of communication to handle the reliable delivery of data. As a connection-oriented protocol, TCP maintains status information about its connections. It uses a connection table to keep track of multiple connections. TCP is also responsible for communicating with the upper layer applications that it is serving. The TCP-to-upper layer protocol is used by TCP to communicate with upper layer applications. The final task that TCP has to perform is to transmit and receive data. The transmission of data is accomplished through the use of the Protocol Data Unit that encapsulates the data.

8.3.1 TCP Connection Establishment

TCP connection is an important function because TCP connection establishment can significantly add to perceived delays. The following is how a TCP connection is established. Suppose a process running in one host wants to initiate a connection with another process in another host. The host that is initiating the connection is called the *client host*, while the other host is called the *server host*. Thus, a TCP connection is established between a client host and a server host. The client application process first informs the client TCP that it wants to establish a connection to a process in the server. The client then proceeds to establish a TCP connection with the server in the following manner:

Step 1: The client-side TCP first sends a special TCP segment to the server-side TCP. This special segment contains no application-layer data. It does, however, have one of the flags bits in the segment's header, the so-called SYN bit, set to 1. For this reason, this special segment is referred to as a *SYN segment*. In addition, the client chooses an *initial sequence number*, for example *x*, and puts this number in the sequence number field of the initial TCP SYN segment. This segment is encapsulated within an IP datagram and sent into the Internet.

Step 2: Once the IP datagram containing the TCP SYN segment arrives at the server host, the server extracts the TCP SYN segment from the datagram, allocates the TCP buffers and variables to the connection, and sends a connection-granted segment to the client TCP. This connection-granted segment also contains no application-layer data. However, it does contain three important pieces of information in the segment header:

(a) The SYN bit is set to 1.
(b) The acknowledgment field of the TCP segment header is set to $x + 1$ to indicate that it is now expecting a segment with sequence number $x + 1$, which shows that the previous segment was received without error.
(c) The server chooses its own *initial sequence number*, for example y, and puts this value in the sequence number field of the TCP segment header.

This connection-granted segment is saying, in effect, "I received your SYN segment to start a connection with your initial sequence number, x. I agree to establish this connection. My own initial sequence number is y and I am expecting a segment with sequence number $x + 1$." The connection-granted segment is sometimes referred to as a *SYNACK segment*.

Step 3: Upon receiving the connection-granted (SYNACK) segment, the client also allocates buffers and variables to the connection. The client host then sends the server yet another segment; this last segment acknowledges the server's connection-granted segment. The client does so by putting the value $y + 1$ in the acknowledgment field of the TCP segment header. The SYN bit is set to 0, since the connection is established. The sequence number of this segment is $x + 1$.

Once the above three steps have been completed, the client and server hosts can send segments containing data to each other. In each of these future segments, the SYN bit will be set to zero.

Note that in order to establish the connection, three packets are sent between the two hosts. For this reason, this connection establishment procedure is often referred to as a *three-way handshake*, which is illustrated in Figure 8.3.

8.3.2 TCP Connection Release

Either of the two processes participating in a TCP connection can end the connection. When a connection ends, the "resources" (i.e., the buffers and variables) in the hosts are deallocated. As an example, suppose the client decides to close the connection. The client application process issues a close command. This causes the client TCP to send a special TCP segment to the server process.

This special segment has a flags bit in the segment's header, the so-called FIN bit, set to 1. When the server receives this segment, it sends the client an acknowledgment segment in return. The server then sends its own shutdown segment, which has the FIN bit set to 1. Finally, the client acknowledges the server's shutdown segment. At this point, all the resources in the two hosts are now deallocated.

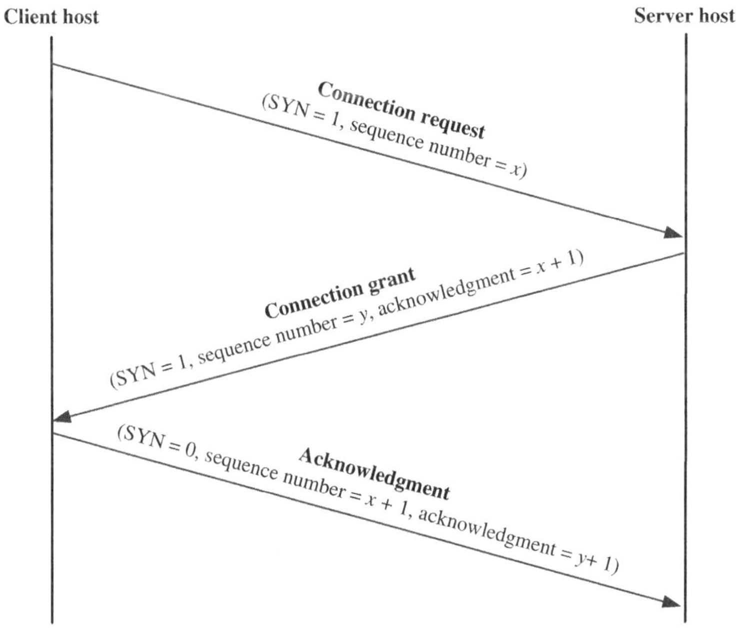

Figure 8.3 The Three-Way Handshake.

8.3.3 TCP Connection Management

During the life of a TCP connection, the TCP running in each host makes transitions through various *TCP states*. Figure 8.4 illustrates a typical sequence of TCP states that are visited by the *client* TCP. The client TCP begins in the closed state. The application on the client side initiates a new TCP connection. This causes TCP in the client to send a SYN segment to TCP in the server. After sending the SYN segment, the client TCP enters the SYN_SENT state. While in the SYN_SENT state, the client TCP waits for a segment from the server TCP that includes an acknowledgment for the client's previous segment as well as the SYN bit set to 1. Once having received such a segment, the client TCP enters the ESTABLISHED state. While in the ESTABLISHED state, the TCP client can send and receive TCP segments containing payload (i.e., application-generated) data.

While in the ESTABLISHED state, the client can decide to close the connection. This causes it to send a TCP segment with the FIN bit set to 1 and to enter the FIN_WAIT_1 state. While in the FIN_WAIT_1 state, the client TCP waits for a TCP segment from the server with an acknowledgment. When it receives this segment, the client TCP enters the FIN_WAIT_2 state. While in the FIN_WAIT_2 state, the client waits for another segment from the server with the FIN bit set to 1; after receiving this segment, the client TCP acknowledges

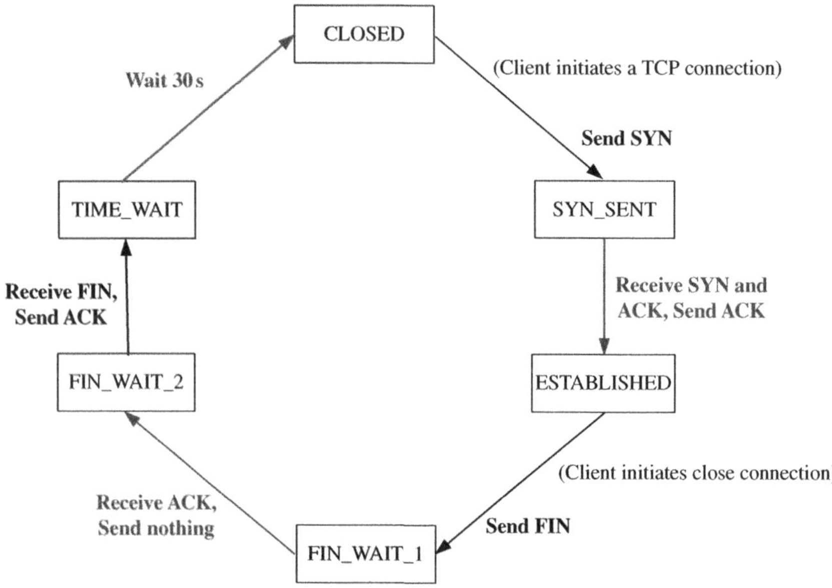

Figure 8.4 Client-Side TCP State Diagram.

the server's segment and enters the TIME_WAIT state. The TIME_WAIT state allows the TCP client to resend the final acknowledgment in the case the ACK is lost. The time spent in the TIME-WAIT state is implementation dependent, but a typical value is 30 s. After the wait, the connection formally closes and all resources on the client side (including port numbers) are released.

The server-side state diagram is shown in Figure 8.5. The server must be in the LISTEN state to be able to know when the client has sent the SYN segment.

8.4 TCP Flow Control

As we discussed earlier, each host on each side of a TCP connection sets aside a receive buffer for the connection. When the TCP connection receives bytes that are correct and in sequence, it places the data in the receive buffer. The associated application process will read data from this buffer, but not necessarily at the instant the data arrives. The receiving application may be busy with some other task and may not even attempt to read the data until long after it has arrived. If the application is relatively slow at reading the data, the sender can very easily overflow the receiver's receive buffer by sending too much data too quickly.

TCP uses a *flow control service* to prevent the possibility of the sender overflowing the receiver's buffer. Flow control is essentially a speed matching

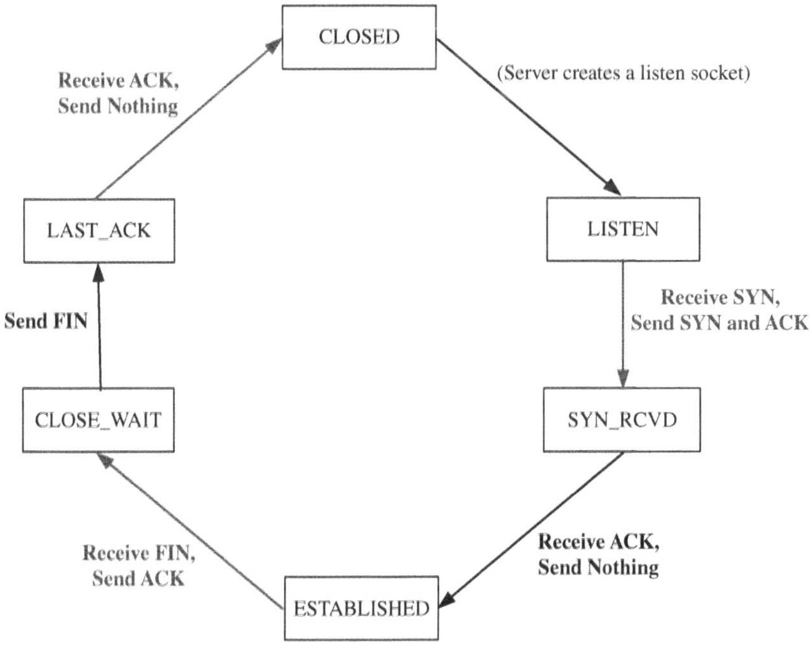

Figure 8.5 Server-Side TCP State Diagram.

service – matching the rate at which the sender is sending to the rate at which the receiving application is reading. A TCP sender can also be throttled due to congestion within the IP network; this form of sender control is referred to as *congestion control*. Flow control is a local phenomenon that is controlled by the receiver and its goal is to ensure that the sender does not overload the receiver.

Network congestion leads to packet loss in the network, which causes the sender to retransmit packets for which no ACK is received within the timeout period. These retransmissions lead to a reduction in the *network throughput*, where network throughput is the amount of data that is moved successfully from one place to another per second. For an already congested network, further retransmission aggravates the problem, which leads to *congestion collapse*. Congestion collapse is the condition that arises when the throughput becomes zero. The goal of flow control is to prevent congestion collapse, which is illustrated in Figure 8.6.

In fixed networks, packet loss is usually due to network congestion. Routers discard packets as soon as their buffers are full. TCP recognizes congestion only indirectly via missing acknowledgments. When a packet is unacknowledged within the timeout period, TCP usually retransmits the packet. If the lack of acknowledgment was due to network congestion, retransmitting the packet will only aggravate the situation because the retransmitted packet will cause more congestion to occur, which, as we stated earlier, can lead to congestion collapse.

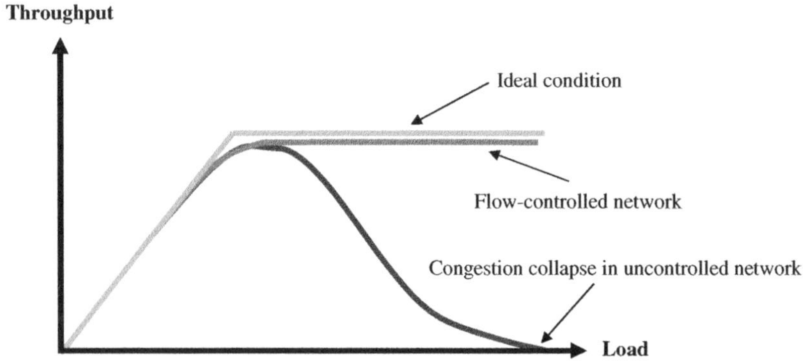

Figure 8.6 Illustration of Congestion Collapse in Uncontrolled Network.

Congestion control is a global phenomenon that ensures that the network can handle the load presented to it. To avoid congestion collapse, several congestion control schemes have been proposed for TCP. These include the following:

- Slow start
- Congestion avoidance
- Fast retransmit
- Fast recovery.

In the rest of this section, we describe each of these schemes. A typical network uses all these schemes.

8.4.1 Slow Start

In old implementations of TCP, a sender could inject multiple segments into the network, up to the window size of the receiver (*rwnd*), which is the value in the window field of the TCP header. This may be OK when the two hosts are on the same LAN, but when the hosts are separated by a slow WAN link there can be problems. Some intermediate routers may not be able to handle the traffic thereby causing packets to be dropped; retransmissions are initiated and the performance is degraded.

The slow-start algorithm avoids this by injecting segments into the network at the same rate that ACKs are returned by the receiver. Slow start adds another window to the sender's TCP called the *congestion window* (*cwnd*), which is set to be equal to one segment when a connection is established and never exceeds the receiver's advertised window. TCP also defines a congestion threshold called the *slow-start threshold* (*ssthresh*), which is a value to be used to determine if the slow-start algorithm should be executed. Specifically, whenever an ACK arrives, if $cwnd < ssthresh$ then slow start is performed; otherwise, congestion avoidance is performed. With slow start, the ACK to the first packet causes the window size to increase to 2. The ACKs for these two packets cause the window size to increase to 4. So, the window size is doubled after each round trip time until eventually *cwnd* exceeds *ssthresh* at which time congestion avoidance kicks in.

Note that *cwnd* is the control imposed by the sender, while *rwnd*, which is the advertised window, is the flow control imposed by the receiver. *cwnd* is based on the sender's assessment of perceived network congestion, while *rwnd* is related to the amount of available buffer space at the receiver for this connection. Note also that if multiple segments are acknowledged by one ACK message, cwnd increases as if one segment were ACKed. That is, an ACK for one segment causes the same change in the value of *cwnd* as an ACK for a group of segments. The serial increase in the number of segment per round trip time is illustrated in Figure 8.7.

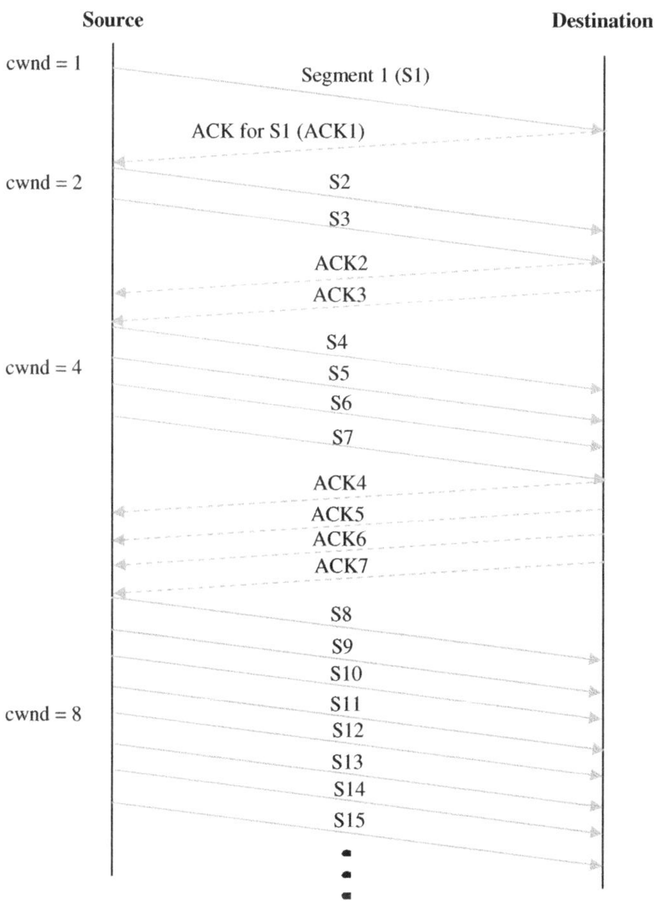

Figure 8.7 Illustration of the *cwnd* Increase.

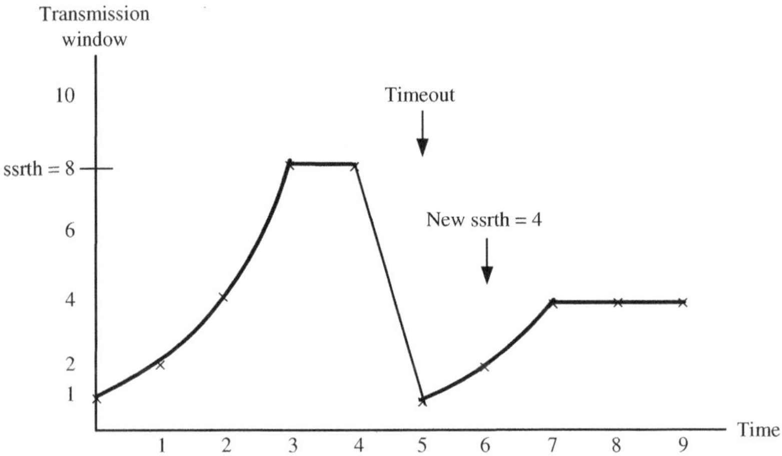

Figure 8.8 Illustration of Slow-Start Behavior When Timeout Occurs.

A missing ACK causes the timer to expire and this leads to a reduction of *ssthresh* to one half of the current *cwnd*. Thereafter, the congestion window starts again with one segment. This is illustrated in Figure 8.8. Note that the window size is bounded by *ssthresh* in the absence of the congestion avoidance scheme that is discussed next.

8.4.2 Congestion Avoidance

Packets are lost for one of two reasons:

(a) They are damaged in transit.
(b) They are dropped by routers as a result of congestion in the network.

In wired networks, packets are lost mainly due to congestion in the network. Congestion avoidance limits the expansion of *cwnd* by restricting it to a linear increase of 1 each round trip time. This means that rather than doubling the window size each round-trip time as slow start does, it increases the window size by 1 each round trip. Thus, congestion avoidance gives a linear growth rather than exponential growth. Also, when congestion occurs (i.e., timer expires or duplicate ACK is received), it sets the *ssthresh* to half the current *cwnd*.

Summary:

a. Congestion avoidance starts when $cwnd \geq ssthresh$ when using the slow start.
b. On each successful ACK the congestion window is updated as follows: $cwnd = cwnd + 1/cwnd$, which means that *cwnd* is increased by one only if

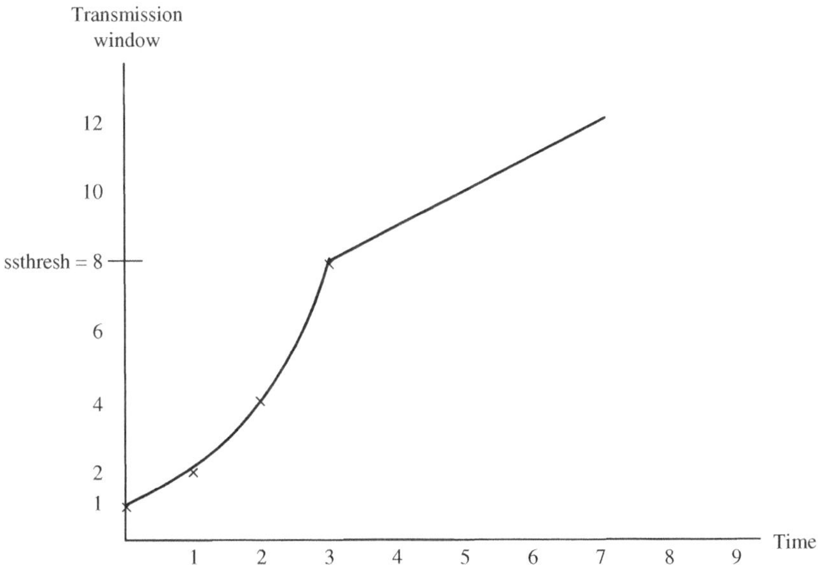

Figure 8.9 The Congestion Avoidance Process.

all *cwnd* segments in one round time have been acknowledged, which leads to a linear growth of *cwnd*.

c. When the timer expires or a duplicate ACK is received, $ssthresh = cwnd/2$ and $cwnd = 1$, which means that slow start resumes.

Figure 8.9 shows the congestion avoidance behavior when no timeout occurs. In the figure, the initial $ssthresh = 8$.

We can summarize the combined slow start and congestion avoidance as follows:

(a) Start with $cwnd = 1$ (slow start).
(b) For each round trip delay double *cwnd*, which implies an exponential growth of *cwnd*.
(c) When $cwnd \geq ssthresh$, enter the congestion avoidance phase where for each round trip delay, $cwnd = cwnd + 1$, which implies linear increase of *cwnd*.

The combined slow start and congestion avoidance is illustrated in Figure 8.10. In the figure, we originally have ssthresh = 8.

8.4.3 Fast Retransmit

TCP generates an immediate ACK (called a *duplicate ACK*) when an out-of-order segment is received. The purpose of the duplicate ACK is to let

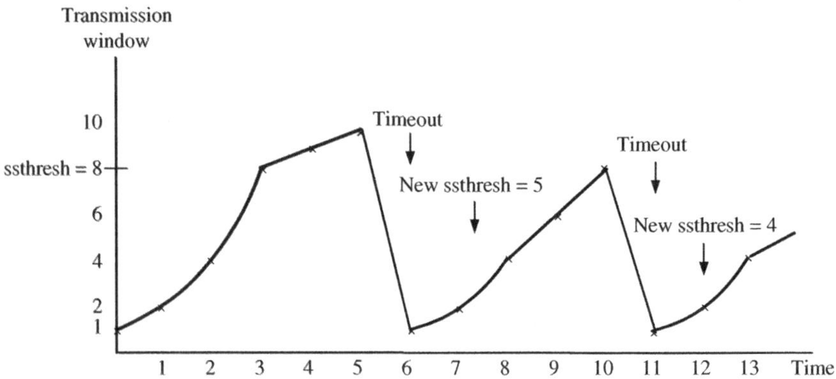

Figure 8.10 Illustration of the Combined Slow Start and Congestion Avoidance.

the other end of the connection know that a segment was received out-of-order and to let it know the sequence number that is expected. Duplicate ACKs are not expected to be delayed. The reception of three or more duplicate ACKs in a row is a strong indication that a segment has been lost. This is what fast retransmit is designed for. It avoids having TCP wait for a timeout before resending lost segments. Thus, a lost segment is retransmitted before the retransmission timer expires if three duplicate ACKs are received before the timer expires.

Figure 8.11 illustrates the fast retransmit scheme. In the figure, segment 3 (S3) is lost. The destination receives S4 and not S3, so it sends ACK2 again (a duplicate ACK) to indicate that it has received a segment (S4) and not S3. On receiving S5, it sends ACK2 again (a second duplicate ACK). On receiving S6, it sends ACK2 again (the third duplicate ACK). At this time, the source has received three duplicate ACKs, so it retransmits S3 and receives ACK6 since S4, S5, and S6 were correctly received. Note that receiving three duplicate ACKs means receiving four ACKs with the same acknowledgment number.

8.4.4 Fast Recovery

The receiver can only generate the duplicate ACK when another segment is received. Thus, the network does not seem to be congested since data is still flowing between the sender and receiver. The fast recovery is designed to be used in conjunction with fast retransmit and the goal is to avoid reducing the flow abruptly by going into slow start as the congestion avoidance calls for.

The algorithm works as follows. After the third duplicate ACK is received, it sets the slow-start threshold (*ssthresh*) to half the congestion window; that is, $ssthresh = cwnd/2$ and it retransmits the missing segment. Then it sets the

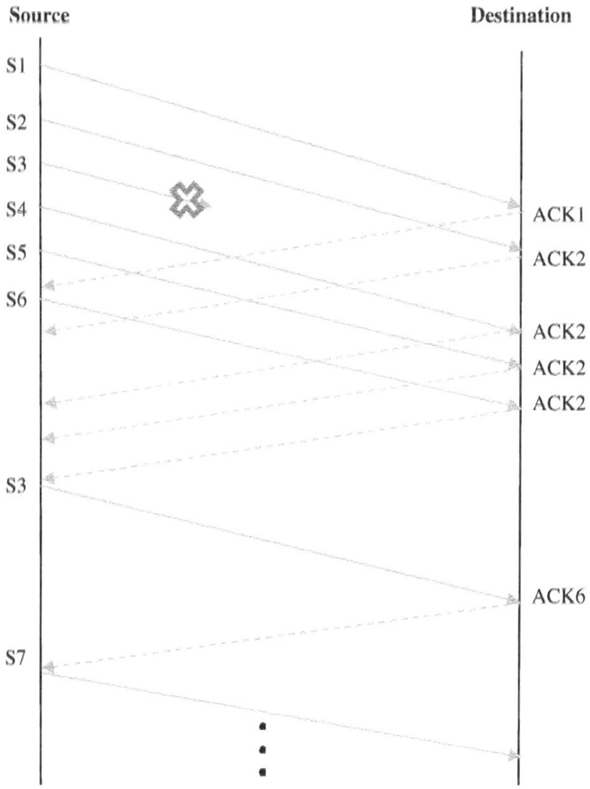

Figure 8.11 Illustration of Duplicate ACKs and Fast Retransmit.

congestion window as follows: $cwnd = ssthresh + 3$ since three segments have been received. If more ACKs are received for the same segment, *cwnd* is incremented by one each time an ACK is received.

8.5 TCP and Explicit Congestion Notification

Routers have traditionally reacted to network congestion by dropping packets that arrive at a router when its buffers are full. This approach to dealing with network congestion is known as the tail drop method, which has been shown not to perform well because it allows queues in routers to remain full for long periods of time thereby increasing delays and causing unnecessary retransmissions. One of the methods that have been used to overcome the problems with tail drop is the *Active Queue Management* (AQM) method. An example of AQM is the random early detection (RED). A RED router maintains two levels of thresholds: Q_{min} and Q_{max}. Let Q denote the average queue length at a RED router. If a packet arrives when $Q < Q_{min}$, it will be accepted and queued. If the packet arrives when $Q > Q_{max}$, it will be dropped. If the packet arrives when $Q_{min} \leq Q \leq Q_{max}$, it will be dropped in a probabilistic manner, where the probability of its being dropped increases as Q approaches Q_{max}.

Dropping packets sends an implicit message to the sender because it will be effectively notified by a subsequent timeout or duplicate ACK. Consequently, it will retransmit the packet. The ECN refines this scheme by marking instead of dropping packets when queue size is between Q_{min} and Q_{max}. As we discussed in Chapter 6, ECN is a mechanism where network nodes can mark IP packets instead of dropping them to indicate congestion to the endpoints. It uses two bits in the DiffServ field of the IP packet header. This leads to four possible ECN codepoints as follows. Codepoint "00" indicates "Not-ECN-Capable Transport (Not-ECT)," which means that the ECN is ignored; this is the default value for the two bits. Codepoints "01" and "10" both indicate "ECT" and are referred to as "ECT(1)" and "ECT(0)," respectively. Codepoint "11" indicates "CE."

When an end-node is ECN-capable, it sets the ECN codepoint in the IP header of the packet to either ECT(1) or ECT(0). A congested router that wants to send congestion notification sets the ECN codepoint to CE, instead of dropping the packet. When a Non-ECT packet experiences congestion at a router, it will be dropped. A packet with ECT marking will not be dropped at a router when it experiences congestion, and this is the difference between the two.

When a TCP receiver receives a CE-marked packet, it sets the ECN-Echo flag in the TCP header in the next packet that it sends back to the sender (e.g., in an ACK). Therefore, as soon as the sender receives the ECN-Echo flagged packet, it will know about the congestion and will react as if a packet has been dropped by halving its send window. The sender will also set the CWR flag in the TCP header of the next outgoing packet to indicate that it has received and reacted to the congestion signal.

What we have described in the previous paragraph is the ideal condition: the receiver reacts to the congestion by setting the ECN-Echo flag in the next packet sent to the sender. However, this message has the potential to reduce the throughput at the receiver. Therefore, there is a possibility that the receiver can lie to the sender by not setting the ECN-Echo bit thereby fooling the sender into continuing to send with the same rate, and in the process the receiver will gain an unfair advantage over other correctly behaving receivers. The nonce-sum (NS) bit in the TCP header is used to deal with this problem. The problem of hiding or concealing the ECN–CE marked packets from the TCP sender can be created deliberately as described; it can also be accidental when one of the CE bits (ECN = 11) is flipped and the sender is not made aware of the situation.

In a very simple case, the nonce sum operates as follows. A random one-bit value (a nonce) is encoded in the two ECT codepoints. The one-bit (exclusive-OR) sum of these nonces is returned in a TCP header flag as the NS bit. If the packet is marked, the marking will erase the nonce value of the ECT codepoints. If a packet arrives without being ECN–CE marked, the sum value will match the value held at the sender; otherwise, the two sum values will differ and the sender is able to verify whether or not the receiver reports correctly the reception of ECN–CE marked packets.

8.6 The SYN Flood DoS Attach

A SYN flood is a form of denial-of-service (DoS) attack in which an attacker sends a succession of SYN requests to a target's system in an attempt to consume enough server resources to make the system unresponsive to legitimate traffic. Normally, when a client attempts to start a TCP connection to a server, it goes through the three-way handshake:

(a) The client requests a connection by sending a SYN segment to the server.
(b) The server *acknowledges* this request by sending SYN-ACK back to the client.
(c) The client responds with an ACK, and the connection is established.

A SYN flood attack occurs when a client sends a great number of SYN requests to the server and when the server returns a SYN ACK, the client simply ignores it and does not return an ACK. This causes the server to retain the resources committed to the client and may not have enough resources for other clients that need a connection later. This is illustrated in Figure 8.12.

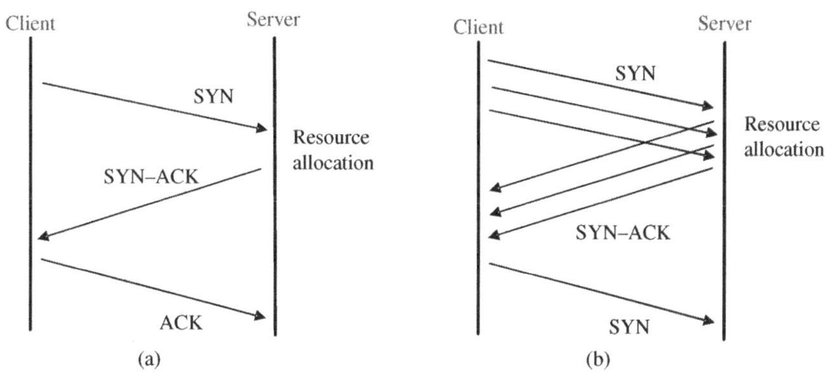

Figure 8.12 Illustration of the SYN Flood Attack. (a) Normal TCP Three-Way Handshake and (b) TCP Handshake with DoS.

8.7 UDP

UDP is another transport-layer protocol that operates in a different manner from TCP. Specifically, it is a simple protocol that implements minimal functions to transport data from one endpoint to another. The difference between TCP and UDP can be summarized as follows:

(a) TCP is connection-oriented, while UDP is connectionless. As discussed earlier in this chapter, TCP uses the three-way handshake to establish a connection prior to data transmission. UDP does not require a connection to be established between a client and server before data can be transmitted.
(b) TCP is a reliable protocol in the sense that data delivery to a client is guaranteed. If a segment is lost in transit, TCP takes steps to recover it. UDP is unreliable with no guarantee of packet delivery. Thus, a UDP datagram may be lost in transit and UDP will not make an effort to recover it.
(c) TCP guarantees order of a message. Segments belonging to the same message are delivered to the client in the same order they were sent into the network, even though they may reach the destination node out of order. TCP will do the resequencing and in-order delivery at the destination by buffering an out-of-order segment until all segments that were transmitted before it have arrived. The segments are then reordered and delivered to the destination host. UDP does not provide ordering of datagrams.
(d) TCP is byte-oriented (or stream-oriented), while UDP is message-oriented. This means that while TCP sends streams of byes with no indication of message boundaries, UDP sends messages individually with definite boundaries so that a read operation at the destination will yield one message as it was sent. Rather than keeping a queue for each message, TCP uses one queue for all messages and adds enough bytes from the queue to generate a segment to be transmitted.
(e) TCP is slow, while UDP is fast. The slowness of TCP arises from its need to create a connection and ensure ordered delivery. UDP does not do either of these and so is suitable for applications where fast delivery is a major issue.
(f) TCP is sometimes called a heavyweight protocol because of the overhead associated with its connection, reliability, and ordered delivery. On the other hand, UDP is considered a lightweight protocol that is not burdened by the need for these functions associated with TCP.
(g) TCP provides flow and congestion control, while UDP does not. Because of this, UDP datagram header does not have a sequence number field or an acknowledgment number field.
(h) Because of TCP's need to do more than UDP, the size of a TCP segment header is at least 20 bytes, while the size of a UDP datagram is fixed at 8 bytes. The TCP header permits optional data of up to 40 bytes, which means that the size of the TCP header lies between 20 bytes and 60 bytes.

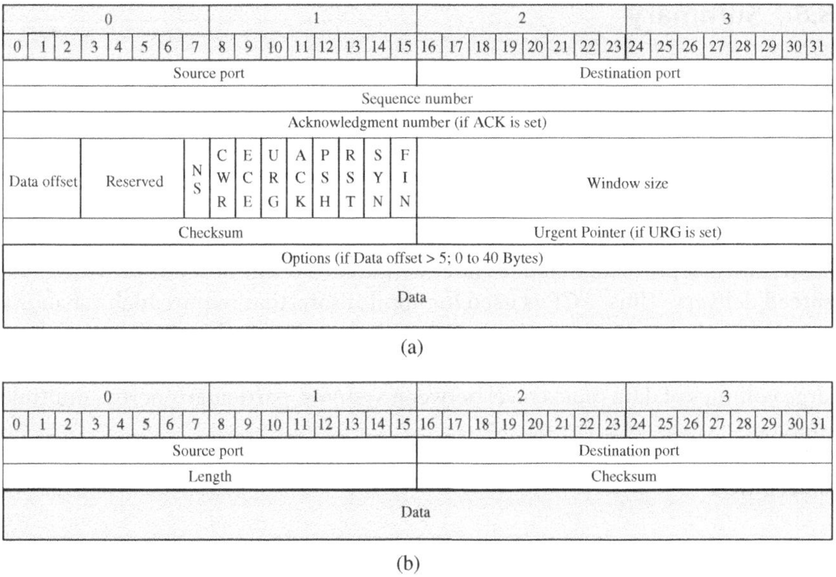

Figure 8.13 Comparison of (a) TCP Segment and (b) UDP Datagram Headers.

The two headers are compared in Figure 8.13. The UDP header is divided into the following four fields:

- Source port number (2 bytes), which is the UDP port number of the sending device; UDP uses the same port concept and numbers as TCP.
- Destination port number (2 bytes), which is the UDP port number of the receiving device
- Length (2 bytes), which represents the total size of a datagram, including both the header and the data
- Checksum (2 bytes), which is used by a receiver to determine if an incoming datagram is error-free or not.

As stated earlier, because UDP does not provide flow control, its datagram header does not have a sequence number field or an acknowledgment number field.

Because of the lack of congestion control, UDP can lead to network congestion collapse if it is used for long-lived applications. Hence, a UDP application can send as much data as it wants, but much of this data might be lost or discarded by the routers because of network congestion. It is designed mainly for applications that do not require the strict service guarantees offered by TCP. UDP is suitable for applications that need fast and efficient transmission, such as games. UDP's stateless nature is also useful for servers that respond to small queries from a large number of clients.

8.8 Summary

This chapter has discussed the transport-layer protocols, which include TCP and UDP. It discussed the TCP flow and congestion control protocols, which include the slow start, congestion avoidance, fast retransmit, and fast response protocols. It also discussed how UDP differs from TCP. Specifically, TCP is a connection-oriented, reliable, and slow protocol that provides guaranteed delivery and preserves order of messages, while UDP is a connectionless, unreliable, and fast protocol that does not guarantee message order or provide guaranteed delivery. Thus, TCP is used for applications that require high reliability but are less time critical, whereas UDP is used for application that are time sensitive but do not require high reliability. Also, TCP is suited for situations where large volumes of data must travel between systems, particularly across multiple routers.

Exercises

1 What is the difference between the transmission control protocol (TCP) and the user datagram protocol (UDP)?

2 What is difference between a well-known TCP port and an ephemeral TCP port?

3 What is the difference between a user port and a dynamic port?

4 What is a TCP socket?

5 What is the name of a TCP data block?

6 List the steps involved in the TCP three-way handshake.

7 What is a SYN segment?

8 What is the name of the network entity that initiates a TCP connection?

9 What does TCP SYN flood mean?

10 What is the difference between the *congestion window* (cwnd) and the *slow-start threshold* (ssthresh)?

11 What does congestion collapse mean?

12 What is the difference between the fast retransmit and the fast recovery?

9

Transport Layer – SCTP and DCCP

9.1 Introduction

In this chapter, we continue our discussion on transport layer protocols. Transmission control protocol (TCP) and user datagram protocol (UDP) that we discussed in Chapter 8 were designed in the early days of the Internet when it was expected that the network would be used primarily for non-real-time traffic. With the increasing use of the Internet for different types of applications most of which generate traffic that needs to be delivered in real time, there is a need to design new transport protocols that can deal with these traffic types. Also, there are new applications that require the simplicity of UDP but they generate a large volume of traffic that needs to be controlled in order to avoid congestion collapse. In this chapter, we discuss two transport protocols that were specifically designed to meet these needs. The first of these protocols is the *Stream Control Transmission Protocol* (SCTP), which, like TCP, is a connection-oriented protocol that provides reliable end-to-end transfer. The second is the *Datagram Congestion Control Protocol* (DCCP), which was designed to inherit some of the features of TCP and some of the features of UDP.

9.2 Stream Control Transmission Protocol

SCTP, which is defined in RFC 3309, is designed to transport SS7 signaling messages over IP networks. As a transport layer protocol, it operates directly on top of IP, at the same level as TCP. Its basic service is the connection-oriented reliable transfer of messages between peer SCTP users. It combines the benefits of the efficient UDP and reliable TCP.

An SCTP packet includes information that allows the combination of several "chunks" of data into each packet. The protocol includes information that allows packets to be routed to avoid congestion so that delay does not cause serious problem for signaling and streaming applications. A packet can

Fundamentals of Data Communication Networks, First Edition. Oliver C. Ibe.

be fragmented if it passes through a network that has a smaller maximum transmission unit (MTU) than the network that originated the SCTP packet. SCTP contains information for sequencing packets.

One of the major differences between SCTP and TCP is the fact that SCTP supports *multihoming* and *multistreaming* and TCP does not. Multihoming is the ability to configure more than one IP address at a given endpoint. This requires a host to have multiple network interfaces each of which has a different IP address. One advantage of multihoming is that in the event that network failure occurs, the use of more than one address could allow rerouting of packets and can also provide an alternate path for retransmissions. During the initiation of a connection, which is called an *association* in SCTP, endpoints exchange the list of addresses that the interfaces will support. One address is designated as the primary address to receive data. A single port number is used across the entire address list at an endpoint for a specific session. SCTP uses the same ports as TCP.

Multistreaming allows for multiple virtual connections on the same physical line. Each user application might be assigned its own stream (virtual connection). This feature allows data to be delivered in multiple independent streams so that if there is data loss in one stream, it will not affect the data delivery for the other streams. The SCTP user can specify at association startup time the number of streams to be supported by the association.

SCTP uses timers of much shorter duration than TCP. It is rate-adaptive, responding to network congestion and throttling back transmission accordingly. It uses an initialization procedure that is based on cookies to prevent denial of service, such as the SYN flood attack that is associated with TCP. Like UDP it is a message-oriented protocol that defines structured frames of data, unlike TCP that imposes no structure on the transmitted stream of bytes.

9.2.1 Motivation for a New Transport Protocol

TCP is a packet-oriented transport protocol for reliable data transfer in IP networks that was defined a long time ago. Unfortunately, its design has imposed several limitations for new emerging applications. Some of the limitations include the following:

(a) TCP provides both reliable data transfer, through acknowledgment mechanism, and strict order of transmission delivery of data, through sequencing mechanism. Some applications need reliable transfer without sequence maintenance, while others would be satisfied with partial ordering of the data. In both of these cases, *the head-of-line* (HOL) blocking caused by TCP adds unnecessary delay.

(b) TCP is *stream oriented*, and this can be also an inconvenience for some applications, since usually they have to include their own marks inside the stream so the beginning and end of their messages can be identified.

(c) TCP was never designed to be *multihomed*. As stated earlier, a multihomed host is one that has several network cards and can make use of a number of IP addresses at the same time. The limited scope of the TCP sockets makes it difficult to design any data transfer mechanism in which a multihomed host could use several network cards at the same time. This would provide high availability that is often needed in some applications.
(d) TCP does not scale well since the maximum number of simultaneous TCP connections is dependent on kernel limitations. This is because TCP is generally implemented at the operating system level.
(e) In TCP, there is no possibility of *timer control*; TCP generally does not allow application control over its initialization, shutdown, and retransmission timers.
(f) TCP is relatively vulnerable to *denial-of-service* (DoS) attacks that attempt to exhaust the resources that legitimate users can use. One example of TCP DoS attack is the *SYN flood attack* that we described in Chapter 8.

9.2.2 Illustration of the HOL Blocking

Consider the example of a voice protocol that carries signaling messages for a phone call establishment over an IP network. Assume that three calls are in progress. If one of them is experiencing data loss, this is not expected to cause a delay in the transmission of messages related to the other two calls.

Assume that these voice messages are transmitted over TCP, and one TCP "pipe" carries all three calls. If the establishment messages for all three calls are received correctly, then the calls will be in progress. If the calls are released in the same order, but the release message for Call 1 is lost, then the release of the other two calls will be delayed while TCP attempts to recover the lost packet. This has the potential to delay the release of all the calls for a number of seconds, which is an unacceptable delay in the telephone network.

Because of multistreaming in SCTP, each call has its own connection and the loss of a packet relating to Call 1 affects only the Call 1 stream of data; Calls 2 and 3 are unaffected. Only the resources relating to Call 1 will be tied up until the transport protocol recovers from the data loss. The HOL problem is illustrated in Figure 9.1.

9.2.3 Summary of Features of SCTP

SCTP is a connection-oriented protocol that supports both ordered and unordered transmission. It supports transport-layer fragmentation. It is a message-oriented protocol that preserves message boundaries. In addition, it supports multihoming by allowing an endpoint to have multiple IP addresses on the same interface, and it supports multistreaming. Finally, it has security features that guard against DoS such as the SYN-flood attack that can occur in TCP.

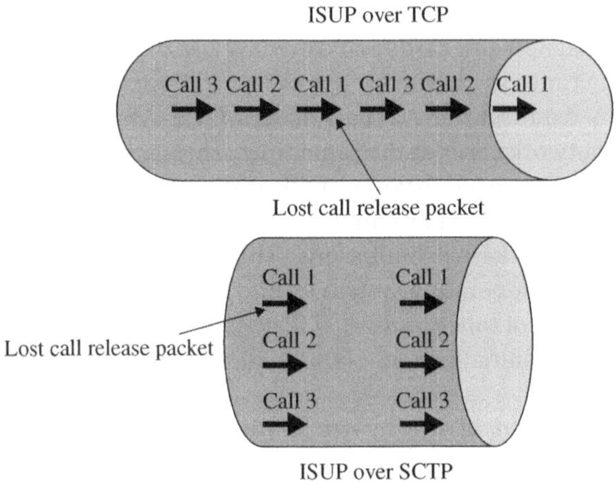

Figure 9.1 Illustration of the HOL Blocking.

9.2.4 SCTP Packet

SCTP has a simpler basic packet structure than TCP. Each packet consists of two basic sections:

(a) The *common header*, which occupies the first 12 octets.
(b) The *data chunks*, which form the remaining portion of the packet.

The structure of the packet is illustrated in Figure 9.2. A chunk contains either control information or user data. In the figure, the first chunk is highlighted in gray and the last of N chunks (Chunk N) is highlighted in light gray.

9.2.5 SCTP Header

All SCTP packets require a three-word (i.e., 96 bits) common header section with the following fields:

(a) *Source port*: A 16-bit field that identifies the sending port.
(b) *Destination port*: A 16-bit field that identifies the receiving port that the destination host uses to route the packet to the appropriate end-point/application.

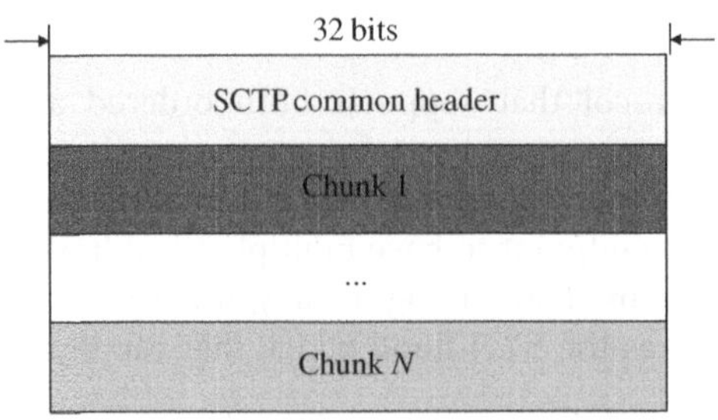

Figure 9.2 SCTP Packet.

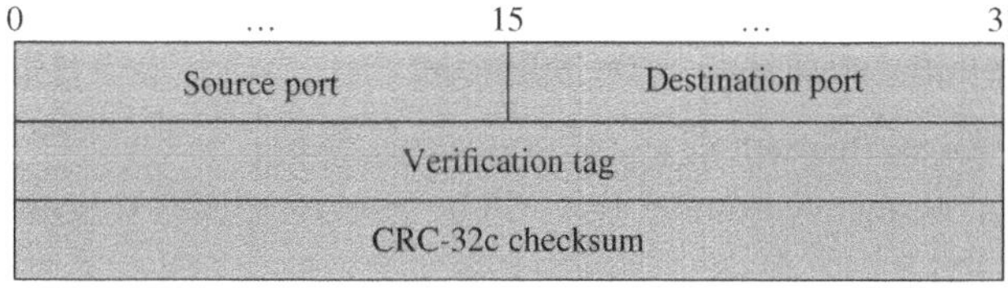

Figure 9.3 SCTP Packet Header.

(c) *Verification tag*: A 32-bit random value created during initialization to distinguish stale packets from a previous connection.
(d) *Checksum*: SCTP uses the CRC-32c algorithm (32-bit CRC).

The header format is shown in Figure 9.3.

9.2.6 Association Establishment

As stated earlier, a connection between an SCTP client and a server is called an *association*. Before data can be exchanged, the two SCTP hosts must exchange the communications state (including the IP addresses involved) using a four-way handshake, as shown in Figure 9.4.

In contrast to TCP's three-way handshake, a four-way handshake eliminates exposure to the TCP SYN flooding attacks. The receiver of the initial (INIT) contact message in a four-way handshake does not need to save any state information or allocate any resources. Instead, it responds with an INIT-ACK message, which includes a state cookie that holds all the information needed by the sender of the INIT-ACK to construct its state. The state cookie is digitally

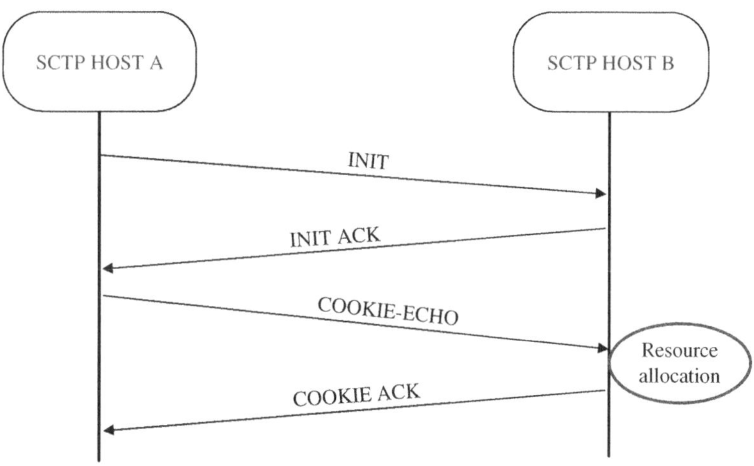

Figure 9.4 SCTP Four-Way Handshake.

signed. Both the INIT and INIT-ACK messages include several parameters used in setting up the initial state such as the following:

(a) A list of all IP addresses that will be a part of the association
(b) An initial transport sequence number that will be used to reliably transfer data
(c) An initiation tag that must be included on every inbound SCTP packet
(d) The number of outbound streams that each side is requesting
(e) The number of inbound streams that each side is capable of supporting.

After exchanging these messages, the sender of the INIT message echoes back the state cookie in the form of a COOKIE-ECHO message that might have user DATA messages bundled onto it as well (subject to path MTU constraints). Upon receiving the COOKIE-ECHO, the receiver fully reconstructs its state and sends back a COOKIE-ACK message to acknowledge that the setup is complete. This COOKIE-ACK can also bundle user DATA messages with it. No other chunk can be carried in a packet that carries an INIT chunk.

9.2.7 Four-Way Handshake and the SYN Flood DoS Attach

Recall that a SYN flood attack occurs when a client sends a great number of SYN requests to the server and when the server returns a SYN ACK, the client simply ignores it and does not return an ACK. This causes the server to retain the resources committed to the client and may not have enough resources for other clients that need a connection later. SCTP protects against DoS attacks using a cookie. The server bundles the cookie in the INIT-ACK from the server to the client, and the server will only commit resources for the connection when it receives the COOKIE-ECHO response from the client. The difference between the TCP three-way handshake and the SCTP four-way handshake is illustrated in Figure 9.5. The figure also includes the TCP handshake with DoS.

9.2.8 Multihoming

One of the most important enhancements in SCTP over traditional transport layer protocols is the multihoming capability. Multihoming is the ability of a single SCTP *endpoint* to contain multiple interfaces with different IP addresses. Thus, a multihomed host can be reached using more than one IP address, usually through more than one network interface. This feature allows an endpoint of an SCTP association to be mapped to multiple IP addresses. The multihoming concept is illustrated in Figure 9.6.

In a single-homed connection, an endpoint contains only one network interface and one IP address. When a network or path failure occurs, the endpoint is completely isolated from the network. Multihoming in SCTP ensures a better chance of survival if a network failure occurs, when compared to TCP. The

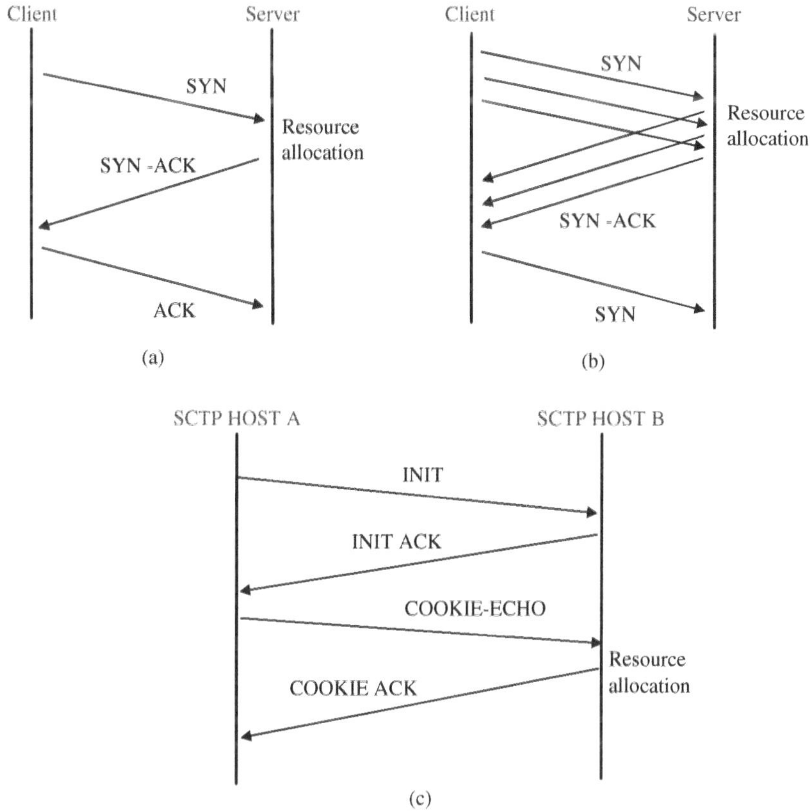

Figure 9.5 Comparison of SCTP Handshake and TCP Handshake. (a) Normal TCP Three-Way Handshake; (b) TCP Handshake with DoS; and (c) SCTP Four-Way Handshake.

built-in support for multihomed hosts in SCTP enables a single SCTP association to run across multiple links or paths, to achieve link or path redundancy. This enables an SCTP association to achieve faster failover from one link or path to another, with minimum interruption in the data transfer service.

Among the possible IP addresses, one is selected as "primary address." The *primary path* is the network path that leads to the primary address. The primary address is used as the destination for all DATA chunks for normal transmission. All the other addresses are considered as alternate IP addresses. SCTP uses these alternate IP addresses to retransmit DATA chunks and to improve the probability of reaching the remote endpoint. Retransmission may occur because of continued failure to send DATA to the primary address. When the primary address fails, all DATA chunks are transmitted to an alternate address until contact with the primary address is reestablished.

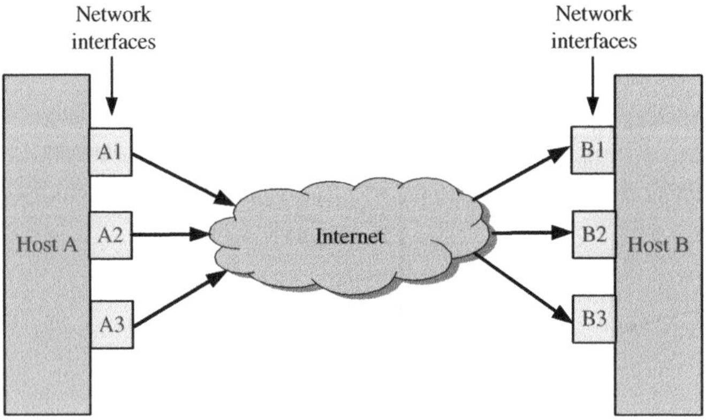

Figure 9.6 Illustration of SCTP Multihoming Feature.

During the initiation of an association, the SCTP endpoints exchange the list of IP addresses so that each endpoint can receive messages from any of the addresses associated with the remote endpoint. For security reasons, SCTP sends response messages to the source address in the message that prompted the response.

9.2.9 Multistreaming

In TCP, only a single data stream is allowed per connection and all of the information must be passed through that one stream. As discussed earlier, TCP transmits data sequentially in the form of bytes in a single stream and ensures that all the bytes are delivered in a particular order. Therefore, a second byte is sent only after the first byte has safely reached the destination. This sequential delivery of data causes delay when a message loss or sequence error occurs within the network. An additional delay occurs when TCP stops sending data until the correct sequencing is restored, either upon receiving an out-of-sequence message or by retransmitting a lost message. The strict preservation of message sequence in TCP poses a limitation for certain applications that require sequencing of messages that affect the same resource (such as the same call or the same channel), so that messages are loosely correlated and delivered without maintaining the overall sequence integrity.

SCTP allows multiple simultaneous data streams within one SCTP connection or association. Each message sent to a data stream can have a different final destination, but each must maintain message boundaries. Thus, all the streams (chunks) within an association are independent but related to the association. In particular, systems cannot send parts of the same message through different streams; one message must go through one stream. This delivery scheme reduces unnecessary HOL blocking between independent

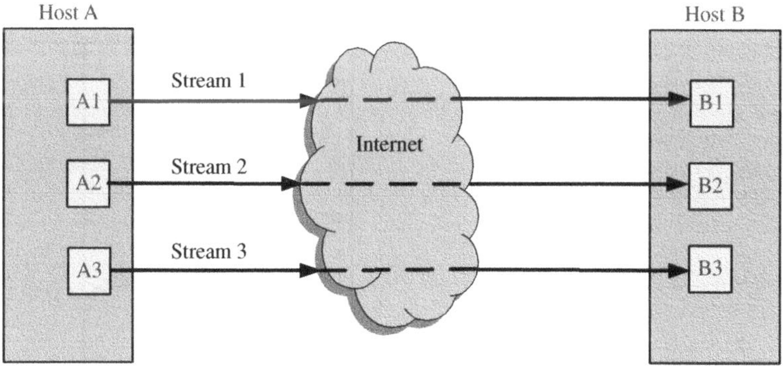

Figure 9.7 Illustration of SCTP Multistreaming Feature.

streams of messages. In other words, a blocked stream does not affect the other streams in an association. Each stream has a stream number that is included inside SCTP packet's chunk headers.

SCTP streams are unidirectional channels, and within each channel the data messages are usually transported in sequence. An application may also request that a message to be delivered unordered, which can reduce blocking effects in case of message loss, since the reordering mechanism of one stream is not affected by that of another stream that may have to wait for a retransmission of a previously lost data chunk. The multistreaming concept is illustrated in Figure 9.7, where there are three streams in the association between hosts A and B.

SCTP multistreaming is particularly effective in situations where there is a need to separate the control channel and the data channels. In TCP, control and data typically share the same connection, which can pose a problem because control packets can be delayed behind data packets. If control and data messages are transmitted in different streams, then control data could be handled in a more timely manner, resulting in a better system management.

9.2.10 SCTP Graceful Shutdown Feature

SCTP does not support a "half-open" connection, which can occur in TCP. In a half-open connection, even though an endpoint indicates that it has no more data to send, the other endpoint can continue to send data indefinitely. In SCTP, when the shutdown procedure begins, both of the endpoints will stop sending new data across the association. SCTP assumes that it needs only to clear up acknowledgments of the previously sent data. The SCTP shutdown feature uses a three-message procedure to gracefully shut down the association, in which each endpoint has confirmed the receipt of the DATA chunks before completing the shutdown process. When an immediate shutdown is required, SCTP

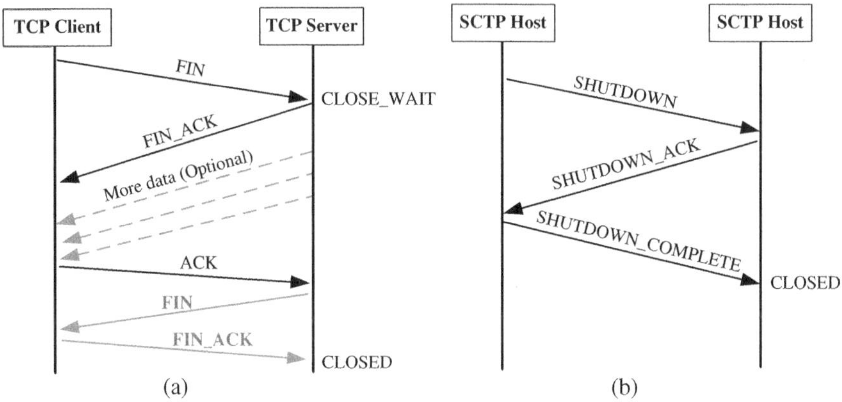

Figure 9.8 Comparison of SCTP and TCP Shutdown Processes. (a) TCP Connection Termination (Half-Open Connection in Light Gray) and (b) SCTP Connection Close.

sends an ABORT message to an endpoint. A comparison of the two procedures is illustrated in Figure 9.8.

9.2.11 Selective Acknowledgments

In standard TCP, every message, or packet of information, must be accounted for, resent as necessary, and processed in the order they were sent. SCTP has the ability to selectively acknowledge receipt of missing, disordered, or duplicated messages.

9.3 Datagram Congestion Control Protocol

Up until recently most Internet unicast traffic has been TCP-based traffic with much less using UDP. Applications that use UDP are mainly short request–response types such as domain name system (DNS) and simple network management protocol (SNMP) applications. UDP is used by those applications that have no need for TCP's three-way handshake, retransmission, and stateful connections. Also, TCP can introduce arbitrary delay because of its reliability and in-order delivery requirements. Recently, there has been an increase in applications that use UDP in a different manner. These applications, which include streaming audio, voice over IP (VoIP) and multiplayer online games are characterized by their timeliness. They are willing to forego the reliability that TCP provides in favor of the simplicity associated with UDP.

As discussed earlier, UDP is a connectionless protocol that does not care about reliable data packet delivery, network congestion control, or packet reordering on the receiver end. Because of the lack of any type of congestion control, UDP has the potential to lead to network congestion collapse. Hence, a UDP application can send data as much as it wants, but much of this data might be lost or discarded by the routers because of network congestion. The increasing growth of long-lived non-congestion-controlled traffic, relative to congestion-controlled traffic, poses a real problem for the Internet. To deal with this critical situation, the IETF has developed a new transport protocol called the DCCP that combines the connection-oriented and congestion-control features of TCP and the unreliable data transmission of UDP.

DCCP is a message-oriented protocol that implements reliable connection setup, teardown, explicit congestion notification (ECN), congestion control, and feature negotiation. It is particularly useful for applications with timing constraints on the delivery of data, which include streaming media, multiplayer online games, and VoIP where old messages quickly become stale so that getting new messages is preferred to resending lost messages. DCCP was designed to have as little overhead as possible both in terms of the size of the packet header and the state and CPU overhead required at the end hosts. Thus, it can be looked at in two ways:

(a) DCCP is UDP plus congestion control, or
(b) DCCP is TCP less reliability and byte-stream semantics.

In the remainder of this chapter, we provide a brief discussion on DCCP.

9.3.1 DCCP Packet Structure

Currently, 10 packet types implement DCCP's protocol functions. These are defined in Table 9.1. From the table, we can see that the first eight packet types are used during the progress of a typical connection: connection initialization, data transfer, and connection termination; and the two remaining packet types are used to resynchronize after bursts of loss.

All DCCP packets begin with a 12- or 16-byte generic header followed by additional fixed-length fields and option field required by the particular packet type. Also, every packet type includes its own fixed-size header data. Thus, different types of packet can have different packet header size. Figure 9.9 illustrates the structure of a DCCP packet.

A special 1-bit field called the X-field (or extended sequence numbers field) determines the form of the generic header of the DCCP packet header. If $X = 1$, the sequence number field is 48 bits long and the generic header takes 16 bytes. If $X = 0$, the sequence number field is 24 bits long, and the generic header is 12 bytes long. These two types of packets are shown in Figure 9.10.

The fields are defined as follows:

- *Source port* (16 bits): Similar to the TCP and UDP source port
- *Destination port* (16 bits): Similar to the TCP and UDP destination port

Table 9.1 Different DCCP Packet Types

Type	Name	Purpose	Use
0	DCCP-Request	Sent by the client to initiate a connection (the first part of the three-way initiation handshake)	Connection initialization
1	DCCP-Response	Sent by the server in response to a DCCP-Request (the second part of the three-way initiation handshake)	
2	DCCP-Ack	Used to transmit pure acknowledgments	Connection initialization and data transfer
3	DCCP-Data	Used to transmit application data	Data transfer
4	DCCP-DataAck	Used to transmit application data with piggybacked acknowledgments	
5	DCCP-CloseReq	Sent by the server to request that the client close the connection	Connection termination
6	DCCP-Close	Used by the client or the server to close the connection; elicits a DCCP-Reset in response	
7	DCCP-Reset	Used to terminate the connection, either normally or abnormally	
8	DCCP-Sync	Used to resynchronize sequence numbers after large bursts of loss	Resynchronization
9	DCCP-SyncAck	Used to resynchronize sequence numbers after large bursts of loss and to respond to a received DCCP-Sync	

- *Data offset* (8 bits): The offset from the start of the packet's DCCP header to the start of its application data area, expressed in 32-bit words
- *CCval* (4 bits): Used for providing the desired connection control ID
- *Checksum coverage* (*CsCov*) (4 bits): Identifies the part of the segment that is checked by the checksum field
- *Checksum* (16 bits): The checksum covers all the 16-bit words in the *DCCP header*, *DCCP options*, a *pseudoheader* taken from the network-layer header, and, depending on the value of the checksum length field, some or all of the payload
- *Type* (4 bits): Identifies the type of DCCP segment, as shown in Table 9.1
- *Extended sequence number* (1 bit): Used to indicate if the sequence number field is 48 bits long and the generic header takes 16 bytes (X = 1), or the sequence number field is 24 bits long and the generic header is 12 bytes long (X = 0)

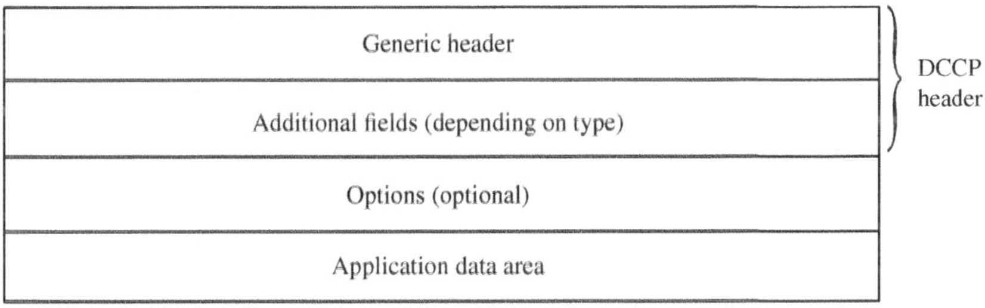

Figure 9.9 Structure of DCCP Packet.

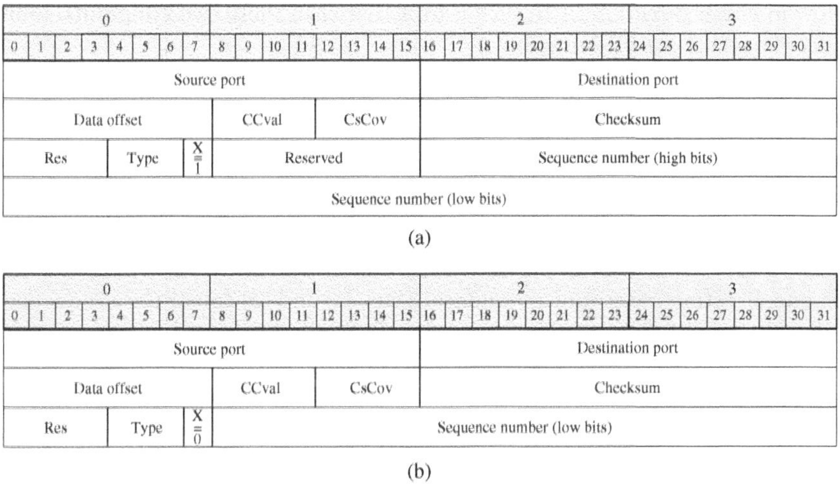

Figure 9.10 The Generic Header Fields. (a) Generic Header with X = 1 and (b) Generic Header with X = 0.

- *Sequence number* (24 bits or 48 bits): The sequence number field is initialized by a *DCCP-Request* or *DCCP-Response* packet, and increases by one (modulo 16777216) with every packet sent. The receiver uses this information to determine whether packet losses have occurred or not.

Every DCCP packet uses a new sequence number; this includes data, acknowledgment, and control packets. Also, acknowledgments are not cumulative; an acknowledgment is for last packet received. There are options for acknowledgment vectors that provide a history of received segment. An example of an Ack vector is as follows:

0:Received; 1:Received ECN; 2:Received; 3:Not yet received

9.3.2 DCCP Connection

There are nine connection states, which are as follows:

- CLOSED: No connection
- LISTEN: Server in passive listening state
- REQUEST: Client is beginning three-way handshake
- RESPONSE: Server responding to request
- PARTOPEN: Client waiting for response from the server
- OPEN: Data transfer (connection established)
- CLOSEREQ: Server asking client to close
- CLOSING: Client waiting for final reset
- TIMEWAIT: 2 maximum segment lifetime wait (at receipt of reset).

Each DCCP connection runs between two endpoints A and B and data can flow in either direction: A to B or B to A. Between these two endpoints, four sets of packets can be defined:

a. Data packets from A to B
b. Acknowledgments from B to A
c. Data packets from B to A
d. Acknowledgments from A to B.

Each of these four sets of packets constitutes what is known as a *subflow*; thus, a subflow is a set of packets sent in one direction. All the packets sent in one direction constitute a sequence. Thus, (a) and (d) constitute a sequence since they are traffic that flows from A to B. Similarly, (b) and (c) constitute a sequence, the traffic that flows from B to A.

A *half-connection* consists of the combination of data packets sent in one direction and the corresponding acknowledgments that flow in the other direction. Thus, (a) and (b) constitute a half-connection, which is the flow "A → B data plus B → A ack." Similarly, (c) and (d) constitute another half-connection, which is the flow "B → A data plus A → B ack." Separating a connection into two half-connections is useful because since traffic is typically asymmetric, using two different routes allows different congestion schemes and connection parameters to be used.

Figure 9.11 illustrates the half-connection concept and how it can be used to piggyback Acks on data packets.

A typical DCCP connection is established as follows:

(a) The client sends a DCCP-Request packet to the server specifying the client and server ports, the service being requested and any features that can be negotiated, such as the CCID.
(b) The server sends a DCCP-Response packet to the client to accept the connection. The response includes any features and options requested by the client that the server agrees to.
(c) The client sends a DCCP-Ack packet that acknowledges the DCCP-Response packet.
(d) A few more packets may be exchanged to finalize the feature negotiations.

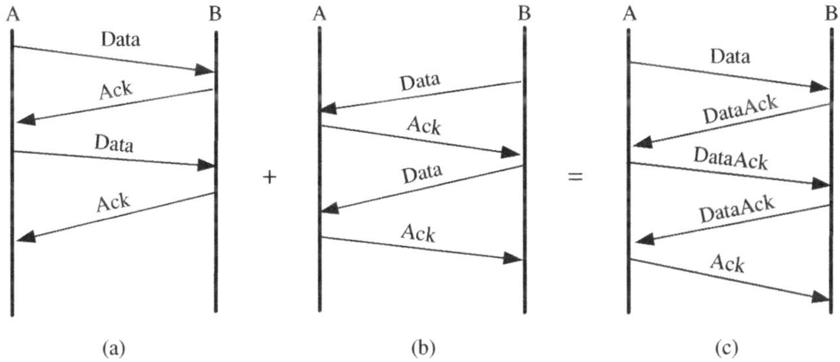

Figure 9.11 Illustration of DCCP Half-Connections. (a) A → B Half-Connection; (b) B → A Half-Connection; and (c) Combined Half-Connections.

(e) The client and server then exchange a series of DCCP-Data packets, DCCP-Ack packets, and DCCP-DataAck packets that contain piggybacked data and acknowledgments.
(f) The client (or the server) sends a DCCP-CloseReq packet requesting that the connection be closed.
(g) The server (or the client) sends a DCCP-Close packet to acknowledge the close.
(h) The client (or the server) sends a DCCP-Reset packet to clear its connection stare.
(i) The server (or the client) receives the DCCP-Reset packet and holds the state long enough to allow any remaining packets to clear the network.

Figure 9.12 illustrates the connection establishment process, and Figure 9.13 shows the DCCP state diagram. Note that to close a connection, one side sends Close and the other responds with Reset. Reset is used for normal close as well as for exceptional conditions. Because whoever sends the Close has to go to the TIMEWAIT state, the server side may send CloseReq to ask the client to send Close.

9.3.3 DCCP Congestion Management

DCCP provides applications a choice of congestion control mechanisms. The choice is made through congestion control IDs (CCIDs) that are negotiated at connection startup. Each congestion control mechanism that is supported by DCCP is assigned a 1-byte CCID: a number from 0 to 255:

- CCID 0 and CCID 1 are reserved.
- CCID 2 is a TCP-like congestion control.
- CCID 3 is a TCP friendly rate control (TFRC).
- CCID 4–255 are reserved.

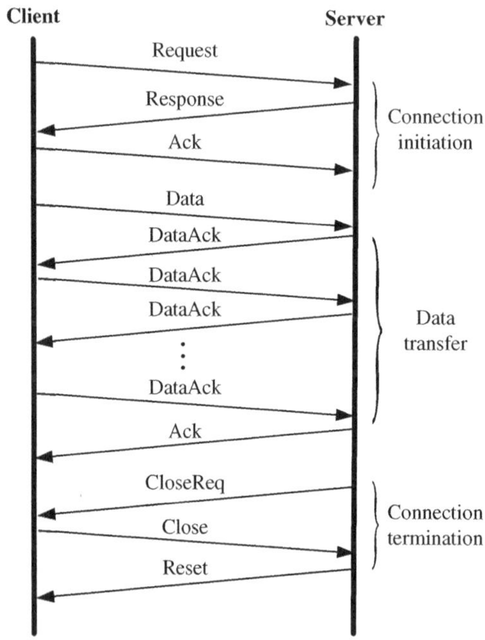

Figure 9.12 Illustration of DCCP Connection Process.

9.3.3.1 CCID 2–TCP-Like Congestion Control

CCID 2 is called TCP-like congestion control mechanism. It is perfect for those applications that can adapt to the changes of congestion control window and that need as much bandwidth as possible in the network. It uses TCP-like congestion control mechanism. Specifically, after a congestion event occurs, CCID 2 reduces its congestion window. A congestion event may be indicated explicitly via ECN marking or via duplicate acknowledgments. For either case, cwnd is halved.

There are some particular features of CCID2 connection:

- Duplicate acknowledgment indicates some loss of data packet.
- The sender has timeout option, which is handled like TCP's retransmission timeout. The sender calculates round-trip time for a window at most once and uses TCP's algorithm for maintaining the round-trip time.

9.3.3.2 CCID 3–TCP Friendly Rate Control

The TCP-friendly rate control (TFRC) is a receiver-based feedback mechanism that uses a sending rate instead of a congestion window to deal with congestion control. Periodically, the receiver sends feedback packets to the sender containing loss event rate. The sender uses these feedback messages to measure the round-trip time (RTT). The loss event rate and RTT are then fed into

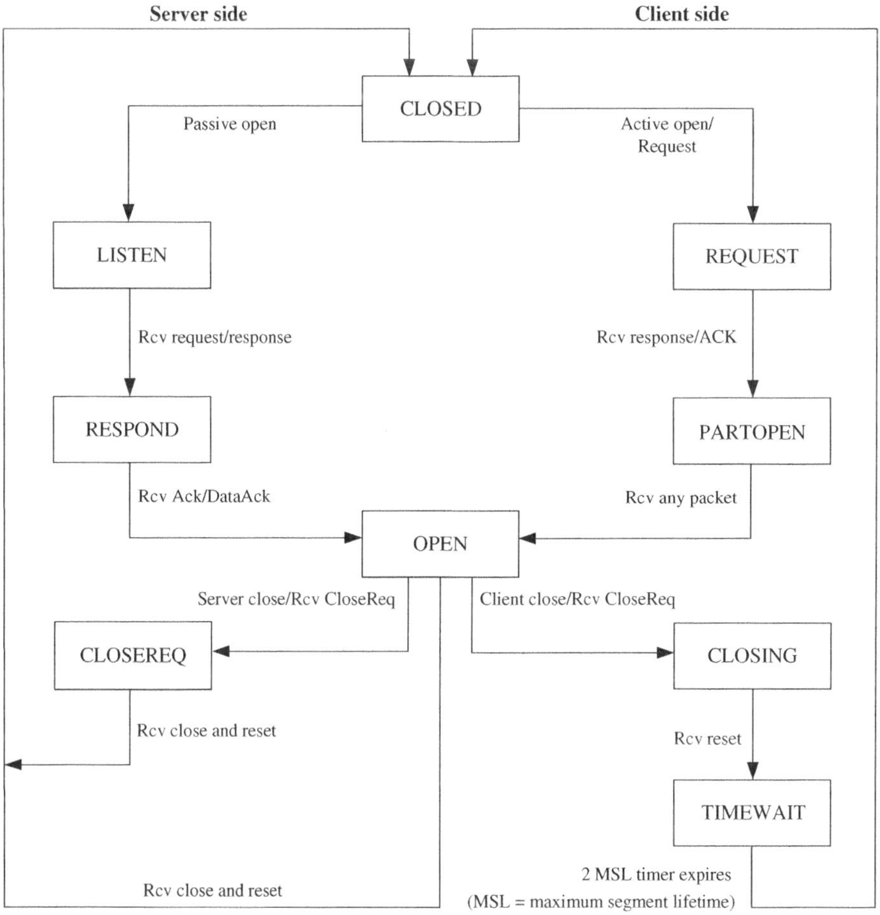

Figure 9.13 DCCP State Diagram.

a throughput equation that gives the acceptable transmit rate. The sender then adjusts its transmit rate to match the calculated rate. In case of no feedback for several RTTs, the sender halves its rate.

9.4 Summary

This chapter has discussed two transport layer protocols that were designed to meet the needs of new applications whose behaviors are contrary to the principles on which TCP and UDP are based. These are the SCTP and the DCCP. Special features of SCTP include multihoming that enables multiple

Table 9.2 Comparison of TCP, UDP, SCTP, and DCCP

Services/features	TCP	UDP	SCTP	DCCP
Connection-oriented	Yes	No	Yes	Yes
Ordered data delivery	Yes	No	Yes	No
Unordered data delivery	No	Yes	Yes	Yes
Reliable data transfer	Yes	No	Yes	No
Congestion control	Yes	No	Yes	Yes
Flow control	Yes	No	Yes	Optional
Byte-oriented	Yes	No	Yes	No
Message-oriented	No	Yes	Yes	Yes
Multistreaming	No	No	Yes	No
Uses selective acknowledgments	Optional	No	Yes	Yes
Multihoming	No	No	Yes	No
Protection against SYN flood attack	No	No	Yes	No
Allows half-closed connections	Yes	No	No	No

IP addresses to be used on one endpoint, and multistreaming that allows multiple parallel streams to exist on one endpoint. Multistreaming solves the problem of the head-of-the-line blocking associated with TCP. Also, SCTP assigns resources in a way that prevents the DoS that can occur in TCP from occurring. Finally, SCTP allows graceful shutdown that does not allow half-open connections to exist, as in TCP.

DCCP is designed for applications that prefer the simplicity of UDP to the congestion-oriented and strict-order-of-delivery features of TCP. These applications do not fit the profile of applications that traditionally use UDP because they are usually long-lived applications with timing constraints on the delivery of data. For these applications, which include streaming media, multiplayer online games, and VoIP, old messages quickly become stale so that getting new messages is preferred to resending lost messages. Table 9.2 is a comparison of the major features of TCP, UDP, SCTP, and DCCP.

Exercises

1 What does "multihoming" mean with respect to SCTP?

2 What does "multistreaming" mean with respect to SCTP?

3 List the steps involved in the SCTP four-way handshake.

4 What does a half-open connection mean?

5 In which transport layer protocol does the half-open connection occur?

6 What does graceful shutdown mean?

7 In which transport layer protocol is the graceful shutdown used?

8 How does SCTP prevent the SYN flood attack?

9 Name one similarity between DCCP and UDP.

10 Name one similarity between DCCP and TCP.

11 What is a half-connection and where is it used?

12 Name two congestion management schemes used in DCCP.

10

Application Layer Services

10.1 Introduction

The application layer is the topmost layer of the protocol hierarchy. It is the layer where actual communication is initiated. It uses the services of the transport layer, the network layer, the data link layer, and the physical layer to transfer data to a remote host. Two remote application processes can be in one of the two different modes:

(a) *Peer-to-peer mode*, which means that the two processes are executing at the same level and they exchange data using some shared resource.
(b) *Client–server mode*, which means that one process acts as a *client* that is requesting some resource from the other process that is acting as the *server* that is providing the service.

Many protocols have been defined for the application and they include the following:

(a) Dynamic host configuration protocol (DHCP) is a protocol that automatically provides an IP address to a host. It also provides all related configuration information, such as subnet mask and default gateway, to the host. It is a client–server protocol that has the port number 546 for the client and 547 for the server.
(b) Domain name system (DNS) is used to resolve human-readable host names into IP addresses. In particular, an e-mail address such as xyz@uml.edu cannot be routed as is; it needs to be mapped to an IP address, and this is the role of the DNS. The port number for DNS is 53.
(c) File transfer protocol (FTP) is a standard network protocol used to transfer computer files between two hosts on a computer network. FTP is built on a client–server model architecture and uses separate control and data connections between the client and the server. The port number for data is 20, and the port number for control is 21.
(d) Hypertext transfer protocol (HTTP) is a protocol for distributed, collaborative, hypermedia information systems. It operates in a client–server

Fundamentals of Data Communication Networks, First Edition. Oliver C. Ibe.

mode, and it is the underlying protocol used by the World Wide Web. It defines how messages are formatted and transmitted, and what actions Web servers and browsers should take in response to various commands. Its port number is 80.

(e) Simple mail transfer protocol (SMTP) is a protocol used to transfer an electronic mail from one user to another. It is client–server type of protocol with the port number 25.

(f) Simple network management protocol (SNMP) is a protocol that is used for collecting information from, and configuring, network devices such as servers, printers, hubs, switches, and routers in an IP network. It requires a *management station*, which is simply the software that collects information from the network. Generally, a management station regularly polls the network for information. Management stations range from the very simple to highly complex. Also, the hardware or software that is to be monitored must have an *agent* running. The agent collects information and then sends it to the management station when polled. An agent can also send notification to the management station without being polled, for example, if an error is detected. The port number is 161.

In the remainder of this chapter, we discuss two of the application layer protocols in greater detail. These application layer protocols are as follows:

a. DHCP
b. DNS.

10.2 Dynamic Host Configuration Protocol

IP address management requires a lot of work. Prior to the advent of the DHCP, in a typical network the network administrator performs the following functions:

- Manually assigns a *unique* IP address to each host
- Assigns a host to a subnet by defining its default gateway
- When the host moves to a new location, the network configuration must be updated.

Since moves, adds, and changes are a constant feature in a typical enterprise network, IP network configuration can be time-consuming. DHCP was developed to solve the IP address management problem.

10.2.1 DHCP Basics

DHCP is defined in RFC 1541 and updated in RFC 2131. It was designed to provide a centralized approach to configuring and maintaining IP addresses.

It allows the dynamic allocation of reusable IP addresses to hosts when they want to connect to the network. It is a client–server system in which the *DHCP client* is an Internet host that uses DHCP to obtain configuration parameters including an IP address and a *DHCP server* is an Internet host that returns configuration parameters to DHCP clients.

When a DHCP client boots up, it broadcasts a DHCP request asking any DHCP servers in the network to provide it with an IP address and other configuration parameters (such as the default gateway and subnet mask). Any authorized DHCP server that receives the broadcast will send an IP address and other configuration parameters as well as the duration of the IP address lease time to the client. From the responses to its broadcasts that it receives, the client chooses a DHCP server and sends an ACK that includes the IP address and configuration parameters; this ACK is sent as a broadcast. Four basic steps are involved in obtaining an IP address:

- Discovery phase
- Offer phase
- Request phase
- Acknowledgment phase.

These phases are often abbreviated as DORA (Discovery, Offer, Request, Acknowledgment).

10.2.2 Discovery Phase

In this phase, the DHCP client is booting up and needs an IP address to connect to the network. It broadcasts a packet containing the DHCPDISCOVER message to the IP address 255.255.255.255 over the UDP port 67, which is the *DHCP server port*. The DHCPDISCOVER message includes the client's MAC address and may include optional parameters that can help the server to determine if it can meet the need.

10.2.3 Offer Phase

In this phase, any DHCP server that receives the DHCPDISCOVER message and can meet the parameter values checks its pool of IP addresses for an available IP address. If the server finds an address, it returns a DHCPOFFER message that contains a valid IP address via UDP port 68, the DHCP client port. The response is sent as a broadcast since the client does not yet have an IP address.

10.2.4 Request Phase

In this phase, the client collects all offers from the responding DHCP servers and selects the most desirable configuration. The client then responds to that offer by broadcasting a DHCPREQUEST message to the server to accept the address. Since it is a broadcast message that is received by all DHCP servers that responded to the request, any server whose offer was rejected will return the offered address to its pool of available IP addresses.

10.2.5 Acknowledgment Phase

In this phase, the DHCP server whose IP address was selected will commit the configuration by responding to the DHCPREQUEST message with a DHCPACK message that contains additional information that the client might have requested. The message is sent as a broadcast and contains the IP address lease time. This phase ends the configuration phase. The client is now able to apply the configuration parameters to its network interface and be able to send and receive unicast data.

10.2.6 Example of Configuration Process Timeline

In this example, there are three DHCP servers: S1, S2, and S3. The client selects the offer from S2. The sequence of events is shown in Figure 10.1.

10.2.7 Address Lease Time

A DHCP server leases IP addresses to clients. The duration of a lease, or the *lease time*, is specified in the DHCPACK message. Before the expiration of the lease time, a DHCP client may request an extension by sending a DHCPREQUEST message to the server. DHCP is a client–server protocol, which means that the server cannot initiate a session with a client. Thus, if a server does not receive a request for a lease extension, it will withdraw the IP address at the end of the lease time. A DHCP client may also relinquish an IP address prior to

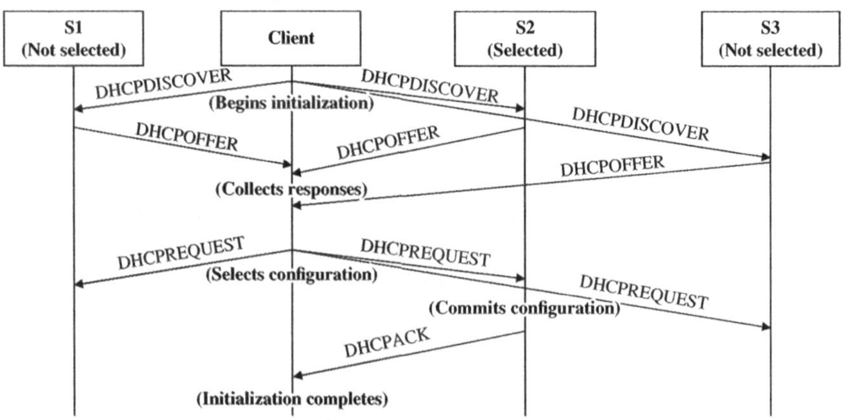

Figure 10.1 Example of the Configuration Process.

the expiration of the lease time by sending a DHCPRELEASE message to the DHCP server that owns the IP address.

10.2.8 Static Addresses

DHCP allows the network administrator to partition the IP address space into a portion that can be dynamically assigned and a portion that can be statically assigned. Static assignment is required in applications where network servers need fixed IP addresses. Web servers, mail servers, and print servers are examples of network devices that must be configured with static IP addresses so the network administrator does not waste time tracking changes. It may be helpful to think of static IP addresses as dynamically assigned IP addresses with very long (ideally infinite) lease times.

10.3 Domain Name System

Typing the information https://www.uml.edu/Academics/colleges.aspx brings you to a website at the University of Massachusetts Lowell with the title "Colleges & Schools." The information is called a universal resource locator (URL), which is a reference to the address of a resource on the Internet. A URL has two main components:

- *Protocol identifier*: For the URL https://www.uml.edu/Academics/colleges.aspx, the protocol identifier is *https*.
- *Resource name*: For the URL https://www.uml.edu/Academics/colleges.aspx, the resource name is "www.uml.edu/Academics/colleges.aspx."

The protocol identifier and the resource name are separated by a colon and two forward slashes. The protocol identifier indicates the name of the protocol to be used to access the resource. Examples of protocols that can be used to access resources on the Internet include the HTTP, HTTP Secure (HTTPS), FTP, Telnet, and trivial file transfer protocol (TFTP).

The prefix "www" of the resource name stands for World Wide Web, which is called the *subdomain*. This portion is not required because typing https://uml.edu/Academics/colleges.aspx still gets you to the website. The next part "uml.edu" is the "domain name" for the website. The last portion of the domain name (i.e., the ".edu") is called the top-level domain (TLD) and is used to identify the type of the website. We often see domain names such as "abc.com" and "xyz.org." The .edu, .com, and .org are examples of what is called "generic top-level domain" (gTLD). For completeness, the part "Academics" is a directory in the website, and "colleges.aspx" is a web page.

The purpose of the preceding discussion is to illustrate the importance of the topic at issue: DNS. When a computer user types a web address directly into the field at the top of their browser window, it initiates a process of locating the page requested. To locate the page, we must know the IP address of the server in which the resource is located. As we have seen in previous chapters, an IP address is a set of numbers that determines the exact location of the device with that IP address. Unfortunately, we cannot memorize the IP address of every resource that we would like to access in the Internet. But the URL is much easier to remember. The DNS enables us to determine the IP address of the domain name part of the URL.

10.3.1 Structure of the DNS

The DNS is a distributed, hierarchical database where authority flows from the top (or root) of the hierarchy downward. It allows hostnames to be mapped to IP addresses, exactly in the same way as a telephone directory maps people's name to phone numbers. When a user wants to send an e-mail to a person with the address abc@uml.edu, the e-mail address must be resolved (or mapped) to an IP address because servers and terminals in a network are located via their IP addresses. Similarly, a website such as www.uml.edu must be resolved to the IP address of the server that is hosting the website. Unfortunately, the IP address of every system in even a small network is not likely to be known. More importantly, since IP addresses are leased and are, therefore, likely to change with time, knowing the IP address at one time may not help the user to easily locate the intended system in the network. The *DNS server* is responsible for this hostname to IP address mapping.

DNS uses a tree (or hierarchical) name structure. At the top of the tree is the root followed by the top level domains (TLDs). These are followed by the domain name and any number of lower levels each separated with a dot. TLDs are classified into two types:

1. gTLD
2. Country code top level domain (ccTLD)

The gTLDs are administered by the Internet Corporation for Assigned Numbers and Names (ICANN) and delegated to a series of accredited registrars. The following are some of the more common gTLDs:

- .com, which is intended for commercial entities; for example, *ibm.com*
- .edu, which is intended for higher-education institutions, such as 4-year colleges and universities; for example, *uml.edu* for the University of Massachusetts Lowell
- .gov, which is intended for use by agencies of the US Federal Government; for example, *uspto.gov* for the United States Patent and Trademark Office
- .mil, which is intended for use by agencies of the US military; for example, *af.mil* for the US Air Force

- .net, which is intended for use by network providers and organizations dedicated to the Internet, such as Internet service providers (ISP); for example, *verizon.net* for the service provider Verizon
- .org, which is intended for nonprofit or noncommercial establishments, such as professional groups, charities, and other such organizations; for example, *redcross.org* for the American Red Cross.

The ccTLDs use a standard two-letter sequence defined by ISO 3166. Some countries with federal governments, such as Canada (ccTLD.ca) and the United States (ccTLD.us), administer their ccTLDs at the national level and delegate to each province or state a two-character province/state code; for example, .ma = Massachusetts and .ny = New York.

Countries such as the United Kingdom, Brazil, and Spain with more centralized governments have a functional segmentation in their delegation models. For example, .co = company, .ac = academic, and so on. Thus, *abc.co.uk* is the domain name of *abc* registered as a company from the UK registration authority. Figure 10.2 illustrates the DNS structure.

The names used in the DNS are called *domain names* and they have a particular structure. Specifically, they are divided in two parts: the *first level domain*

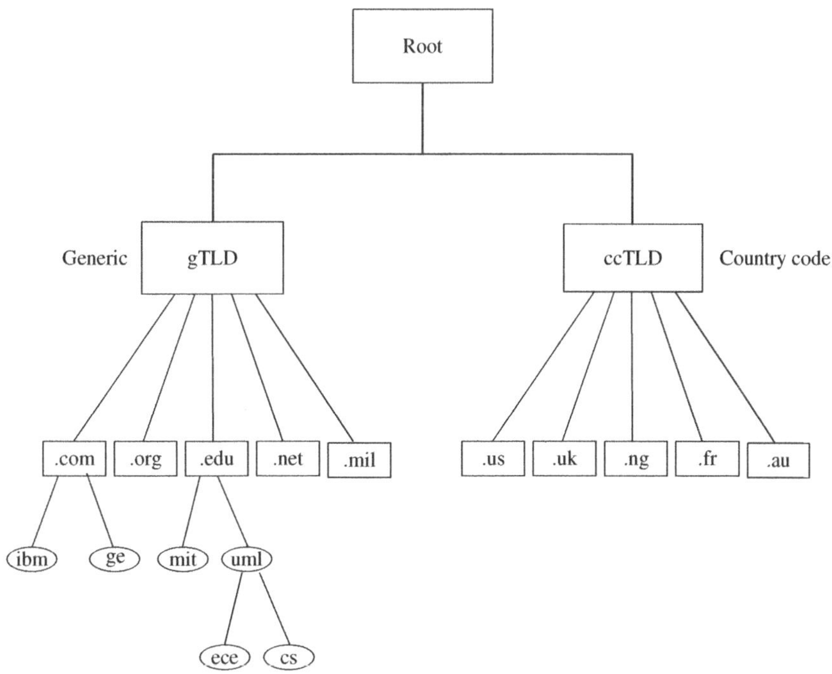

Figure 10.2 Structure of the DNS.

and the *second level domain*. For example, in uml.edu the first level domain is ".edu" and the second level domain is "uml."

As we have discussed, the DNS is used to translate a hostname to an IP address. The host's name must be a *fully qualified domain name* (FQDN), which means that the name must list the host's precise location in the DNS hierarchy. An example of a fully qualified name is the following:

solutionece43.ece.uml.edu

This DNS name represents the host solutionece43 in the subdomain *ece*, which is a department in the subdomain *uml* that is the name of the organization that has registered the domain is in the TLD.*edu*. Thus, an FQDN is a domain name that includes all domains between the host and the root of DNS.

10.3.2 DNS Queries

A *DNS resolver* is any system that has been configured with one or more DNS server IP addresses and that performs queries against these DNS servers. There are two types of queries that a resolver can request from a DNS server:

(a) A recursive query
(b) An iterative query.

Before we discuss the two systems, we define the concept of *DNS forwarding*. DNS forwarding is the process by which a DNS query that cannot be handled locally is forwarded to another DNS server outside the network that can resolve the query. Thus, a *DNS forwarder* sends name queries that cannot be resolved within its local networks to a remote DNS server outside of its local network for resolution.

A recursive query is one in which the DNS client requires the DNS server to respond to the client with either the requested resource record or an error message that indicates that the record does not exist. In this scheme, if the DNS server is not able to resolve the requested query, then it forwards the query to another DNS server. If the second DNS server cannot resolve the query, the original servers keeps trying other DNS servers until it either gets an answer or the query fails because no DNS server can resolve it. Thus, the DNS will do all it can to provide a resolution to the query. For this to happen, the DNS server must be a DNS forwarder.

In an iterative query, recursion is disabled. This means that the query cannot be forwarded to another DNS server if the DNS server to which the DNS client sent the query cannot resolve it. Thus, when a DNS client sends a query to a DNS server, the latter will return an answer if it can resolve the query. If the DNS server cannot resolve the query, it will refer the client to another DNS server that might be able to resolve the query.

Thus, unlike the recursive query where the DNS server returns an answer to the query after possibly using the services of other DNS servers, the iterative query requires the DNS server to either answer the query itself or refer the DNS client to another DNS server that can resolve the query. This means that in iterative query the answer to the query must come from the DNS server that provides the answer, while in the recursive query it must come from the DNS server to which the query was first sent. Alternatively, in the recursive query, how the query is resolved is transparent to the DNS client but in the iterative query it is not.

10.3.3 Name-to-Address Resolution Process

Domain names are used to identify one or more IP addresses. For example, the domain name *uml.edu* represents several IP addresses. Thus, a domain name can represent a personal computer used to access the Internet or a server computer hosting a website that is connected to the Internet.

Consider a user who enters the domain name www.uml.edu. The process of converting the domain name into an IP address takes numerous queries as follows:

(a) The computer sends a request to the DNS server of the ISP for the user's network. This request essentially says "what is the IP address of the server that is hosting uml.edu?"
(b) If the ISP's DNS server has cached this address, it will return it to the computer; otherwise, since the TLD is *.edu*, it will send a request to the DNS root server for the IP address of the *.edu* domain server. In this example, we assume that the IP address is not cached, which means that ISP's DNS server will send a request to a DNS root server for the IP address of the *.edu* domain.
(c) Upon receiving a response about the IP address of the *.edu* domain from the DNS root server, the ISP's DNS server will request the IP address of the authoritative DNS server for the uml.edu domain.
(d) Upon receiving this information from the DNS server for the *.edu* domain, the ISP's DNS server sends a request to the DNS server for the uml.edu domain requesting the IP address of the server that is hosting the uml.edu website.
(e) Upon receiving a response from the DNS server for the *.edu* domain, the ISP's DNS server returns the answer to the user's computer, which then uses the IP address to connect to the server hosting the *uml.edu* website.

The above sequence of activities is illustrated in Figure 10.3.

If the uml.edu Web address was previously requested by one of the ISP's customers, its IP address is cached in the DNS server and the server returns the IP address immediately thereby eliminating the long process shown earlier. Also, a user's computer typically caches IP addresses, which can eliminate the DNS query altogether. This is illustrated in Figure 10.4.

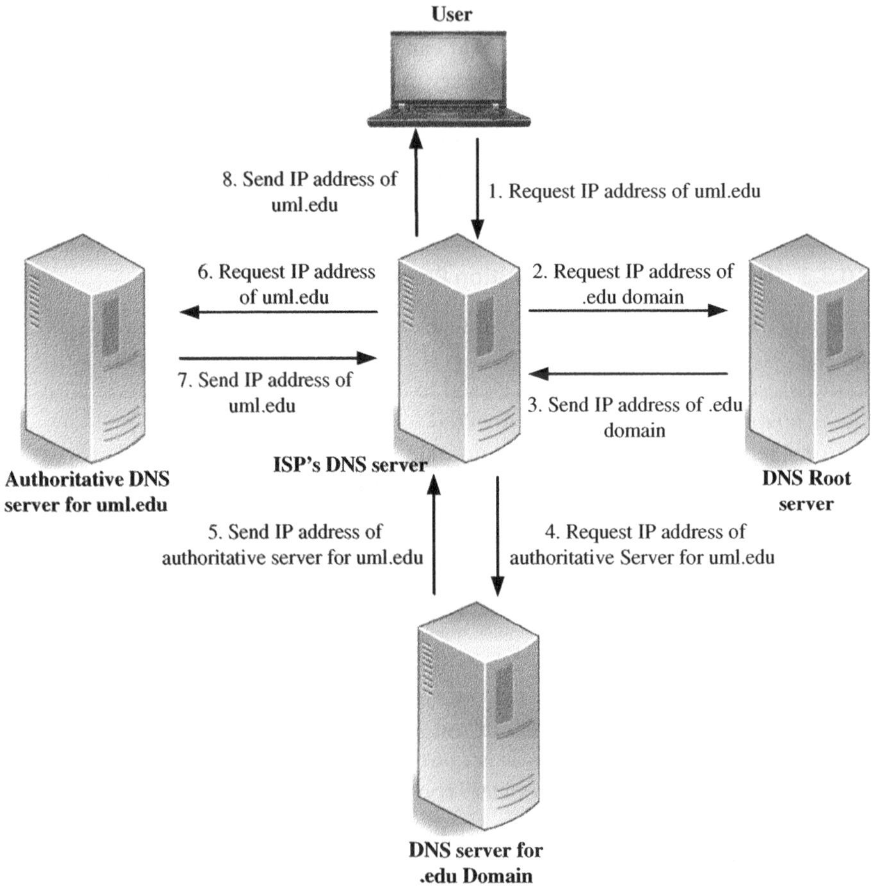

Figure 10.3 Name-to-IP Address Resolution Process.

10.3.4 DNS Zones

A DNS zone is the contiguous portion of the DNS domain name space over which a DNS server has authority. A DNS zone can contain one or more contiguous domains. A DNS server can be authoritative for multiple DNS zones. A noncontiguous namespace cannot be a DNS zone.

There are different types of zones and they include the following:

- Primary zone
- Secondary zone.

A *primary zone* is the original source of the data for all domains in the zone. Thus, it is the only zone type that can be edited or updated. Updates made to the primary zone are made by the DNS server that is authoritative for the

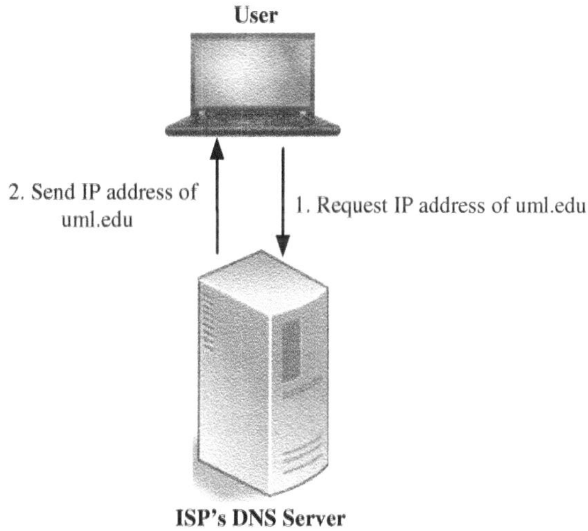

Figure 10.4 Name-to-IP Address Resolution When IP Address is Cached.

specific primary zone. Users can also back up data from a primary zone to a secondary zone.

A *secondary zone* is a read-only copy of the zone that was copied from the master server during zone transfer. In fact, a secondary zone can only be updated through zone transfer.

10.3.5 DNS Zone Updates

DNS allows changes to be propagated by means of zone transfer. Thus, a zone transfer is the process that occurs when the zone's resource records (RRs) on the primary DNS server are copied to secondary DNS servers. It enables a secondary DNS server to continue handling queries if the primary DNS server fails. There are two basic zone transfer methods: full zone transfer (AXFR) and incremental zone transfer (IXFR). Another process that is associated with transfer is Notify.

10.3.5.1 Full Zone Transfer

When the user configures a secondary DNS server for a zone and starts the secondary DNS server, the secondary DNS server requests a full copy of the zone from the primary DNS server. A full transfer of all the zone information is performed using TCP on port 53. AXFRs tend to be resource intensive. This disadvantage of full transfers has led to the development of IXFRs.

10.3.5.2 Incremental Zone Transfer

In an IXFR, only those RRs that have since changed in a zone are transferred to the secondary DNS servers. Every period called the refresh period the secondary DNS server sends a query to the primary DNS server to see if there is a change in the DNS database. The primary DNS server compares the data in its database with that in the secondary DNS server's database. If the primary and secondary DNS servers' databases are the same, zone transfer does not take place. If the DNS data of the two servers are different, transfer of the delta RRs starts. For IXFR to occur, the primary DNS server has to record incremental changes to its DNS database. IXFRs require less bandwidth than AXFRs.

10.3.5.3 Notify

DNS Notify is a mechanism that enables a primary DNS server to inform secondary DNS servers when its database has been updated. Fundamentally, it informs the secondary DNS servers when they need to initiate a zone transfer request so that the updates of the primary DNS server can be replicated to them. When a secondary DNS server receives the notification from the primary DNS server, it can start an IXFR or an AXFR to pull zone changes from the primary DNS servers.

10.3.6 Dynamic Update

Dynamic update is a standard that provides a means of dynamically updating zone data on a zone's primary server. Originally, DNS was designed to support only static changes to a zone database. Because of the design limitations of static DNS, the ability to add, remove, or modify RRs could only be performed manually by a DNS system administrator. For example, a DNS system administrator would edit records on a zone's primary server and the revised zone database is then propagated to secondary servers during zone transfer. This design is workable when the number of changes is small and updates occur infrequently but can otherwise become unmanageable.

With dynamic update, on the other hand, the primary server for the zone can also be configured to support updates that are initiated by another computer or device that supports dynamic update. For example, it can receive updates from DHCP servers. Updates are sent using a standard UPDATE message format and can include the addition or deletion of individual RRs or sets of resource records (RRsets).

In order for a request for a dynamic update to be performed, several prerequisite conditions can also be identified. Where prerequisites are set, all such conditions must be met before an update is allowed. Some examples of prerequisites that can be set are as follows:

(a) A required RR or RRset already exists or is in use prior to an update.
(b) A required RR or RRset does not exist or is not in use prior to an update.

A requester is permitted to initiate an update of a specified RR or RRset. Each prerequisite must be satisfied in order for an update to occur. After all prerequisites are met, the zone's primary server can then proceed with an update of its

Table 10.1 List of Root Servers

Hostname	IP Addresses	Manager
a.root-servers.net	198.41.0.4, 2001:503:ba3e::2:30	VeriSign, Inc.
b.root-servers.net	192.228.79.201, 2001:500:84::b	University of Southern California (ISI)
c.root-servers.net	192.33.4.12, 2001:500:2::c	Cogent Communications
d.root-servers.net	199.7.91.13, 2001:500:2d::d	University of Maryland
e.root-servers.net	192.203.230.10, 2001:500:a8::e	NASA (Ames Research Center)
f.root-servers.net	192.5.5.241, 2001:500:2f::f	Internet Systems Consortium, Inc.
g.root-servers.net	192.112.36.4, 2001:500:12::d0d	US Department of Defense (NIC)
h.root-servers.net	198.97.190.53, 2001:500:1::53	US Army (Research Lab)
i.root-servers.net	192.36.148.17, 2001:7fe::53	Netnod
j.root-servers.net	192.58.128.30, 2001:503:c27::2:30	VeriSign, Inc.
k.root-servers.net	193.0.14.129, 2001:7fd::1	RIPE NCC
l.root-servers.net	199.7.83.42, 2001:500:9f::42	ICANN
m.root-servers.net	202.12.27.33, 2001:dc3::35	WIDE Project

local zones. Multiple updates can be processed concurrently only if one update does not depend on the final result of another update.

10.3.7 Root Servers

Root servers are the authoritative name servers that serve the DNS root zone. These servers form a network of hundreds of servers in many countries around the world. They are configured in the DNS root zone as 13 named authorities, as shown in Table 10.1.

While there are 13 designated DNS root server names, none of these names represents a single computer but rather a server cluster consisting of many computers. This was done to increase the reliability of DNS without any negative effect on its performance.

10.4 Summary

The application layer is the topmost layer of the OSI protocol hierarchy and it is where actual communication is initiated. This chapter has briefly discussed some of the application layer protocols. These include the DHCP, the DNS, the FTP, the HTTP, the SMTP, and the SNMP. A more detailed discussion is made for DHCP and DNS.

Exercises

1 What is the dynamic host configuration protocol (DHCP) used for?

2 Name the four steps used in DHCP.

3 Which network element is responsible for initiating a DHCP request?

4 Which network element is responsible for responding to a DHCP request?

5 What is the domain name system (DNS) used for?

6 What is a fully qualified DNS domain name?

7 Give two types of top-level DNS domains and give an example of each type.

8 What is a DNS zone?

9 Name the two types of DNS zone transfers

10 What is DNS update?

11

Introduction to Mobile Communication Networks

11.1 Introduction

By all accounts, wireless communication was introduced with the invention of the wireless telegraph in 1896 by Marconi. He was able to transmit encoded alphanumeric characters as telegraphic signals across the Atlantic Ocean. Since then, there has been a rapid growth in wireless communication as can be seen in the development of radio, television, and satellite communications. More recently, the development of different broadband wireless technologies has led to higher data rates than were previously available.

In this chapter, we consider the architecture of modern wireless communication networks; specifically, the cellular wireless communication network. The specific topics to be discussed include the following:

1. Introduction to radio communication
2. Cellular network architecture
3. 2G Networks
4. 3G Networks
5. 4G Networks
6. 5G Networks.

11.2 Radio Communication Basics

The electromagnetic (EM) spectrum is the fundamental resource of a wireless communication system. The EM spectrum is divided into several frequency bands that are used for specific applications. The frequency band assigned to an application determines the propagation characteristics of the channel, including the attenuation and other impairments.

The radio spectrum (or radio waves) is a part of the EM spectrum that extends from below the low frequencies used for modern radio communication to gamma radiation at the short wavelength (high-frequency) end. Thus,

Fundamentals of Data Communication Networks, First Edition. Oliver C. Ibe.

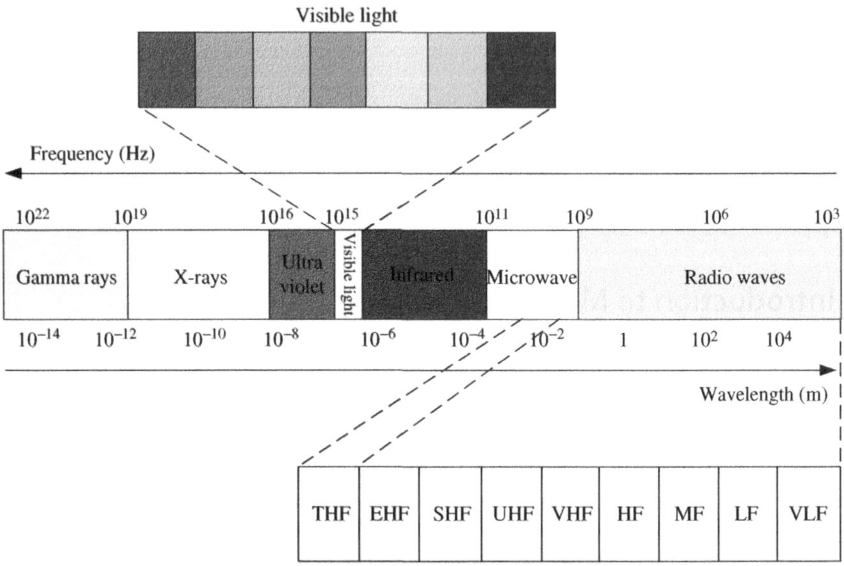

Figure 11.1 Expanded Electromagnetic Spectrum.

the EM spectrum covers wavelengths from thousands of kilometers down to a fraction of the size of an atom. The visible light lies toward the shorter end, with wavelengths from 400 to 700 nm. The ultraviolet, visible light, and infrared regions constitute the *optical spectrum*. Figure 11.1 is an expanded version of Figure 2.43; it includes the different radio frequency (RF) bands.

Table 11.1 is a summary of the different RF bands along with the typical applications for which they are used. It is an expanded version of Table 2.2.

11.3 Model of Radio Communication System

The components of a radio communication system include the following:

(a) *Transmitter*, which converts information to be transmitted into RF signal
(b) *Transmitting antenna*, which converts the RF signal into an EM wave
(c) *Transmitting medium*, which is free space
(d) *Receiving antenna*, which intercepts the EM wave and converts it back to RF signal
(e) *Receiver*, which converts the received RF signal back to a form that the intended recipient can use.

These components are illustrated in Figure 11.2.

An RF signal is characterized by its frequency f and its wavelength λ. The parameters f and λ are related by $f\lambda = v$, where v is the velocity of the

Table 11.1 The Radio-Frequency Spectrum with Applications.

	Band name	Frequency range	Wavelength range	Examples of use
1	Very low frequency (VLF)	3–30 kHz	100–10 km	Navigation, time signals, and submarine communication
2	Low frequency (LF)	30–300 kHz	10–1 km	Navigation clock time signals, RFID, and amateur radio
3	Medium frequency (MF)	0.3–3 MHz	1 km–100 m	AM (medium-wave) broadcasts
4	High frequency (HF)	3–30 MHz	100–10 m	Shortwave broadcasts, citizens' band radio, amateur radio, and RFID
5	Very high frequency (VHF)	30–300 MHz	10–1 m	FM, TV broadcasts, land mobile and maritime mobile communications, and weather radio
6	Ultrahigh frequency (UHF)	0.3–3 GHz	1 m–10 cm	Television broadcasts, microwave oven, mobile phones, wireless LAN, Bluetooth, ZigBee, and GPS
7	Super high frequency (SHF)	3–30 GHz	10–1 cm	Radio astronomy, microwave devices/communications, and wireless LAN
8	Extra high frequency (EHF)	30–300 GHz	1 cm–1 mm	Radio astronomy, high-frequency microwave radio relay, microwave remote sensing, and amateur radio
9	Tremendously high frequency (THF)	0.3–3 THz	1–0.1 mm	No applications have yet been identified

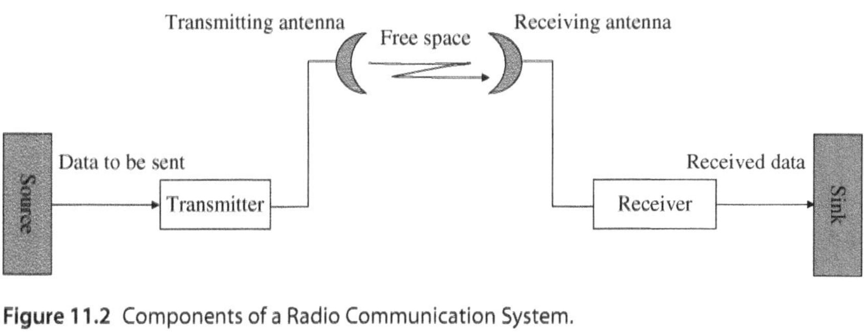

Figure 11.2 Components of a Radio Communication System.

wave in the medium through which it propagates. For free space (or air), $v = c = 3 \times 10^8$ m/s. Frequency is measured in Hertz (Hz) and wavelength in meters.

11.4 Radio Wave Propagation

Radio propagation is defined as the way radio waves travel or propagate when they are transmitted from one point to another. Radio waves are affected by the medium in which they travel. There are four fundamental ways that radio waves are propagated. These are as follows:

(a) Free-space propagation
(b) Reflection
(c) Diffraction
(d) Scattering.

11.4.1 Free-Space Propagation

Free-space propagation deals with how radio waves travel in free space without the influence of any objects along the propagation trajectory. The only thing that affects the signal strength is the distance from the source. Under this propagation scheme, the RF waves leaving a transmitting antenna reach the receiving antenna in one of the many ways:

(a) *Ground waves* travel along the ground and follow the terrain. They are used mostly for communication in the very low frequency (VLF) and low frequency (LF) bands. When signals travel via the ground wave, they are modified by the ground or terrain over which they travel. As the distance between the transmitter and the receiver increases, the radio signal strength decreases. The medium over which the wave travels has significant resistivity, which means that when the radio wave comes in contact with the medium, the energy from the wave is converted into heat, leading to loss in signal strength. Thus, ground waves do not usually travel very far from the transmitter.
(b) *Sky waves* can travel up to 4000 km from transmitter and are used for communication in the HF bands. In this case, the radio signals are modified and influenced by a region high in the earth's atmosphere known as the *ionosphere*. The ionosphere is at a distance of 30–620 miles from the surface of the earth. It is ionized by solar radiation, which varies according to the time of the day and season of the year. Radio waves are reflected in the ionosphere, depending on their frequencies. The ionosphere is divided into three layers: D, E, and F. Layer D is the lowest layer and it absorbs high-frequency radio waves. Also, it exists mainly during the day and disappears at night. When layer D disappears, high-frequency radio waves

penetrate into the higher layers where they are reflected back to earth. This is why AM radio signals from distant radio stations can be received at night but not during the day. Thus, with sky waves, radio stations can be heard from the other side of the globe depending on many factors that include the radio frequencies used, the time of day, the season of the year, and the geographical location of the transmitter.

(c) *Direct waves* travel in a line-of-sight (LOS) manner from the transmitter to the receiver. LOS transmission is used when the transmitter is visible from the receiver and is the most common of the radio propagation modes at very high frequency (VHF) and higher frequencies.

Sometimes ground waves are reflected, causing the receiving antenna to receive both ground-reflected waves and direct waves; the combination of these two types of waves is called *space wave*. These propagation modes are illustrated in Figure 11.3.

11.4.2 Reflection

Reflection occurs when an EM wave falls on an object that has very large dimensions compared to the wavelength of the propagating wave. For example, such objects can be the earth, buildings, and walls.

If a radio wave is initially traveling in free space and comes into contact with an object that has different electrical properties from the free space in which

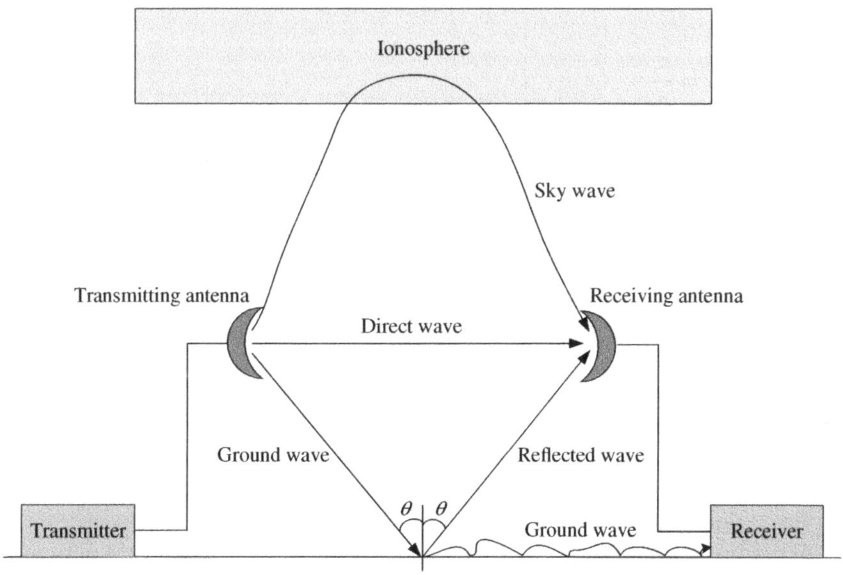

Figure 11.3 Free-Space Propagation Modes of Radio Waves.

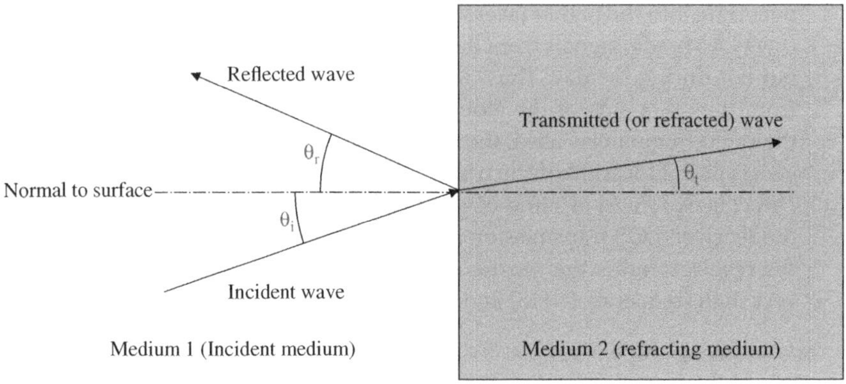

Figure 11.4 Wave Reflection and Refraction at a Plane Surface of Solid Medium.

it is traveling, such as a perfect dielectric, the wave is partially reflected and partially refracted (or transmitted through the object).

Consider Figure 11.4 where a wave is incident on a plane surface at an angle θ_i to the normal of the surface. Part of the wave is reflected at an angle θ_r to the normal and part of the wave is refracted at an angle θ_t to the normal. From the law of reflection, we know that the angle of incidence is equal to the angle of reflection; that is, $\theta_i = \theta_r$.

Let n_i be the refractive index of the incident medium and let n_t be the refractive index of the refracting medium. Then, according to Snell's law, we have that

$$\frac{\sin\theta_i}{\sin\theta_t} = \frac{n_t}{n_i}$$

11.4.3 Diffraction

Diffraction is a phenomenon whereby radio waves bend around obstacles that obstruct them. This phenomenon can be explained by the Huygens' principle, which states that every point on a wave front acts as point sources for the production of secondary wavelets, and they combine to produce a new wave front in the direction of propagation. The propagation of secondary wavelets in the shadowed region results in diffraction. The field in the shadow region is the vector sum of the electric field components of all the secondary wavelets that are received by the receiver.

As an illustration, assume that we have a uniform wave front along the A_1–A_2 line in Figure 11.5. It is natural to expect a zero field strength behind the obstacle in the region below the C_1–C_2 line.

According to the Huygens' principle, each point on a wave front acts as a source of a secondary wave front called *wavelet*. These wavelets do not radiate equally in all directions: For a direction that makes an angle ϕ with the

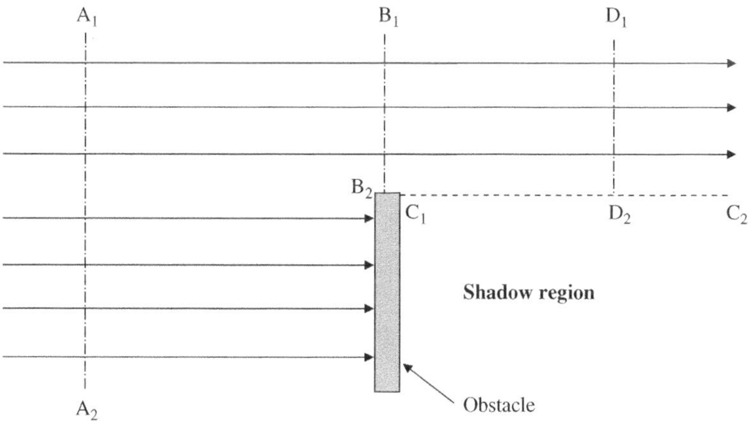

Figure 11.5 Illustration of an Obstacle that Creates Shadow Region.

direction of propagation, the amplitude of the field generated by a wave front is proportional to $(1 + \cos\phi)$. Thus, the amplitude in the reverse direction is zero since $\cos\pi = -1$; the amplitude is greatest in the direction of propagation since $\cos 0 = 1$. The Huygens' principle is illustrated in Figure 11.6.

11.4.4 Scattering

Scattering occurs when the medium through which the radio wave propagates contains a large number of objects that are smaller in size than the wavelength. The wave is scattered in all directions. Objects that can cause scattering include

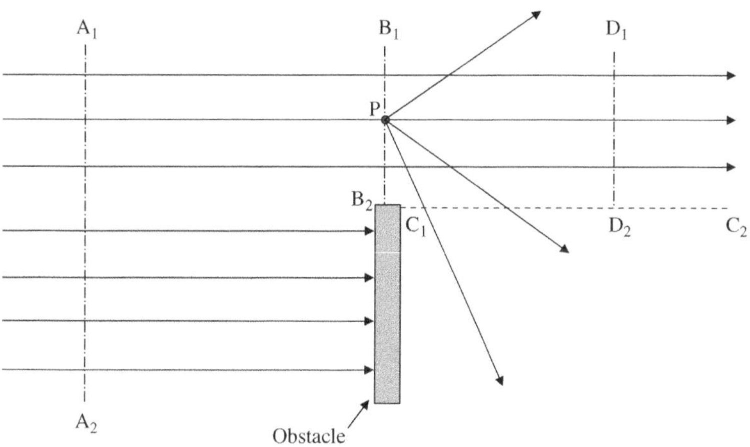

Figure 11.6 Illustration of the Huygens' Principle.

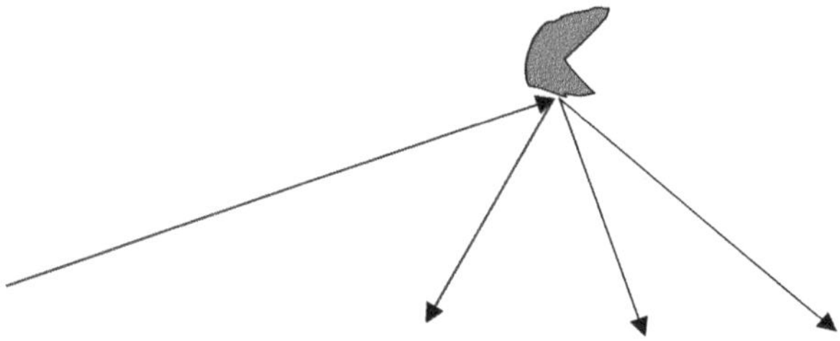

Figure 11.7 Illustration of Wave Scattering.

foliage, lamp posts, snowflakes, street signs, and so on. This is illustrated in Figure 11.7.

11.5 Multipath Fading

The different paths that a signal from a transmitter can be received at a device are illustrated in Figure 11.8. The figure is an illustration of the concept of *multipath*.

Multipath occurs when a radio signal is received directly by an antenna, and later the same signal is received again after it has been reflected, diffracted, or scattered from a building, a mountain, or any object. At some locations, the signals traveling by different paths may arrive in-phase and thus add up to make

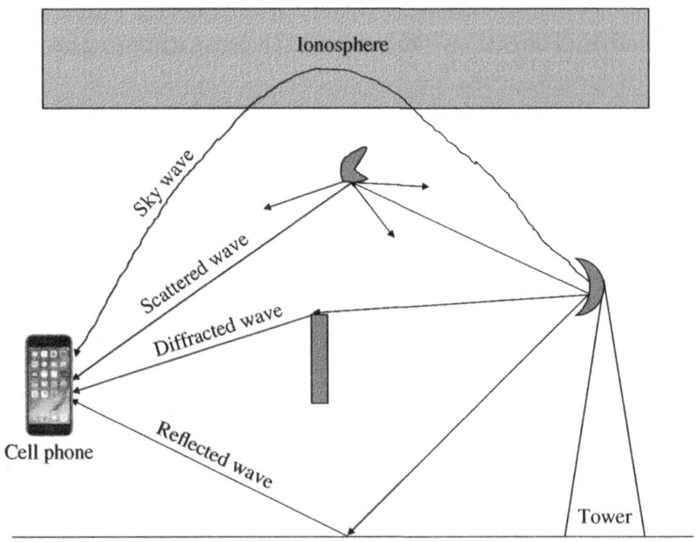

Figure 11.8 Different Paths from Transmitter to Receiver.

the signal stronger. At other locations, the signals may arrive out-of-phase and thus cancel one another, causing the signal to fade (or lose its strength). The effect whereby out-of-phase components combine at the receiver to cancel each other is referred to as *multipath fading.*

There are two types of multipath fading:

(a) In *Rician fading*, there is a strong, constant component to the signal that results from a direct unobstructed path between the two stations, in addition to the multiple reflected, diffracted, and/or scattered components. This occurs in LOS systems, as illustrated in Figure 11.9.
(b) In *Rayleigh fading*, there is no direct unobstructed component; all signals arrive after suffering one or more reflections, diffractions, and/or scattering. This is used in non-line-of-sight (NLOS) systems as illustrated in Figure 11.10.

Multipath poses particular problems for digital transmission systems. In the time domain, the receiver sees multiple copies of the signal with different time delays. The time difference between two paths often means that different symbols can overlap or smear into each other and create *intersymbol interference* (ISI).

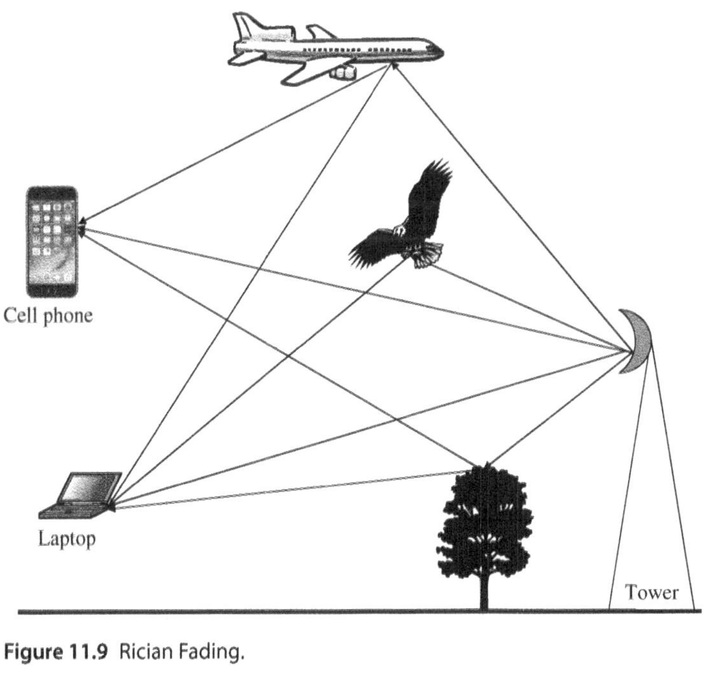

Figure 11.9 Rician Fading.

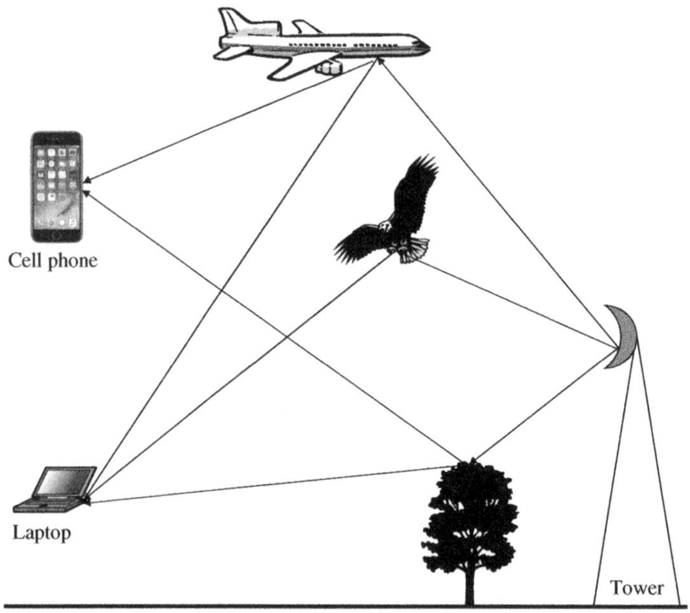

Figure 11.10 Rayleigh Fading.

11.6 Introduction to Cellular Communication

In television broadcasting, one very powerful transmitter is located at the highest spot in an area and broadcasts in a radius of up to 50 km. Mobile communication service operates in a slightly different way: It uses the cellular concept in which many low-power transmitters are placed throughout a coverage area instead of using one powerful transmitter. By dividing a service area into different areas called *cells*, each with a low-power transmitter, the service provider can increase the system capacity by many orders of magnitude.

11.6.1 Frequency Reuse

If all the transmitters transmit on the same channel, then there will be interference problems caused by mobile units using the same channel in adjacent areas. One of the solutions to the interference problem is to use different channels in different cells. Since the radio energy dissipates over distance, the signal strength from one transmitter can be reduced to zero after some distance from the transmitter. This means that the same channel can be reused in another cell whose transmitter is located at a distance that is so far from the current transmitter that the signals from that transmitter are not received at that location.

This practice is called *frequency reuse*. It means that the same channel is systematically used several times in a service area without causing interference in the different cells where the channel is used.

11.6.2 Cellular System Architecture

When it comes to partitioning a network into cells, the geometric feature that comes to mind is the circle. However, when we consider three circles, we find that there is an area that is not covered by any of the three. This means that a cell that is circular in design will leave some areas uncovered. Any attempt to close the gap results in at least three areas of overlap that will result in interference between at least one other cell. This is illustrated in Figure 11.11.

A cellular system is usually depicted as an area that is totally covered by radio, without any gaps or overlaps. This property of covering a surface with a pattern of flat shapes so that there are no overlaps or gaps is called *tessellation*. A *regular* tessellation is a pattern made by repeating one regular polygon. There are only three regular tessellations:

- Square
- Equilateral triangle
- Hexagon.

These polygons are shown in Figure 11.12.

The tessellation patterns of these polygons are shown in Figure 11.13.

In deciding which of the three polygons is the best choice there is one thing to consider: the area of the cell. The coverage area of each object is specified with respect to a circle that encloses these three polygons. Figure 11.14 illustrates

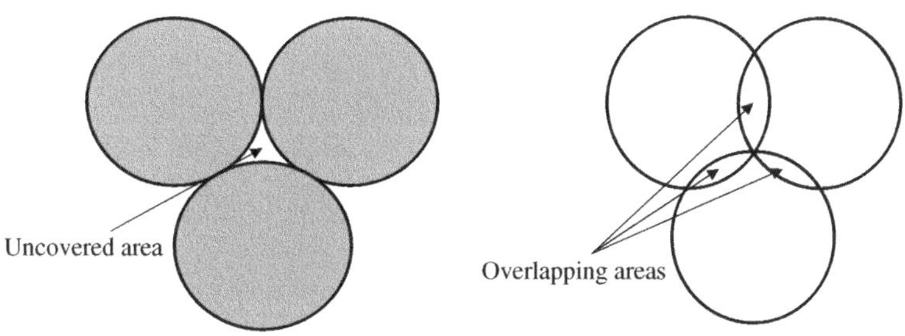

Figure 11.11 Illustration of Problems with Circular Cells.

Figure 11.12 The Regular Polygons for Regular Tessellations.

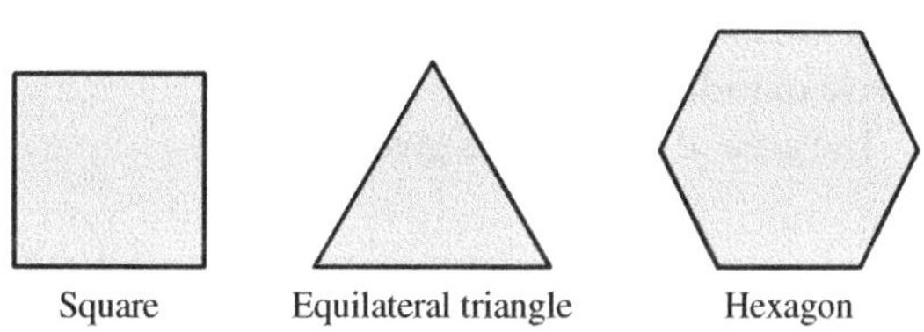

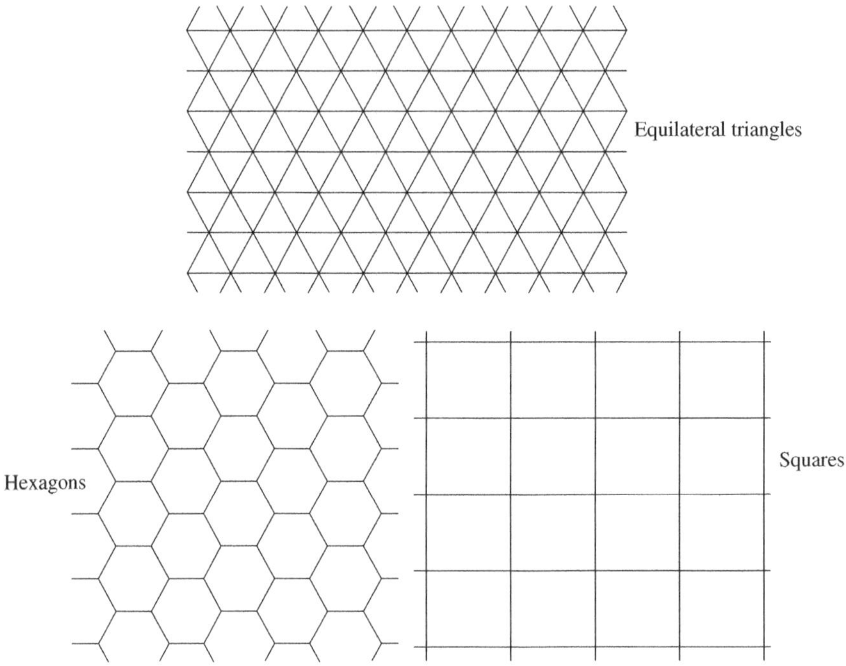

Figure 11.13 The Three Regular Tessellations.

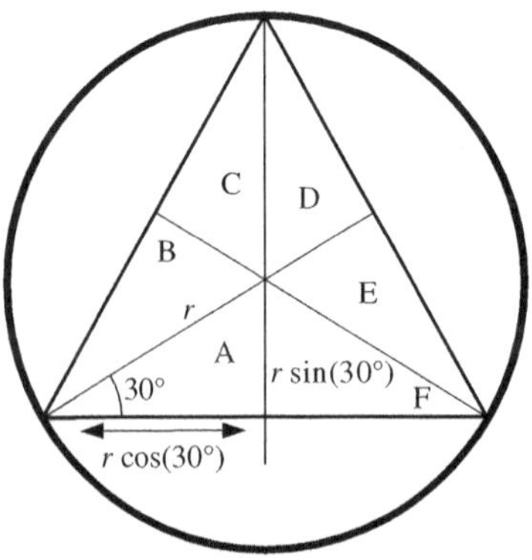

Figure 11.14 An Equilateral Triangle Subtended by Circle.

how we can compute the area of an equilateral triangle that is enclosed in a circle of radius r.

The area of region A is given by

$$A_{\mathrm{T}} = \frac{1}{2}\{r\cos(30^{\circ})\}\{r\sin(30^{\circ})\} = \frac{1}{2}\left(\frac{r\sqrt{3}}{2}\right)\left(\frac{r}{2}\right) = \frac{r^2\sqrt{3}}{8}$$

Since the six regions have the same area, we have that the ratio of the area of an equilateral triangle to the area of a circle of radius r (with area $A_C = \pi r^2$) that encloses the triangle is given by

$$\frac{A_T}{A_C} = \frac{6\sqrt{3}}{8\pi} = \frac{3\sqrt{3}}{4\pi} = 0.4135$$

For the square that is subtended by a circle of radius r as shown in Figure 11.15, we know that the diagonal of the square is the diameter of the circle.

Thus, if the length of each side of the square is a, then from the Pythagorean theorem we have that

$$2a^2 = (2r)^2 = 4r^2 \Rightarrow A_S = a^2 = 2r^2$$

Therefore, the ratio of the area of a square to the area of the circle that encloses it is given by

$$\frac{A_S}{A_C} = \frac{2r^2}{\pi r^2} = \frac{2}{\pi} = 0.6366$$

For the hexagon that is subtended by a circle of radius r as shown in Figure 11.16, we know that there are six equilateral triangles, as shown in the figure.

Figure 11.15 A Square Subtended by a Circle.

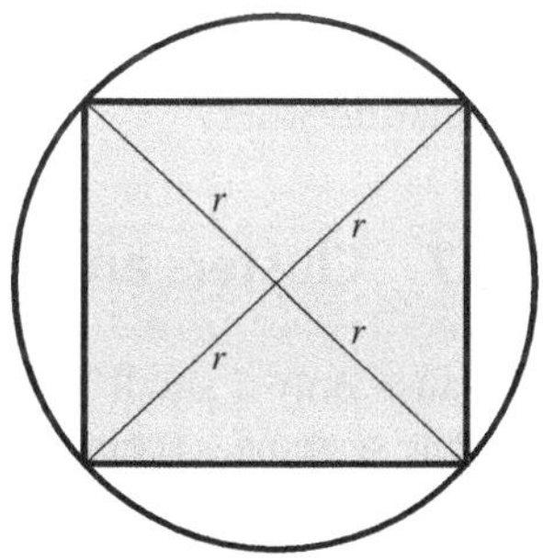

Figure 11.16 A Hexagon Subtended by a Circle.

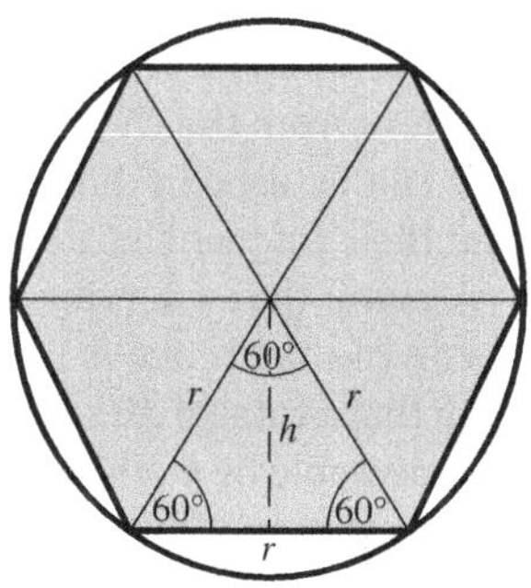

The height of each triangle is $h = r\sin(60^{\circ}) = \frac{r\sqrt{3}}{2}$. Thus, the area of each triangle is

$$A_1 = \frac{1}{2}rh = \frac{1}{2}(r)\left(\frac{r\sqrt{3}}{2}\right) = \frac{r^2\sqrt{3}}{4}$$

This means that the area of the hexagon is

$$A_{\text{H}} = 6A_1 = \frac{3r^2\sqrt{3}}{2}$$

The ratio of the area of the hexagon to the area of the circle that encloses it is

$$\frac{A_{\text{H}}}{A_{\text{C}}} = \frac{3r^2\sqrt{3}}{2\pi r^2} = \frac{3\sqrt{3}}{2\pi} = 0.8270$$

From these computations, we have the following observations for the three regular tessellations:

(a) The area of an equilateral triangle is approximately 41.35% of that of a circle enclosing it.
(b) The area of a square is approximately 63.70% of that of a circle enclosing it.
(c) The area of a hexagon is approximately 83% of that of a circle enclosing it.

This means that the hexagon has the highest coverage area after a circle among the three. Thus, the hexagon not only tessellates like the equilateral triangle and the square but also covers more area than the other two, coming very close to the circle in coverage area. This is why a cell is hexagonal in cellular network.

11.7 Clusters and Frequency Reuse

Because only a small number of radio channel frequencies are available for mobile systems, a scheme was developed to reuse radio channels to allow many conversations to take place at the same time. The solution the industry adopted is called frequency reuse, as we discussed earlier. The concept of frequency reuse is based on assigning to each cell a group of radio channels used within a small geographic area. A cell is assigned a group of channels that is completely different from the channels used in neighboring cells. The same group of channels can be used in different cells that are far enough away from each other so that their frequencies do not interfere.

Assume that a service provider has N channels available in a service area. A common practice is to divide these N channels among k neighboring cells such that a total of N/k channels are available in each cell, where we assume that N is divisible by k. This group of cells where frequency reuse is not practiced is called a *cluster*. Thus, all the available channels are used within one cluster with different channels used in the different cells of the cluster. The rate at which the

Figure 11.17 Illustration of a Seven-Cell Cluster.

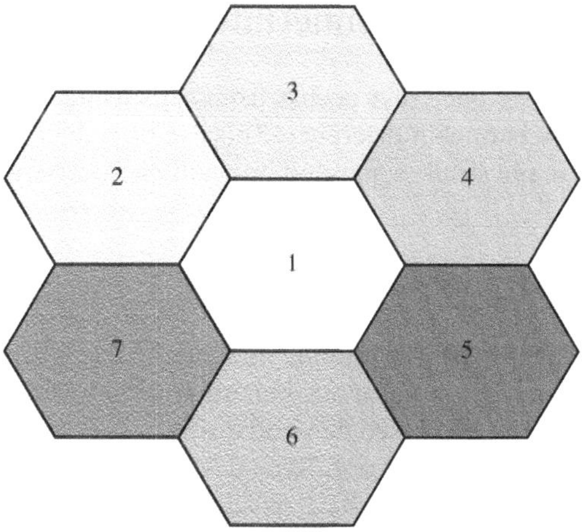

same frequency can be used in the network is called the *frequency reuse factor* and is equal to 1 divided by the number of cells in a cluster. Thus, the reuse factor of the cluster in Figure 11.17 is 1/7, which means that there are seven cells in a cluster and each cell is using 1/7 of the total number of the available channels.

The cluster is the basis for frequency reuse. Specifically, all the available channels are assigned to one cluster and the cluster is duplicated everywhere. Figure 11.18 shows two clusters where the cells with the same number use the same set of channels.

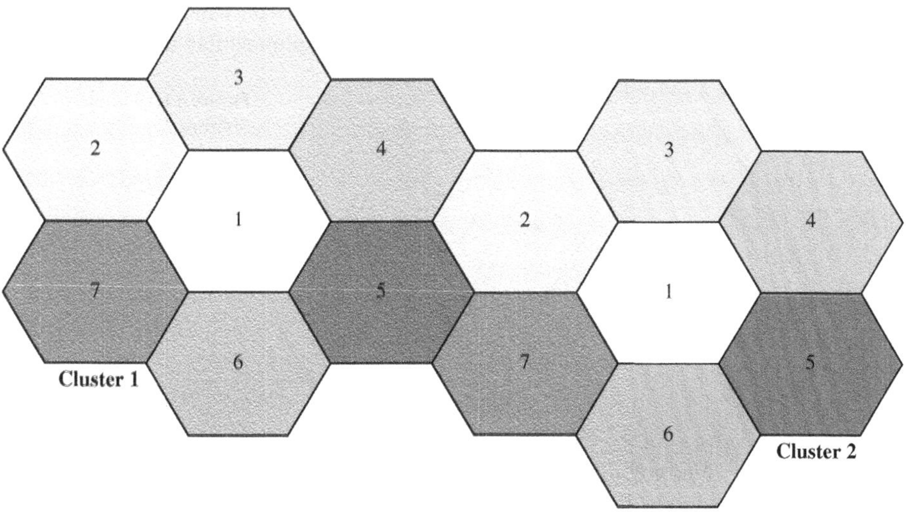

Figure 11.18 Illustration of Frequency Reuse in Two Adjacent Clusters.

11.8 Co-Channel Interference

Interference that occurs from cells using the same set of frequencies is called co-channel interference. This can be reduced by ensuring that co-channel cells are physically separated by a sufficient distance to provide sufficient isolation.

11.9 Cell Splitting

The demand for mobile network access is usually not the same everywhere; it is more in some areas than others. For example, areas around a sports stadium have an increased demand during a sports event compared to when there is no ongoing sport event. To deal with this imbalance in demand for network access, service operators have developed the idea of cell splitting.

As a service area becomes full of users, cell splitting is used to split a single area into multiple smaller ones. In this way, urban centers can be split into as many areas as necessary to provide acceptable service levels in heavy-traffic regions, while larger, less expensive cells can be used to cover remote rural regions. The splitting creates a hierarchy such that the splitting can be done at any level of the hierarchy. Thus, we can have *microcells* within macrocells and *picocells* within microcells as shown in Figure 11.19.

11.10 Introduction to Mobile Cellular Networks

There are different types of mobile communication networks in the world today. These networks are distinguished by the technologies on which they are based. In North America, the first widely deployed mobile cellular network was an

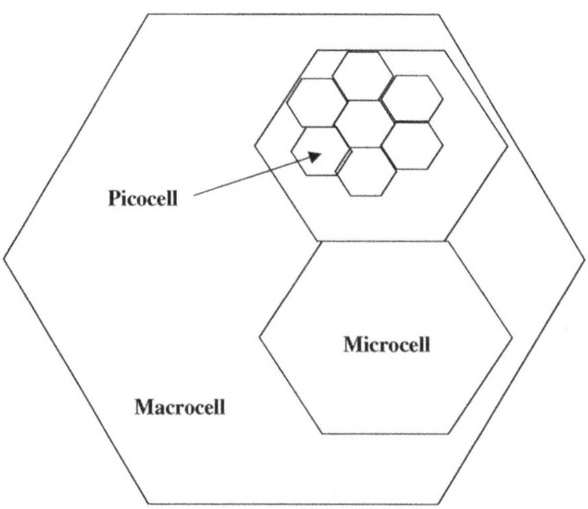

Figure 11.19 Cell Splitting.

analog system that was based on the Advanced Mobile Phone System (AMPS) protocol in which mobile stations (MSs) used frequencies 824–849 MHz to talk to the base stations (BSs) and the BSs used 869–894 MHz to talk to the MS. The two frequency bands are separated by 45 MHz.

AMPS is generally referred to as a first-generation network and its successors are referred to as second-generation (or 2G) networks and are mainly digital systems, though some are hybrid analog and digital systems. 2G networks include the following:

(a) GSM networks
(b) IS-54/IS-136 TDMA networks
(c) IS-95 code-division multiple access (CDMA) networks.

Newer generations of networks include the 2.5G, 2.75G, 3G, and 4G networks. These networks are discussed later in the chapter. A newer type of network called the 5G network is currently under development at the time of writing.

11.11 Mobile Cellular Network Architecture

The cellular communications system consists of the following four major components that work together to provide mobile service to subscribers:

1. Public switched telephone network (PSTN)
2. Mobile switching center (MSC)
3. Cell site with antenna system
4. Mobile subscriber unit (MS).

The PSTN is made up of local telephone networks, the interexchange telephone networks, and the long-haul network that interconnect telephones and other communication devices on a worldwide basis. The MSC switches calls from cell sites to wireline central offices in the PSTN; specifically, as follows:

- It gathers traffic from dozens of cells and passes it on to the PSTN via a central office switch.
- It controls calls, tracks billing information, and locates cellular subscribers.

The term cell site is usually called a BS and is used to refer to the physical location of the radio equipment that provides coverage within a cell. A BS communicates with the MSs in its cell over the air interface. Sometimes a *Base Station Controller* (BSC) is used to control multiple BSs. The MS unit (such as a cell phone) consists of a transceiver that transmits and receives radio transmissions to and from a BS. Two other network elements are located at the MSC; these are as follows:

(a) Home location register (HLR)
(b) Visitor location register (VLR).

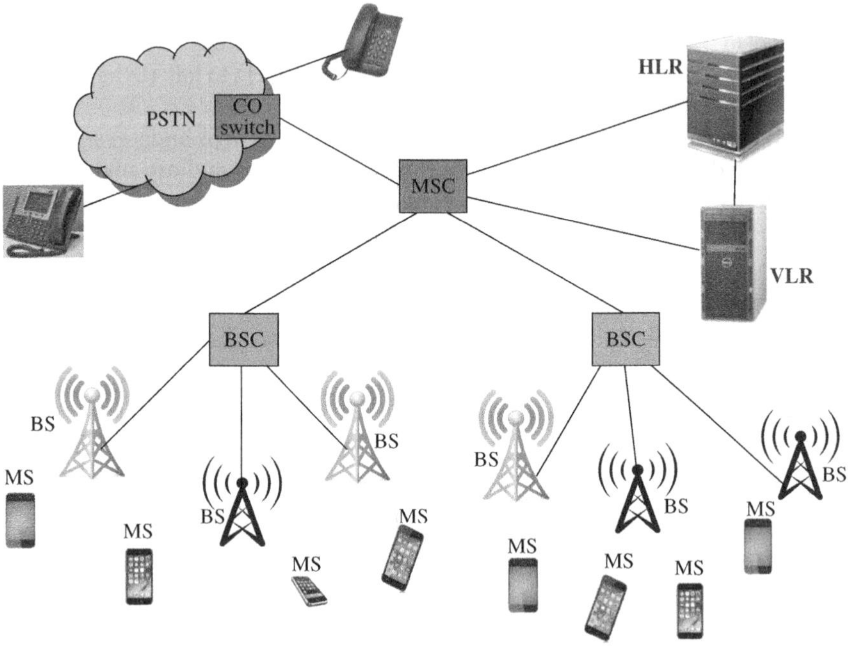

Figure 11.20 Architecture of Mobile Communication Networks.

The HLR stores complete local information. It is the main database that contains such information as the mobile equipment identity number, the subscriber's telephone number, the class of service, subscriber's current city, and their last known "location area" or the place they last used their mobile.

The VLR contains roamer information and the details of all the subscribers that are currently in the service provider's service area. As a person passes through another carrier's network, and the visited network detects their mobile unit, the new carrier's VLR queries the roaming person's assigned HLR. The VLR makes sure the person is a valid subscriber, then retrieves just enough information from their own carrier's HLR to manage their call.

Figure 11.20 is an illustration of the architecture of a cellular communication network.

11.12 Mobility Management: Handoff

Adjacent cells do not use the same radio channels, which means that a call must either be dropped or transferred from one radio channel to another when a user crosses the line between adjacent cells. *Handoff* is a scheme that was developed to minimize the dropping of calls. It occurs when the mobile network automatically transfers a call from a radio channel in one cell to another radio channel in another cell as an MS crosses adjacent cells.

When the MS moves out of the coverage area of a given cell site, the signal strength becomes weak. At this point the serving cell site requests a handoff from the system controller. The system controller transparently switches the call to a stronger-frequency channel in a new cell without interrupting the call.

11.12.1 Handoff Schemes

As discussed earlier, handoff is a mechanism that permits an MS to roam from cell to cell without interrupting a conversation in progress. Three issues are involved in a handoff:

1. Measurements of signal strength
2. Decision of where and when the MS will be handed off
3. Execution that transfers traffic and control to another BS.

Both the measurement and execution depend on the type of decision that is used. There are three types of handoff decisions:

(a) Network-initiated handoff, where the BS takes the measurements and MSC decides where and when to handoff
(b) Mobile-initiated handoff, where mobile device measures signal strength and determines when and where to handoff
(c) Mobile-assisted handoff (MAHO), where mobile device takes the measurements and reports back to network that ultimately makes the handoff decision.

11.12.2 Hard Handoff versus Soft Handoff

Regardless of which decision method is used to initiate handoff, it can take one of two forms:

(a) Hard handoff, called "break-before-make," where the old connection is broken as soon as handoff decision is made and generally before the new connection is established
(b) Soft handoff, called "make-before-break," where connections to new and previous base stations are kept active for some time until the mobile station has fully settled in the new cell when the old connection is then broken.

11.13 Generations of Mobile Communication Networks

As discussed earlier, the original mobile communication networks were called the first-generation mobile networks. They were replaced by second-generation networks, which themselves have been superseded by the third-generation networks. Similarly, third-generation networks have been superseded by fourth-generation networks, and there are plans for fifth-generation networks. In this section, we discuss the architecture of each of these generations of networks.

11.13.1 First-Generation Networks

The first-generation systems provided voice communication using frequencies around 900 MHz and analog modulation. One important thing that differentiated the 1G mobile communications technologies from previous technologies was the cellular technology. The mobile communication technologies before the 1G era focused on developing a powerful BS system that could send signals to cover as large an area as possible. The coverage of a single BS was about 50 miles or more, which was enough to cover most metropolitan regions at the time. The number of channels that subscribers could use at the same time was small, which meant that the rate at which calls were blocked was very high. The use of cellular system in 1G wireless telecommunication technology resulted in a more efficient spectrum usage.

1G wireless telecommunication technology used analog transmission techniques that were basically used for transmitting voice signals. The three most popular 1G systems are as follows:

(a) Nordic Mobile Telephone System (NMTS-450 and NMTS-900), which was the first multinational cellular system that was developed in 1981 within the Scandinavian countries of Denmark, Sweden, Finland, and Norway. It was used initially in the 450 MHz frequency band and later in the 900 MHz frequency band. In addition to the Scandinavian countries, it was also used in more than 30 other countries.
(b) AMPS, which was developed in the United States and operated in the 800 MHz frequency band (specifically between 824 and 894 MHz) and used frequency division multiple access (FDMA) technology for transferring information
(c) Total Access Communication System (TACS), which was developed in the United Kingdom and was based upon the AMPS standards of using FDMA analog technology and was within the 800 MHz frequency band and the 900 MHz frequency band.

11.13.2 Second-Generation Networks

2G networks are voice-centric; data is "secondary." Thus, it uses the capacity available from the PSTN to provide a data rate of up to 64 kbps, which was an improvement over the 2.4 kbps that 1G networks provided. This was made possible by the fact that 2G network uses digital technology as opposed to the

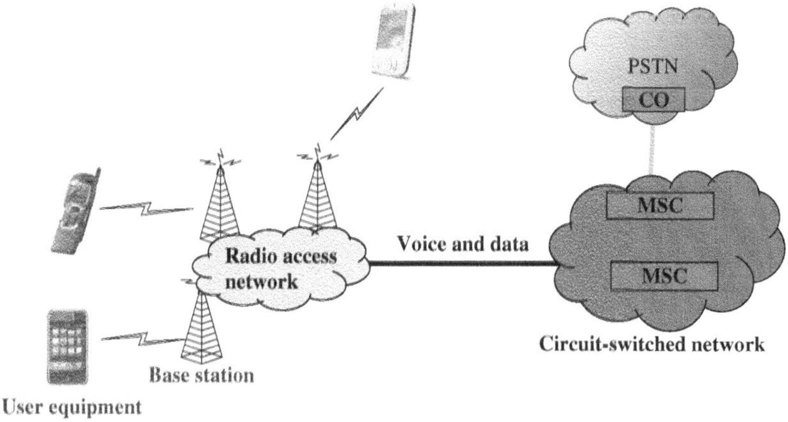

Figure 11.21 Architecture of 2G Networks.

1G network that uses analog technology. The architecture of the network is shown in Figure 11.21.

Two examples of 2G networks are the Global System for Mobile Communications (GSM) network and the CDMA network. These networks generally have devices that work only on one of the networks. Thus, a device made for a GSM network may not work in a CDMA network. In the United States, Sprint and Verizon operate CDMA networks while AT&T and T-Mobile operate GSM networks.

11.13.3 Introduction to the GSM Network

The GSM network is the most pervasive 2G network in the world. It was developed as a pan-European digital cellular network and predates the American version, the IS-136 network. It was the first network to introduce the short message service (SMS). It is a TDMA-based system in which frequency spectrum is partitioned into two 25-MHz subbands in the original version:

(a) 890–915 MHz used for mobile device to BS communication (i.e., the uplink)
(b) 935–960 MHz for the downlink (from BS to mobile device).

A newer system is called the GSM-1800 (or digital cellular system, DCS, 1800) and uses two 75-MHz subbands, which are as follows:

(a) 1710–1785 MHz for uplink
(b) 1805–1880 MHz for downlink.

The key features of GSM include the fact that it has international roaming capability, which means that a single subscriber number can be used in different

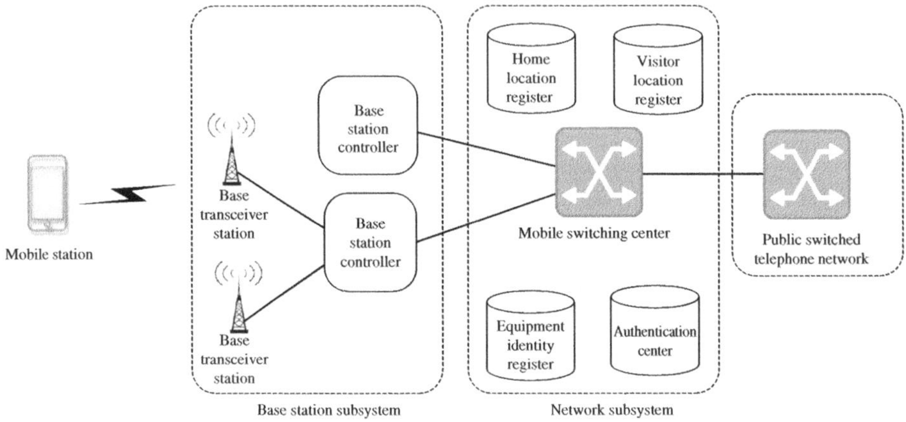

Figure 11.22 GSM Architecture.

parts of the world. Also, it provides a high level of security. It introduced the concept of universal and inexpensive mobile handsets.

The architecture of the network is shown in Figure 11.22. Every GSM mobile phone has a subscriber identity module (SIM). The SIM is a smart card that provides the mobile phone with a unique identity through the use of the international mobile subscriber identity (IMSI). The SIM is like a key without which the mobile phone cannot function.

The base station subsystem (BSS) connects the user on a mobile phone with other landline or mobile users. The base transceiver station (BTS) is in direct contact with the mobile phones via the air interface. The BSC controls several BTSs. It monitors each call and decides when to handover the call from one BTS to another, as well as manages radio frequencies allocated for the calls through the BTS.

The network subsystem (NSS) routes calls from a fixed network via the BSC and BTS to an individual MS. The MSC interconnects the cellular network with the PSTN. The MSC also serves to coordinate setting up calls to and from GSM users. NSS includes the HLR, the VLR, the equipment identity register (EIR), and the authentication center (AuC).

The HLR stores information of all subscribers belonging to an area served by an MSC. It stores permanent data such as the IMSI, services subscribed by the user, subscriber's number from a public network, a security key called K_I and some other temporary data. The HLR provides the MSC with all the necessary information to handle a call that is coming from a public network.

The VLR contains relevant information for all mobile devices currently served by an MSC. It stores the temporary mobile subscriber identity (TMSI), which is used for limited intervals to prevent the transmission of the IMSI via the air interface. The VLR supports the MSC during call establishment and authentication when the call originates from a mobile device.

The EIR stores all the international mobile equipment identities (IMEI) of mobile equipment and their rights on the network. It maintains a white, gray, and black list. Those on the white list are permitted on the network while those on the black list are blocked from the network. The gray list consists of faulty equipment that may pose a problem on the network but are still permitted to participate in the network. The IMEI reveals the serial number of the mobile device, manufacturer, type of approval, and country of production.

The AuC is a protective database that houses the K_I, the A3 authentication algorithm, the A5 ciphering algorithm, and the A8 ciphering key generating algorithm. It is responsible for user authentication and data encryption.

11.13.4 GSM Channels

Each of the two subbands contains 124 frequency carriers with appropriate guard bands. Each frequency carrier is 200 kHz wide and divided into eight TDMA time slots, each of which is essentially a channel. Thus, each channel is identified by its frequency carrier and its slot number within the carrier. This means that the system uses a combination of FDMA and TDMA. Two classes of logical channels are defined:

(a) *Traffic channels* are used for user data.
(b) *Signaling channels* are used for network control and management.

There are three types of signaling channels:

(a) Broadcast channels
(b) Common control channels
(c) Dedicated control channels.

Broadcast channels carry downlink information from the BS and include the following:

(a) *Frequency correction channel* allows a mobile terminal to synchronize its own frequency to that of the transmitting base site.
(b) *Synchronization channel* carries the information to enable the MS to synchronize to the TDMA frame structure and know the timing of the individual timeslots.
(c) *Broadcast control channel* is used by all MSs to measure their signal strength.

Common channels are used for conveying information from network to the MSs and provide access to the MSs. They are divided into three groups of channels:

(a) *Paging channel*, a downlink channel used to alert a mobile device when an incoming call needs to be delivered to it
(b) *Access grant channel*, used to direct an MS to a bidirectional standalone dedicated control channel that the subscriber can use for its communication

(c) *Random access channel*, which is used by a mobile device to initiate a call setup or respond to a paging; it is used in a random access manner. MSs use the *slotted Aloha scheme* to access this channel.

A dedicated control channel is used to coordinate and control specific mobile devices in a wireless system. It is divided into three groups of channels, which include the following:

(a) *Stand-alone dedicated control channel*, which is used for registration, authentication, call setup, and location updating; also supports SMS
(b) *Slow associated control channels*, used to transmit measurement samples, control the power, and maintain and correct the timing alignment in MSs
(c) *Fast associated control channels*, bidirectional channel used to exchange information more quickly than the slow associated control channel. For example, when handoff occurs, FACCH steals a traffic channel to transmit power and handoff signaling messages.

11.13.5 Power Control

While BSs perform the timing measurements, they also perform measurements on the power level of the different MSs. These power levels are adjusted so that the strengths of the signals reaching the BS are nearly the same for all MSs. The MS measures the strength and the quality of the signal between itself and the BS. If the MS does not receive the signal from the BS correctly, the BS changes its power level.

11.13.6 Overview of IS-136 TDMA Networks

The IS-136 TDMA network operates in a manner similar to GSM. It can be considered the American version of the GSM. The major differences are as follows:

(a) Each carrier frequency is 30 kHz wide and is divided into three TDMA time slots called digital traffic channels (DTCs) that are assigned to voice calls.
(b) It is a TDMA-based system in which frequency spectrum is partitioned into two 25 MHz subbands:
    i. 824–849 MHz used for MS to BS communication (i.e., the uplink)
    ii. 869–894 MHz for the downlink

The other features are essentially minor variations of those of the GSM network.

11.13.7 Overview of IS-95 CDMA Networks

CDMA is a spread spectrum scheme that spreads a narrowband signal over a much wider bandwidth. In the IS-95 CDMA, the RF spectrum is divided into carrier frequencies that are 1.25 MHz wide. Each call is assigned a different code, which acts as the digital carrier, and the codes are ideally mutually orthogonal (or mutually uncorrelated). Each code constitutes a row of the 64×64 Walsh–Hadamard matrix. The 64 codes are also referred to as Walsh functions.

Walsh–Hadamard matrices are characterized as follows:

(a) There is single row of 0s.
(b) In each of the other rows, there is an equal number of 0s and 1s.
(c) The matrices have N row (and N columns) such that $N = 2^n$, where n is an integer.

They can be recursively constructed as follows:

$$H_1 = [0]$$

$$H_2 = \begin{bmatrix} H_1 & H_1 \\ H_1 & \overline{H}_1 \end{bmatrix} = \begin{bmatrix} 0 & 0 \\ 0 & 1 \end{bmatrix}$$

$$H_{2N} = \begin{bmatrix} H_N & H_N \\ H_N & \overline{H}_N \end{bmatrix}$$

where the elements of $\overline{H}_N$ are the binary complements of the corresponding elements of H_N. For example, H_4 represents the following matrix:

$$H_4 = \begin{bmatrix} H_2 & H_2 \\ H_2 & \overline{H}_2 \end{bmatrix} = \begin{bmatrix} 0 & 0 & 0 & 0 \\ 0 & 1 & 0 & 1 \\ 0 & 0 & 1 & 1 \\ 0 & 1 & 1 & 0 \end{bmatrix}$$

One implication of the orthogonality feature of the codes is that with IS-95 CDMA networks there is no frequency planning among the cells because the system uses the one-cell frequency reuse plan. That is, the same carrier frequencies used in one cell are used in all cells. This is illustrated in Figure 11.23, where it is compared with the TDMA network frequency reuse scheme.

IS-95 CDMA uses soft handoff and a call can be in a soft handoff condition with up to three cells at the same time. This reduces the chances of dropping a call during a handoff. Unlike other technologies, CDMA uses different transmission techniques in different directions. The forward direction includes the following:

(a) One *pilot channel*, used by MSs to acquire carrier phase and timing information from the BS
(b) One *synch channel*, which contains repeatedly transmitted system information, such as BS's ID and data rate of the cell's paging channels; information that is important to the mobile device
(c) Up to seven *paging channels*, which carry information to the MSs that do not have calls in progress. Information conveyed is a notification of incoming call. Data rate is either 4.8 or 9.6 kbps.
(d) At least 55 *forward traffic channels*, used by BS to send information to MSs with calls in progress.

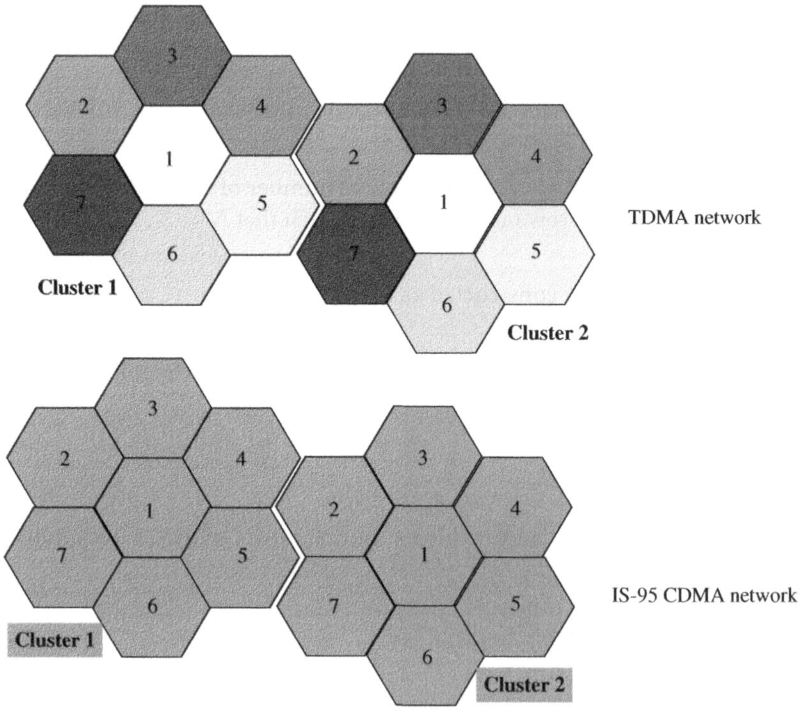

Figure 11.23 Frequency Reuse in TDMA Versus CDMA Networks.

The reverse direction includes two sets of channels:

(a) *Access channels*: The BS maintains up to 32 of these channels, which are used by idle MSs to originate calls, respond to paging messages, or register their locations. An MS chooses one of them at random when it needs to request a channel. MSs use the slotted Aloha scheme to transmit in these channels.
(b) *Traffic channels*: These are used by the MSs with calls in progress to send information to the BS.

IS-95 performs power control: BSs receive transmission from all MSs at the same power level, which is the solution to the so-called *near–far problem*. Specifically, if uncontrolled, stations near the BS will transmit with high power, which introduces noise in the channel thereby reducing capacity. Every 1.25 ms the BS instructs MSs to raise or lower their power levels. CDMA provides 84 power levels in increments of 1 dB. Power level is important because since all MSs use the same frequency spectrum in each cell, transmission in one cell may be affected by transmissions in adjacent cells. But power control can minimize the interference.

11.13.8 Third-Generation Networks

3G networks enable advanced services such as real-time video, high-speed multimedia, and mobile Internet access in addition to voice. In Europe, 3G systems are referred to as the Universal Mobile Telecommunications System (UMTS). According to the ITU-T definition, a network qualifies to be called a 3G network if it provides the following data rates:

(a) 2 Mbps in fixed or in-building environments
(b) 384 kbps in low mobility or pedestrian users
(c) 144 kbps in high mobility or vehicular users.

Thus, there is a trade-off between speed and data rate. The ITU has approved the following five International Mobile Telecommunications-2000 (IMT-2000) terrestrial radio interfaces, most of which are based on CDMA:

(a) IMT-MC: multi-carrier, which is also called multi-carrier CDMA (MC-CDMA) or CDMA2000
(b) IMT-DS: direct spread, which is also called wideband CDMA (W-CDMA), UTRA-FDD, UMTS-FDD, where UTRA stands for *UMTS terrestrial radio access.*
(c) IMT-TC: time code, which is also called UTRA-TDD and China's TD-SCDMA
(d) IMT-SC: single carrier, which is also called UWC-136/EDGE, SC-TDMA, where UWC stands for *universal wireless communications*
(e) IMT-FT: frequency time, which is also called the digital enhanced cordless telecommunications (DECT) or FDMA/TDMA.

The 3G radio interfaces can be summarized as shown in Figure 11.24.

Architecturally, 3G networks have separate paths for voice and data; the radio access network (RAN) performs the traffic separation. The architecture of 3G networks is shown in Figure 11.25.

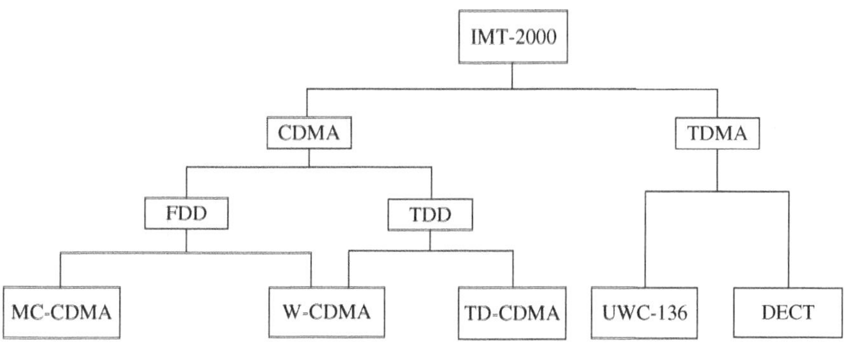

Figure 11.24 Summary of the 3G Radio Interface.

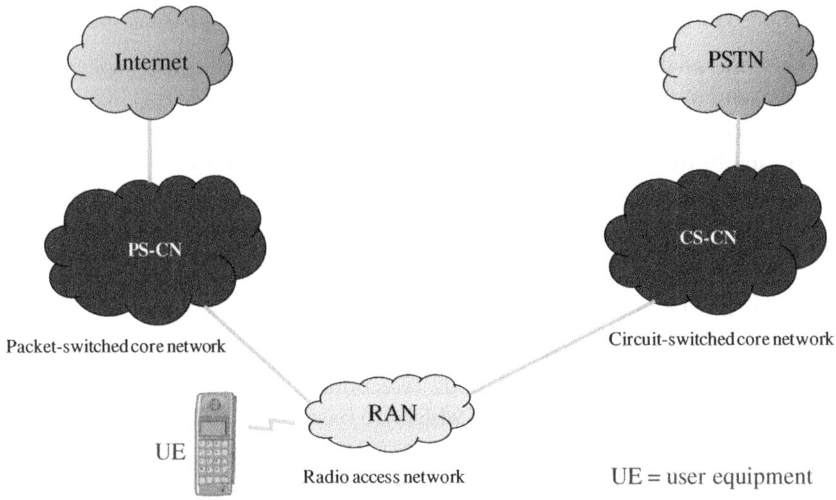

Figure 11.25 Architecture of 3G Networks.

The features of the network include the following:

- Data transmission speed increased from 144 kbps to 2 Mbps
- Providing faster communication than 2G
- Send/receive large e-mail messages
- High-speed web/more security
- Support for video conferencing and 3D gaming
- Provides TV streaming/mobile TV.

11.13.9 Fourth-Generation Networks

As defined in the international mobile telecommunications advanced (IMT-advanced) by ITU-R, 4G is an IMT-advanced cellular system that fulfills the following requirements:

- It is based on an all-IP packet switched network.
- It has peak data rates of up to approximately 100 Mbits/s for high mobility such as mobile access and up to approximately 1 Gbits/s for low mobility such as nomadic/local wireless access.
- It is able to dynamically share and use the network resources to support more simultaneous users per cell.
- It uses scalable channel bandwidths of 5–20 MHz, optionally up to 40 MHz.
- It has a peak link spectral efficiency of 15-bit/s/Hz in the downlink, and 6.75-bit/s/Hz in the uplink (meaning that 1 Gbit/s in the downlink should be possible over less than 67 MHz bandwidth).

- It has an indoor system spectral efficiency of 3-bit/s/Hz/cell for downlink and 2.25-bit/s/Hz/cell for uplink.
- It ensures smooth handoffs across heterogeneous networks.
- It has the ability to offer high quality of service (QoS) for next-generation multimedia support.

Thus, 4G network is a data-centric all-IP network that supports VoIP. It is also capable of providing 100Mbps – 1Gbps speed and supports mobile multimedia and customized personal services. In addition, it supports high QoS, high security, and ubiquitous service.

4G networks are generally implemented using the Long-term Evolution (LTE). Another method called WiMAX was considered but has now been dropped by most service providers. In the LTE network, a BS is called an Evolved Node B, which is abbreviated as *eNodeB*. The architecture of a 4G network is shown in Figure 11.26.

Table 11.2 is a summary and comparison of the different generations of networks. It details the multiple access scheme used for each generation, switching scheme used, the frequency band used and other special features.

11.13.10 Fifth-Generation Networks

5G is a new network system that has much higher speeds and capacity, and much lower latency, than existing cellular systems. The technologies to be used in 5G are still being defined at the time of writing, but there are some general themes everyone agrees on. A set of eight requirements for 5G has been identified:

(a) 1–10 Gbps connections to end points in the field (i.e., not theoretical maximum)
(b) 1 ms end-to-end round trip delay (latency)
(c) 1000 times bandwidth per unit area
(d) 10–100 times the number of connected devices compared to current technologies; it is being prepped for the Internet of things (IoT).

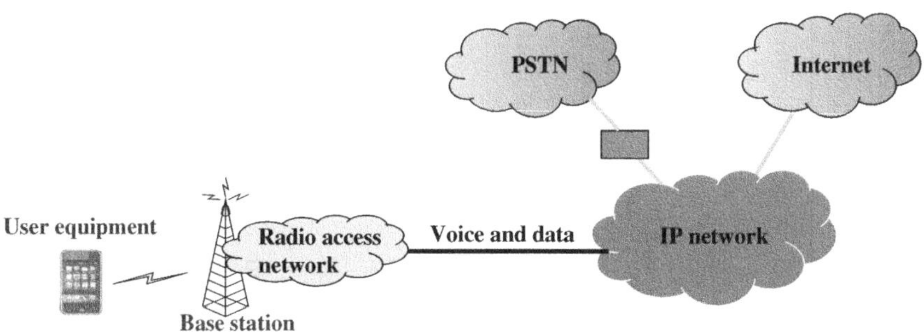

Figure 11.26 Architecture of 4G Networks.

Table 11.2 Comparison of Generations of Mobile Communication Systems.

Parameters	1G	2G	3G	4G
Name	1st generation mobile network	2nd generation mobile network	3rd generation mobile network	4th generation mobile network
Technology	AMPS, NMT, TACS	IS-95, GSM	IMT2000, WCDMA	LTE, WiMAX
Multiple address/access system	FDMA	TDMA, CDMA	CDMA	CDMA
Switching type	Circuit switching	Circuit switching for voice and packet switching for data	Packet switching except for air interface	Packet switching
Speed (data rates)	2.4–14.4 kbps	14.4 kbps	3.1 Mbps	100 Mbps
Special characteristic	First wireless communication	Digital version of 1G technology	Digital broadband, speed increments	Very high speeds, all IP
Features	Voice only	Multiple users on single channel	Multimedia features, video call	High speed, real time streaming
Supports	Voice only	Voice and data	Voice and data	Voice and data
Internet service	No Internet	Narrowband	Broadband	Ultra broadband
Bandwidth	Analog	25 MHz	25 MHz	100 MHz
Operating frequencies	800 MHz	GSM: 900 MHZ, 1800 MHz CDMA: 800 MHz	2100 MHz	850 MHz, 1800 MHz
Band (frequency) type	Narrow band	Narrow band	Wide band	Ultrawide band

Carrier frequency	30 kHZ	200 kHz	5 MHz	15 MHz
Advantage	Simpler (less complex) network elements	Multimedia features (SMS, MMS), Internet access, and SIM introduced	High security, international roaming	Speed, high-speed handoffs, MIMO technology, global mobility
Disadvantages	Limited capacity, not secure, poor battery life, large phone size, background interference	Low network range, slow data rates	High power consumption, low network coverage, high cost of spectrum license	Hard to implement, complicated hardware required
Applications	Voice calls	Voice calls, short messages, browsing (partial)	Video conferencing, mobile TV, GPS	High-speed applications, mobile TV, wearable devices

(e) 99.999% availability
(f) 100% coverage
(g) 90% reduction in network energy usage
(h) Up to 10 year battery life for low-power, machine-type devices.

Currently, it is difficult to conceive of a single technology that could meet all of these conditions simultaneously.

11.14 A Note on Internet-of-Things

The IoT is a system of interrelated computing devices, mechanical and digital machines, objects, animals, or people that are provided with unique identifiers and the ability to transfer data over a network without requiring human-to-human or human-to-computer interaction. A thing, in IoT, can be a person with a heart monitor implant, a farm animal with a biochip transponder, an automobile that has built-in sensors to alert the driver when tire pressure is low, or any other natural or man-made object that can be assigned an IP address and provided with the ability to transfer data over a network. IoT has evolved from the convergence of wireless technologies, microelectromechanical systems (MEMS), microservices, and the Internet.

IoT will not be possible without the support of IPv6. Thus, the global adoption of IPv6 will be critical for the successful development of the IoT because every "thing" must have an IP address to be able to connect to the network and IPv6 has many addresses to give out. In addition to sensing things, IoT systems could also perform actions. In the IoT, sensors and actuators embedded in physical objects are linked through wired and wireless networks that are connected to the Internet. When objects can both sense the environment and communicate, they become tools for making important decisions. For example, intelligent shopping systems could monitor specific users' purchasing habits in a store by tracking their specific mobile phones and using the information to extend to them special offers on their favorite products.

11.15 Summary

This chapter has been concerned with the basics of mobile communication networks. It has introduced the reader to the fundamentals of radio communication, cellular communication, and the different generations of mobile cellular communication networks. The goal is to expose the reader to the ubiquitous mobile communication network from which almost all services can now be accessed.

Exercises

1 Name two methods of radio propagation.

2 What does multipath fading mean?

3 What is the difference between Rician fading and Rayleigh fading?

4 What does co-channel interference mean?

5 What does frequency reuse mean in cellular networks?

6 What does a frequency reuse factor of 1/12 mean?

7 What is the difference between the frequency reuse scheme in a GSM/TDMA network and that used in a CDMA network?

8 What does hard handoff mean in a cellular network?

9 What does soft handoff mean in a cellular network?

10 What is the function of the home location register (HLR)?

11 What is the function of the visitor location register (VLR)?

12 What type of medium access control scheme is generally used by mobile stations to request transmission slots from the base station?

13 On what type of channel would a mobile station send a request to the base station for a transmission slot?

14 What is the paging channel used for in a mobile communication network?

15 What transmission rates qualify a mobile communication network to be a 3G network?

16 What kind of handoff scheme is used in the IS-95 CDMA network?

17 How far apart in kHz are the frequency carriers in the GSM network separated from each other?

18 How far apart in kHz are the frequency carriers in the IS-136 TDMA network separated from each other?

19 How many time slots are used in one frequency carrier of the IS-136 TDMA network?

20 How many time slots are used in one frequency carrier of the GSM network?

12

Introduction to Network Security

12.1 Introduction

Security concerns are very important in IP networks because such networks are inherently nonsecure. In this lecture, we first examine the types of attacks that can be launched on IP networks. We then examine the security mechanisms used in IP networks. Finally, we discuss the IP security (IPSec) protocol.

12.2 Types of Network Attacks

A network attack is an intrusion on a network infrastructure. The attacker first analyzes the environment and collects information in order to exploit the existing open ports or vulnerabilities. In some cases, the purpose of attack is only to learn and get some information from the system without altering or disabling it; this is referred to as a *passive attack*. In other cases called *active attacks*, the attacker accesses network resources to alter, disable, or destroy them. An attack can be performed either from outside of the organization by an unauthorized entity or from within the company by an "insider" that already has some access to the network.

There are different types of attacks that can be launched in a network. The following are some of the common ones:

(a) *Network sniffing* (*packet sniffing*) is a process of capturing the data packets traveling in the network. It is used by IT professionals to analyze and monitor the traffic to find such things as unexpected suspicious traffic. It is also used by attackers to collect data sent in clear text that is easily readable. In this case, the intent is to gather login names and passwords used to access the network. The information is then passed on to someone who can use it for any intended purpose.

(b) *Spoofing* is a process by which an intruder masquerades as a trusted user in order to gain unauthorized access to a secure environment. It

Fundamentals of Data Communication Networks, First Edition. Oliver C. Ibe.

is particularly used in source routing since the sender can specify the return path of the reply to a message. One of the purposes of spoofing in a corporate environment is to be able to conduct unauthorized business with another company's clients. The common types of spoofing include the following:

(i) *IP address spoofing* is a process of creating IP packets with forged source IP address to impersonate a legitimate system. This kind of spoofing is often used in denial-of-service (DoS) attacks.

(ii) *ARP spoofing* is a process of sending fake ARP messages in the network. The purpose of this type of spoofing is to associate the MAC address with the IP address of another legitimate host causing traffic redirection to the attacker's system. It is often used in man-in-the-middle attacks.

(iii) *DNS spoofing* is an attack where the wrong data is inserted into DNS server cache, causing the DNS server to divert the traffic by returning wrong IP addresses as the results for client queries.

(c) *Man-in-the-middle* (MITM) *attack* is an attack that involves the surreptitious placement of a software agent between the client and server ends of a communication session. With neither party being aware of the presence of the malicious agent, the agent simply relays the data transmissions between client and server as though nothing is happening. In parallel with this process, the agent is also recording the data as it is passed through. This results in a third party having access to a variety of different types of data, from login and password credentials to proprietary and confidential information. The agent can also modify data "on the fly" causing severe problems for the victim. Man-in-middle attacks have increased considerably since the introduction of wireless networking. This is because there is now no need for the attacker to connect to a wire as data can simply be intercepted from anywhere within range of the wireless signal. A simple example of MITM is illustrated in Figure 12.1 where the attacker shown as MITM changes the destination of the deposit from Chris' account to his own account.

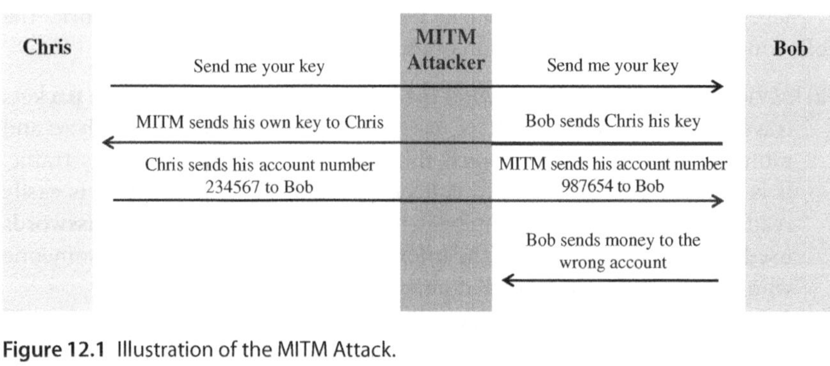

Figure 12.1 Illustration of the MITM Attack.

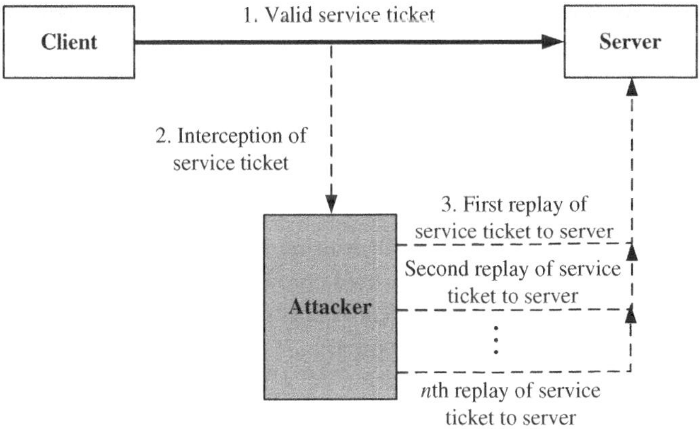

Figure 12.2 Illustration of the Replay Attack.

(d) *Replay attack* is a variation on the man-in-the-middle attack. In this case, an agent is once again placed within the client–server line of communication where it records the transaction data for the express purpose of allowing the data to be modified and replayed to the server at a later time for evil purposes. For example, a replay attack might record the entire process of a user logging into a banking web site and performing transactions. The recorded transcript may then be replayed to repeat the login sequence for the purposes of stealing money from the account. Replay attack is illustrated in Figure 12.2, where an action can be replayed multiple times to the detriment of the client.

(e) *DoS* is an attack that is aimed at preventing unauthorized users from accessing services on the network. A DoS attack can be in the form of flooding the network with invalid data until traffic from authorized network users cannot be processed. It can also be in the form of flooding the network with invalid network service requests until the host providing that particular service cannot process requests from authorized network users. The network would eventually become overloaded. It can also be in the form of disrupting communication between hosts and clients through the modification of system configurations. Finally, it can be in the form of causing physical network destruction, such as crashing a server or router in the network. An example of the DoS is the SYN flood that we discussed in Chapter 8. An attacker can initiate a DoS attack from multiple computers or systems. This type of attack is called a *distributed denial-of-service attack* (DDoS), which is more difficult to deal with than an attack that is initiated from one system. Generally, with DDoS multiple infected systems flood a particular host with traffic simultaneously.

(f) *Trojan horse* is a program that installs malicious software while under the guise of doing something else. The term is derived from the ancient Greek story of the wooden horse that was used to help Greek troops invade the city of Troy by trickery. Similar to the mythical Trojan horse, the malicious code is hidden in a computer program or other computer file that may appear to be useful, interesting, or at the very least harmless to an unsuspecting user. When this computer program or file is executed by the unsuspecting user, the malicious code is also executed resulting in the installation of the malicious Trojan horse program. Trojans are generally spread by some form of social engineering, for example, where a user is duped into executing an e-mail attachment disguised to be unsuspicious.

(g) *Session hijacking* refers to the exploitation of a valid computer to gain unauthorized access to information or services in a computer system. In particular, it is used to refer to the theft of a magic cookie used to authenticate a user to a remote server.

(h) *Phishing* is an attack in which the attacker attempts to fraudulently acquire sensitive information, such as usernames, passwords, and credit card details, by masquerading as a trustworthy entity in a communication session. It is typically carried out by e-mail or instant messaging and often directs users to give details at a website.

12.3 Security Services

There are different types of security services that can be offered. The following is a list of some of these services:

(a) *Authentication* is the process of verifying that the user is exactly who they claim to be. It is usually done through the use of passwords and/or user IDs, if *single-factor authentication* is used. However, sometimes a *two-factor authentication* is used. A two-factor authentication is a two-step verification that provides an extra layer of security beyond user ID and password. In addition to what is used in a single-factor authentication, it requires some information that the user, and only the user, should know or have or be. Things such as place of birth, favorite book, and pet's name are examples of the second factor.

(b) *Data Integrity* is the art of ensuring that data is transmitted from source to destination without alteration. It is usually accomplished via the use of a *digital signature*.

(c) *Confidentiality* is the art of ensuring that data is kept private and accessed only by intended recipient. It is accomplished through *encryption*.

(d) *Nonrepudiation* is the art of ensuring that parties cannot deny their electronic actions. It is usually accomplished through the use of a two-factor authentication and digital signature.

12.4 Data Encryption Terminology

Encryption transforms readable text, called *plaintext* or *cleartext*, into an unintelligible form, called *ciphertext*, using an encryption algorithm. The purpose of an encryption algorithm is to scramble a message so that it remains secure even if the ciphertext is transmitted over a nonsecure medium. The process of recovering a plaintext from its ciphertext is called *decryption*. A system that encrypts and decrypts information is called a *cryptosystem*. The art of creating and using cryptosystems is called *cryptography*. The art of breaking encrypted messages (usually by intruders) is called *cryptanalysis*. Finally, the study of cryptography and cryptanalysis is called *cryptology*.

12.5 Cryptographic Systems

Both encryption and decryption use a *key*. A key in the cryptographic sense is a long string of characters that permits a cryptosystem to encrypt or decrypt information in a distinct way. All modern cryptosystems are key-based systems. Sometimes the same key is used for both encryption and decryption, and sometimes different keys are used for each operation. Based on this observation, there are two classes of key-based cryptosystems:

(a) Symmetric or secret-key cryptosystems
(b) Asymmetric or public-key cryptosystems.

12.5.1 Symmetric Cryptosystems

In a symmetric cryptosystem, the same key is used for encryption and decryption. Thus, both the originator and the recipient of a message must know the key, which is either known to the recipient through some prior arrangement or communicated in parallel with the ciphertext. The main problem with the system is how to send the key to the recipient in a secure manner. An example of a symmetric cryptosystem is the data encryption standard (DES), which has been replaced by advanced encryption standard (AES).

12.5.2 Public-Key Cryptosystems

A public-key cryptosystem uses one key for encryption and another key for decryption. Each user is assigned a pair of unique and mathematically related keys: a public key and a private key. The private key is a secret key that is available only to the owner, and the public key is published. What is encrypted by one key can be decrypted by the other, and vice versa. Though the keys are mathematically related, it is difficult to derive the private key from the knowledge of the public key. Figure 12.3 illustrates the architecture of the

Figure 12.3 Architecture of the Public-Key Cryptosystem.

public-key cryptosystem. To send a message to a user A, we use A's published key to encrypt the message, and A uses their private key to decrypt it upon receiving the message.

The *cryptographic strength* of a public-key cryptosystem depends on how difficult it is for anyone to compute a user's private key given only the corresponding public key. Because of the uniqueness of pair of keys assigned to a user, no other private key can decrypt data encrypted by a given public key than its mate.

Private-key cryptosystems generally use either the Diffie–Hellman algorithm (named after its inventors: Diffie and Hellman) or the RSA algorithm (named for its inventors: Rivest, Shamir, and Adleman). The cryptographic strength of the Diffie–Hellman cryptosystem depends on the difficulty of computing discrete logarithms in a finite field generated by a large prime number. Similarly, the cryptographic strength of RSA cryptosystem depends on the difficulty in finding the prime factors of a large integer. A major difference between the two schemes is the manner in which the keys are assigned: The Diffie–Hellman cryptosystem requires a dynamic key exchange between sender and receiver for each session. The RSA cryptosystem requires each user to have a permanent key pair, one of which is posted in a public place and the other is known only to the user; in this case, keys are never exchanged. Thus, the Diffie–Hellman cryptosystem is susceptible to interception.

12.5.3 Comparing Symmetric and Public-Key Cryptosystems

Public-key cryptosystems provide more security services than symmetric cryptosystems. However, public-key cryptosystems have certain disadvantages over symmetric cryptosystems:

(a) Public-key cryptosystems are not suited for encrypting long messages because they take longer time than symmetric cryptosystems.
(b) Public-key algorithms can generate ciphertexts that are much larger than the corresponding plaintexts.
(c) A ciphertext generated by a public-key cryptosystem cannot be sent to more than one user due to the uniqueness of the keys, but a ciphertext

generated by a symmetric system can be sent to multiple users, provided the encryption key can be sent securely to each recipient.

12.5.4 A Hybrid Encryption Scheme

As stated earlier, symmetric cryptosystems are not as secure as asymmetric cryptosystems because of the problems associated with transmitting the encryption key to the recipient of the ciphertext. However, they are faster than asymmetric cryptosystems because they generate shorter ciphertexts. A hybrid scheme that combines the strengths of both schemes works as follows. A message is encrypted using the symmetric key encryption and sent to the destination. The encryption key, which is usually shorter than the actual message, is then sent in a separate message that is encrypted using the public-key cryptosystem. This means that when the recipient receives the encrypted message, they wait for the key to arrive in another message. When the second message arrives, the recipient uses their private key to recover the attached public key, which they can use to decrypt the first message that contains the hidden information.

This process is similar to how a new credit card can be activated. Usually, the credit card company mails the user a new card. In a separate mail the activation code is enclosed. Before the recipient can use the card, they must use the activation code to validate the card over the phone with the card issuing company or using an ATM machine.

12.6 Technical Summary of Public-Key Cryptography

In this section, we present an introduction to the RSA algorithm. Cryptographic algorithms are based on number theory. Thus, our discussion on the RSA algorithm requires an understanding of prime numbers and modulo-n arithmetic. We start this section by providing a brief introduction to number theory.

12.6.1 Introduction to Number Theory

Integers are whole numbers: $\ldots, -3, -2, -1, 0, 1, 2, 3, \ldots$ *Natural numbers* are positive integers: $1, 2, 3, \ldots$ Number theory deals with the study of natural numbers. An integer d is called a *divisor* of an integer a if there exists an integer k such that $a = kd$. In this case, we write $d|a$, which reads "d divides a with no remainder." For example, $3|15$, and $4|8$.

An integer d is a *common divisor* of integers a and b if there exist integers k and l such that $a = kd$ and $b = ld$. The *greatest common divisor* of two integers a and b that are not both 0 is a common divisor $d > 0$ of a and b such that all other common divisors of a and b divide d. We denote the greatest common divisor of a and b by $\gcd(a, b)$.

An integer $a > 1$ is defined to be *prime* if its only divisors are ± 1 and $\pm a$. Two integers a and b are *coprime* (or *relatively prime*) if they have no positive divisors in common except for 1; that is, a and b are coprime if $\gcd(a, b) = 1$.

Thus, a number is prime if and only if the only positive numbers that divide it evenly are 1 and the number itself. Examples of prime numbers include 1, 2, 3, 5, 7, 11, 13, 17, 19, and 23. But 12 is not a prime number because 2, 3, 4, and 6 divide it. Coprime extends this idea to two numbers. For example, 9 and 15 are not coprime because 3 divides both of them. On the other hand, 4 and 7 are coprime because the only number that divides both of them is 1.

12.6.2 Congruences

If a, b, and n are integers such that n divides $b - a$, then a is said to be *congruent* to b modulo n, and we write $a \equiv b\text{mod}(n)$; that is, $n|(b - a)$. Congruence relation partitions the integers into disjoint subsets such that two integers are in the same subset if and only if their difference is divisible by n. These subsets are called *residue classes* modulo n, and the residue class that contains a is denoted by $[a]$. For example, if $n = 5$, then there are five different residue classes, as follows:

$$[0] = \{0,\ 5, -5,\ 10, -10,\ 15,\ \ldots\}$$
$$[1] = \{1,\ 6, -4,\ 11, -9,\ 16,\ \ldots\}$$
$$[2] = \{2,\ 7, -3,\ 12, -8,\ 17,\ \ldots\}$$
$$[3] = \{3,\ 8, -2,\ 13, -7,\ 18,\ \ldots\}$$
$$[4] = \{4,\ 9, -1,\ 14, -6,\ 19,\ \ldots\}$$

In general, if $n \neq 0$, then there are only $|n|$ different residue classes modulo n. The general framework of congruence relation is usually called *modular arithmetic*. An important property of modular arithmetic is that if a and b are integers, then

$$ab \bmod n = (a \bmod n)(b \bmod n)\bmod n$$

For example, to evaluate $(7 \times 25)\text{mod}13$ we proceed in one of two ways:

(a) Direct method: $(7 \times 25)\text{mod}13 = 175\text{mod}13 = (169 + 6)\text{mod}13 = (\{13 \times 13\} + 6)\text{mod}13 = 6$

(b) $7\text{mod}13 = 7,\quad 25\text{mod}13 = 12 \Rightarrow (7\text{mod}13)(25\text{mod}13) = 7 \times 12 = 84$
$84\text{mod}13 = (78 + 6)\text{mod}13 = (\{6 \times 13\} + 6)\text{mod}13 = 6$

12.6.3 The Square and Multiply Algorithm

Using the preceding property, we can perform the modular arithmetic of larger integers that can be factored into much smaller integers. One application is the evaluation of $x^a \bmod n$. To obtain the result, we use the so-called "square and multiply" algorithm, which can be stated as follows:

(a) Calculate $y = x^{\lfloor a/2 \rfloor} \mathrm{mod} n$, where $\lfloor b \rfloor$ is the largest integer that is smaller than b.
(b) Square y and reduce mod n.
(c) If a is odd, then multiply y by x and reduce mod n.

For example, we apply the method to the evaluation of $3^{11} \bmod 13$. Let

$$y = 3^{\lfloor 11/2 \rfloor} \mathrm{mod} 13 = 3^5 \mathrm{mod} 13 = 243 \mathrm{mod} 13 = (18(13) + 9) \mathrm{mod} 13 = 9$$

Then we have that

$$\begin{aligned} y^2 &= (3^5 \mathrm{mod} 13)(3^5 \mathrm{mod} 13) \mathrm{mod} 13 = (9 \times 9) \mathrm{mod} 13 = 81 \mathrm{mod} 13 \\ &= (6(13) + 3) \mathrm{mod} 13 = 3 \end{aligned}$$

Thus,

$$3^{11} \mathrm{mod} 13 = y^2 (3 \mathrm{mod} 13) = 9$$

Check: $3^{11} \mathrm{mod} 13 = 177147 \mathrm{mod} 13 = (177138 + 9) \mathrm{mod} 13 = (\{13 \times 13626\} + 9) \mathrm{mod} 13 = 9$

12.6.4 Euclid's Algorithm

The Euclid's algorithm is used to calculate the greatest common divisor of two positive integers a and b. The algorithm is based on the observation that a common divisor d of the integers a and b also divides the difference $a - b$. To see this, let $a = kd$ and let $b = ld$, for some integers k and l, and assume that $a > b$. Then $a - b = (k - l)d$; thus, d divides $a - b$. Thus, we can reduce the original problem of finding $\gcd(a, b)$ to a smaller problem of finding $\gcd(a, a - b)$. This means that we can obtain the solution to the problem by a series of division and remainder.

More formally, if a and b are positive integers such that $a > b$, there exist unique nonnegative integers q and r such that

$$a = qb + r \quad 0 \le r < b$$

where q is the quotient and r is the remainder. From the above discussion, $\gcd(a, b) = \gcd(b, r)$. For example, assume that we want to compute $\gcd(36, 15)$. We start by dividing 36 by 15 to obtain

$$36 = 2 \times 15 + 6$$

This means that $\gcd(36, 15) = \gcd(15, 6)$. Repeating the procedure, we obtain

$$15 = 2 \times 6 + 3$$

Thus, gcd(15, 6) = gcd(6, 3). Finally, we obtain

$$6 = 2 \times 3 + 0$$

Because 6 is a perfect multiple of 3, gcd(6, 3) = 3 = gcd(36, 15). Generally, the procedure is arranged as follows:

$$36 = 2 \times 15 + 6$$
$$15 = 2 \times 6 + 3$$
$$6 = 2 \times 3 + 0$$

12.6.5 Extended Euclid's Algorithm

Let $\gcd(a, b) = k$. It can be shown that there exist integers p and s such that

$$pa + sb = k$$

By reversing the steps in the Euclid's algorithm, it is possible to find the integers p and s. We illustrate this with the previous example. From the second to the last line, we have that

$$3 = 15 - 2(6)$$

Substituting for 6 in the first line, we obtain

$$3 = 15 - 2(36 - 2(15)) = -2(36) + 5(15)$$

Thus, we obtain $p = -2$ and $s = 5$. This procedure is known as the extended Euclid's algorithm.

The extended Euclid's algorithm can be used to obtain the inverse of a number modulo n. The inverse modulo n of an integer a is the integer a^{-1} such that

$$aa^{-1} \equiv 1 \text{mod} n$$

A modular inverse can be obtained via the extended Euclid's algorithm. This inverse exists only if $\gcd(a, n) = 1$. In this case, we have that

$$pa + sn = 1$$

This means that

$$pa \equiv 1 \text{mod} n$$

In the previous example, $k \neq 1$, which means that 36 has no inverse modulo 15. However, assume that we want to obtain the inverse modulo 72 of 5. We compute gcd(72, 5) using the Euclid's algorithm as follows:

$$72 = 14(5) + 2$$
$$5 = 2(2) + 1$$

Using the extended Euclid's algorithm, we obtain

$$1 = 5 - 2(2) = 5 - 2(72 - 14(5)) = 29(5) - 2(72)$$

Thus, 5 and 29 are inverses modulo 72. Observe that $5 \times 29 = 145 = (2)(72) + 1$. Note that 4 and 7 are inverses modulo 9 because $4 \times 7 = 28 = (3)(9) + 1$, which means that

$$(4 \times 7) \bmod 9 = 28 \bmod 9 = (27 + 1) \bmod 9 = 1 \Rightarrow 9|(28 - 1)$$

12.6.6 Euler's Phi Function (Euler's Totient Function)

The *Euler phi function*, also known as the *Euler totient function*, of the integer n is defined as the number of natural numbers that are less than or equal to n that are coprime (i.e., relatively prime) to n. It is denoted by $\phi(n)$. From this definition, we have $\phi(n) = n - 1$, if n is prime. For example, $\phi(10) = 4$ since 1, 3, 7, and 9 are the only positive integers that are coprime to 10.

If n is the product of two primes p and q, then $\phi(n) = \phi(p)\phi(q) = (p - 1)(q - 1)$. For example, $15 = 3 \times 5$, which means that $\phi(15) = \phi(3)\phi(5) = (2)(4) = 8$. These numbers are 1, 2, 4, 7, 8, 11, 13, 14.

Note that $\phi(10) = \phi(2)\phi(5) = (2 - 1)(5 - 1) = (1)(4) = 4$, as we obtained earlier.

12.6.7 The RSA Algorithm

The basic idea behind the RSA algorithm is to find three integers d, e, and n such that d and e are inverses modulo $\phi(n)$, where n is the product of two prime numbers p and q; that is, $n = pq$. If such numbers exist, then (e, n) can be the encryption (i.e., public) key that is published and (d, n) will be the corresponding decryption (i.e., private) key that is not published. Let M be the message (or plaintext) to be encrypted, and C be the corresponding ciphertext. If d and e are inverses modulo $\phi(n)$, then it can be shown that if C is generated by the operation

$$\mathrm{C} = E[M] = M^e \bmod n$$

then M can be recovered by the operation

$$M = D[\mathrm{C}] = \mathrm{C}^d \bmod n$$

As stated earlier, the keys are (e, n) and (d, n). $\phi(n)$ is a secret that is not disclosed to the public. Similarly, the private key is not published and is only calculated using the congruence relationship $d \times e \equiv 1 \bmod n$. Only the public key is published.

Example 12.1 In practice p and q are very large numbers, but for illustration purposes we choose small values of p and q. In particular, let $p = 2$ and $q = 11$.

Then $n = pq = 2 \times 11 = 22$. Now, $\phi(22) = 10$ because the numbers 1, 3, 5, 7, 9, 13, 15, 17, 19, and 21 are relatively prime to 22. Note that we can obtain the same result from $\phi(22) = \phi(2 \times 11) = \phi(2)\phi(11) = 1 \times 10 = 10$. Since 3 is relatively prime to 10, we assume that $e = 3$; then d is obtained from the fact that we need the condition $(d \times e) \bmod 10 = 1$ to be satisfied. This means that $10|\{d \times e - 1\}$. The value of d that satisfies this condition for the specified value of d is $d = 7$ because $3 \times 7 = 21$ and $21 \bmod 10 = (20 + 1) \bmod 10 = 1$. Let $M = 5$; then the ciphertext becomes

$$\begin{aligned} C &= E[\mathrm{M}] = M^3 \bmod 22 = 5^3 \bmod 22 = 125 \bmod 22 \\ &= (110 + 15) \bmod 22 = (5 \times 22 + 15) \bmod 22 \\ &= 15 \end{aligned}$$

The recipient of C recovers M by performing the operation

$$\begin{aligned} M &= D[C] = C^7 \bmod 22 = 15^7 \bmod 22 = 170859375 \bmod 22 \\ &= (170859370 + 5) \bmod 22 = (7766335 \times 22 + 5) \bmod 22 \\ &= 5 \end{aligned}$$

Thus, we are able to recover M. Note that we can also use $e = 7$ and $d = 3$ and still recover the message since the private and public keys are interchangeable. C will be different in this case, but we still recover M as follows:

$$C = M^7 \bmod 22 = 5^7 \bmod 22 = 78{,}125 \bmod 22 = (78{,}122 + 3) \bmod 22 = 3$$

$$C^3 \bmod 22 = 3^3 \bmod 22 = 27 \bmod 22 = (22 + 5) \bmod 22 = 5 = M$$

In computing d, we essentially found the solution to the problem:

$$\frac{ed - 1}{10} = k, \ k = 1, 2, 3, \ldots \Rightarrow ed - 1 = 10k \Rightarrow ed = 1 + 10k \Rightarrow d = \frac{1 + 10k}{e}$$

We start with a small value of e; let $e = 3$, since 3 is relatively prime to 10. We find that there is no solution for $k = 1$. For $k = 2$, we obtained $d = 7$. We can also use the Euclid's algorithm and the Euclid's extended algorithm, which are very useful in large values of $\phi(n)$. In this case, we have that the gcd(10, 3) is given by

$$10 = 3(3) + 1$$

Using the extended Euclid's algorithm, we have that

$$1 = 10 - 3(3)$$

Thus, 3 and −3 are congruent modulo 10. But $-3 \bmod 10 = 10 - 3 = 7$. This gives $d = 7$, as we obtained earlier. Note that there are other solutions to the problem

$$d = \frac{1 + 10k}{e} = \frac{1 + 10k}{3}$$

These are $d = 17$, which is obtained when $k = 5$; $d = 27$, which is obtained when $k = 8$; $d = 37$, which is obtained when $k = 11$; and so on.

Example 12.2 As another example, let $p = 3$ and $q = 13$. Then $n = p \times q = 39$, and the totient function is $\phi(39) = \phi(3) \times \phi(13) = 2 \times 12 = 24$. Let $e = 5$. Then d is obtained as follows:

$$d = \frac{1 + k\phi(39)}{e} = \frac{1 + 24k}{5}$$

One solution to the equation is obtained when $k = 6$, which gives

$$d = \frac{1 + (24)(6)}{5} = \frac{145}{5} = 29$$

Another solution is obtained when $k = 16$; in this case, we obtain $d = 77$. Other solutions are obtained when $k = 26, 36, 46, \ldots$. In general, there are many solutions to the problem, which is one of the strengths of the scheme.

12.7 Digital Signatures

The RSA algorithm can be used to generate a *digital signature* that can be attached to a message. A digital signature is a data value that is computed as a checksum of a message and the user's private key. Because public-key cryptosystems are inefficient for encrypting long messages, it is more common to use them to sign a condensed version of the message rather than encrypt the entire message. The condensed version is generated by a *hash function.*

A hash function (or *message-digest function*) is an algorithm that operates on an input string of an arbitrary length and generates an output string of fixed length called the *hash value* or *message digest*. The length of message digest is usually much smaller than that of original message. A message digest is substantially unique to the message and is similar to the fingerprint of the message because it is practically impossible for two distinct messages to have identical message digests using the same hash function. The importance of a message digest lies in the fact that any change in the message will produce a different message digest when the same hash function is used. Thus, digital signatures are used to ensure data integrity.

12.7.1 Generating a Digital Signature

At the source, a hash function operates on the message and generates the message digest. The private key is used to transform the message digest into the digital signature. The digital signature is attached to the message and the combination of the message and the digital signature is sent to recipient. This is illustrated in Figure 12.4.

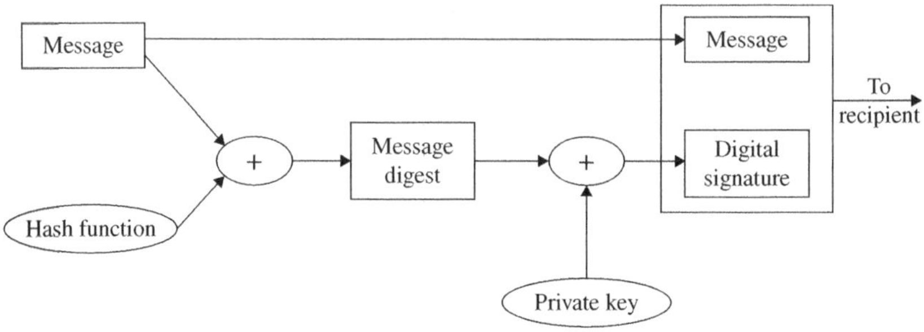

Figure 12.4 Generation of a Digital Signature.

12.7.2 Verifying a Digital Signature

At the destination, the message and the digital signature are separated. The message digest of the received message is generated with the same hash function used at the source. The public key of the source is used on the digital signature to recover the message digest generated at the source. The two message digests are compared. The message is accepted if the two message digests are identical; otherwise, it is rejected. The process at the destination is illustrated in Figure 12.5.

The US government standard for creating digital signatures is the digital signature standard (DSS), which specifies a *digital signature algorithm* (DSA). DSA uses a special hash function called the *secure hash algorithm* (SHA), which generates a 160-bit message digest for each message. DSA then uses the message digest and the private key to generate a 320-bit digital signature. Note that DSA is for signatures only and is not an encryption algorithm. The key size is variable from 512 to 1024 bits.

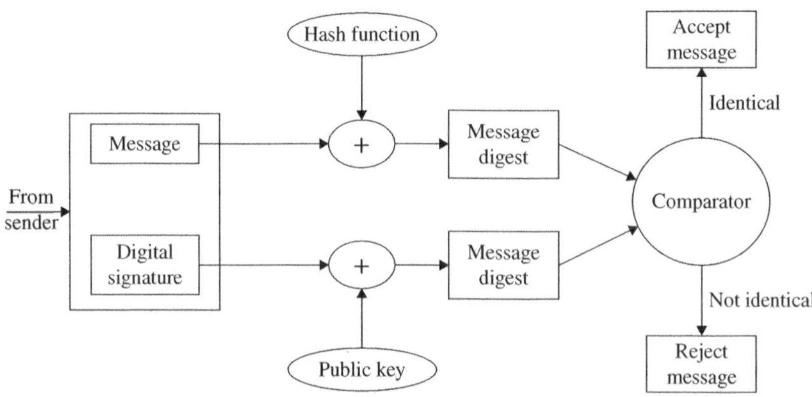

Figure 12.5 Verification of the Message.

12.8 IP Security Protocols

IPSec is a suite of IETF-defined protocols for IP networks. It has three basic components:

(a) Authentication header (AH)
(b) Encapsulating security payload (ESP)
(c) Internet key exchange (IKE).

AH and ESP are used to protect IP traffic. They use cryptographic techniques to provide data confidentiality and use digital signatures to authenticate the source of the data. IKE is a key management protocol that allows users to agree on which keys to use, permits keys to be exchanged securely, and manages the keys after they have been agreed upon and exchanged. IPSec implementation is mandatory in IPv6 but optional in IPv4.

12.8.1 IPSec Modes

IPSec operates in two modes: the *tunnel mode* and the *transport mode*. In tunnel mode, IPSec encapsulates the original IP packet into an IPSec packet with new IP headers. In this mode, it effectively hides the original IP packet from view. The tunnel mode is used mainly between two routers that are the default gateways for their networks. This is illustrated in Figure 12.6.

In the transport mode, IPSec applies IPSec protocols to an IP packet and leaves the original IP headers visible. It is used between hosts. This is illustrated in Figure 12.7.

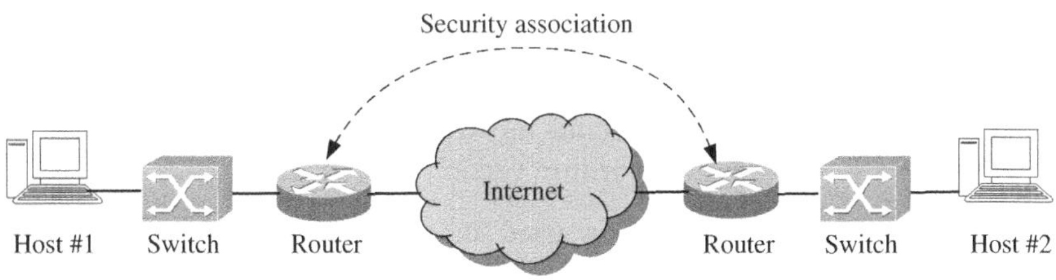

Figure 12.6 IPSec in Tunnel Mode.

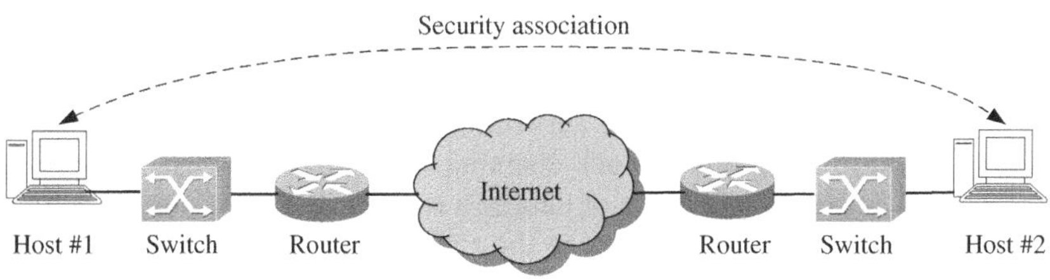

Figure 12.7 IPSec in Transport Mode.

12.8.2 Security Association

For authentication and encryption to work properly, the sender and the receiver must agree on a number of things before hand:

(a) Key and its lifetime
(b) Authentication algorithm
(c) Mode of operation.

This definition of protocols, algorithms, and key validity time periods is called *security association* (SA) between the sender and receiver. Each IPSec connection has two SAs: one in each direction. An SA has three identifiers:

(a) A unique number called *security parameter index* (SPI), which is assigned by the destination to the SA and is 32 bits long
(b) Destination IP address
(c) Protocol (AH or ESP) identifier.

12.8.3 Authentication Header

The AH provides data integrity and protection against *replays* by adding authentication information to an IP packet. As stated earlier, a replay is an attack that attempts to trick the system by retransmitting a legitimate message. The authentication information is a cryptographic checksum calculated with the IP fields that do not change in transit and a secret key. Thus, it ensures that fields that do not change in transit have not been tampered with. If an attacker attempts to compute another checksum without knowledge of the secret key, the computation fails.

AH is not as widely used as the ESP. It provides authentication via a message digest and uses one of two default hash functions:

(a) Hash message authentication code (HMAC) with MD-5
(b) HMAC with secure hash algorithm version 1 (SHA-1).

Figure 12.8 shows how AH is used in the tunnel mode and in the transport mode.

12.8.4 Encapsulating Security Payload

The ESP provides data integrity and confidentiality (or privacy) as well as authentication. Unlike AH, it encrypts the entire payload; and like AH, it uses HMAC with MD5 or SHA-1 authentication. It uses the data encryption standard (DES) in the cipher block chaining (CBC) mode for encryption. A block cipher encrypts one block of data at a time. In the CBC mode, a block cipher adds a feedback mechanism to the encryption scheme: the plaintext is X-ORed with the previous ciphertext block prior to encryption, which means that two identical blocks of plaintext never encrypt to the same ciphertext.

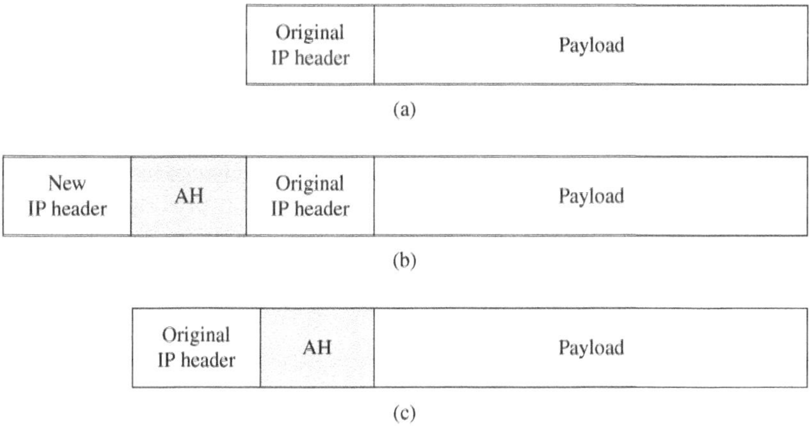

Figure 12.8 AH in the Tunnel and Transport Modes. (a) Original IP Packet, (b) Authentication Header in Tunnel Mode, and (c) Authentication Header in Transport Mode.

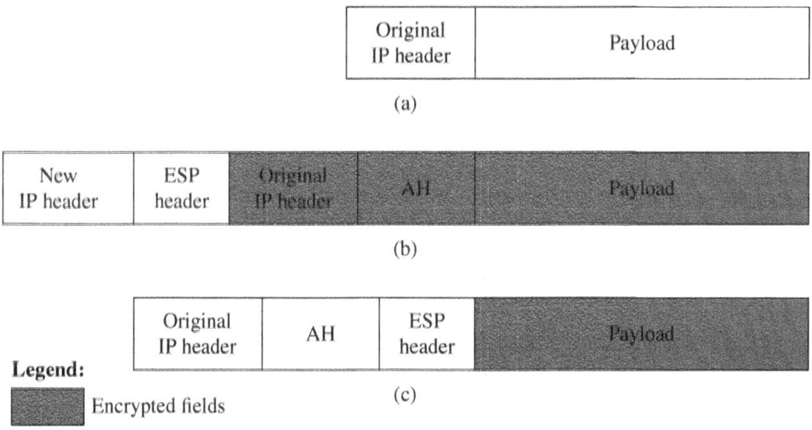

Figure 12.9 ESP in the Tunnel and Transport Modes. (a) Original IP Packet, (b) ESP in Tunnel Mode, and (c) ESP in Transport Mode.

Sometimes, ESP and AH are combined to provide strong security. Figure 12.9 illustrates the use of ESP in the tunnel and transport modes.

Note that when IPSec is implemented in transport mode, the security mechanism is enforced at each host. Similarly, when it is implemented in tunneling mode, security is enforced at security gateways at the end points of the SA.

12.8.5 Key Distribution

The IKE is the IPSec key negotiation and management protocol. It is a hybrid protocol that consists of three components:

(a) Internet security association and key management protocol (ISAKMP) is a protocol that specifies the framework for key exchange.
(b) Oakley is a protocol used to carry out the key exchange negotiation process for both parties. It uses the Diffie–Hellman algorithm to determine a shared key and achieves perfect forward security using the shared.
(c) Secure key exchange mechanism (SKEME) is key exchange technique that provides anonymity, privacy, and key refreshment. It provides modes to perform fast and frequent key refreshment.

12.9 Summary

This chapter has discussed basic features of network security. It discussed the types of security attacks that can be launched in a network as well as the different security services. It discussed the two types of key-based security protocols; namely, the Diffie–Hellman protocol and the RSA protocol. Finally, it discussed the Internet security protocol (IPSec). This is intended to provide a basic introduction to data communication network security and does not go into details on any of the more advanced topics on the subject.

Exercises

1 What does authentication mean in a network?

2 What does confidentiality mean in a network?

3 What does nonrepudiation mean in a network?

4 What is the difference between a symmetric (or secret-key) cryptosystem and an asymmetric (or public-key) cryptosystem?

5 What is a digital signature?

6 State one difference between the authentication header and the encapsulating security payload protocols of the IPSec.

7 Name the two modes of operation of the IPSec protocols.

8 Consider an RSA system with $p = 7$ and $q = 19$.

a. What is the Euler's totient function of the system?
b. Give the steps that are used to generate the encryption key e and the decryption key d.
c. Find one set of values of e and d. (Note that e is usually a small number that is relatively prime to $\phi(n) = \phi(pq) = \phi(133)$.)
d. How would the message $M = 8$ be encrypted with key e?

Bibliography

The following books cover the topics discussed in this book, albeit at an advanced level:

1. Agbo, S.O. and M.N.O. Sadiku, *Principles of Modern Communication Systems*, Cambridge University Press, 2017.
2. Duck, M. and R. Read, *Data Communications and Computer Networks: For Computer Scientists and Engineers*, 2nd Edition, Prentice-Hall, 2003.
3. Easttom, C., *Computer Security Fundamentals*, 3rd Edition, Pearson, 2016.
4. Forouzan, B.A., *Data Communications and Networking*, 5th Edition, McGraw-Hill, 2012.
5. Gallager, R.G., *Principles of Digital Communication*, Cambridge University Press, 2008.
6. Goldsmith, A., *Wireless Communications*, Cambridge University Press, 2005.
7. Gupta, P.C., *Data Communications and Computer Networks*, 2nd Edition, PHI Learning Private Limited, Delhi, India, 2014.
8. Held, G., *Understanding Data Communications*, 7th Edition, Addison-Wesley, 2002.
9. Ibe, O.C., *Fixed Broadband Wireless Access Networks and Services*, Wiley, 2002.
10. Kurose, J. and K. Ross, *Computer Networking: A Top-Down Approach*, 7th Edition, Pearson, 2016.
11. Mir, N.F., *Computer and Communication Networks*, Prentice-Hall, 2007.
12. Moussavi, M., *Data Communication and Networking: A Practical Approach*, Delmar Cengage Learning, 2011.
13. Oppenheim, A.V., Willsky A.S. and Hamid, S. *Signals and Systems*, 2nd Edition, Pearson, 1996.
14. Pal, A., *Data Communication and Computer Networks*, PHI Learning Private Ltd, Delhi, India, 2014.
15. Schiller, J., *Mobile Communications*, 2nd Edition, Pearson, 2003.
16. Schwartz, M., *Mobile Wireless Communications*, Cambridge University Press, 2005.

Fundamentals of Data Communication Networks, First Edition. Oliver C. Ibe.

17 Shay, W.A., *Understanding Data Communications and Networks,* Cengage Learning, 2003.

18 Singh, B., *Data Communications and Computer Networks,* 4th Edition, PHI Learning Private Limited, Delhi, India, 2014.

19 Stallings, W., *Data and Computer Communications,* 10th Edition, Pearson, 2013.

20 Stuber, G., *Principles of Mobile Communication,* 3rd Edition, Springer, 2012.

21 Tomasi, W., *Introduction to Data Communication Networking,* Pearson, 2005.

22 Tripathi, N. and J.H. Reed, *Cellular Communications: A Comprehensive and Practical Guide,* Wiley-IEEE Press, 2014.

23 Tse, D. and P. Viswanath, *Fundamentals of Wireless Communication,* Cambridge University Press, 2005.

24 White, C., *Data Communications and Computer Networks: A Business User's Approach,* 8th Edition, Cengage Learning, 2015.

Index

Fundamentals of Data Communication Networks, First Edition. Oliver C. Ibe.

d

e

Printed and bound by CPI Group (UK) Ltd, Croydon, CR0 4YY
17/09/2026
14964317-0002